D1610195

IEE Electromagnetic Waves Series 17
Series Editors: Professor P.J.B. Clarricoats
E.D.R. Shearman and J.R. Wait

Surveillance radar performance prediction

Previous volumes in this series

Volume 1	Geometrical theory of diffraction for electromagnetic waves Graeme L. James
Volume 2	Electromagnetic waves and curved structures Leonard Lewin, David C. Chang and Edward F. Kuester
Volume 3	Microwave homodyne systems Ray J. King
Volume 4	Radio direction-finding P. J. D. Gething
Volume 5	ELF communications antennas Michael L. Burrows
Volume 6	Waveguide tapers, transitions and couplers F. Sporleder and H.G. Unger
Volume 7	Reflector antenna analysis and design P. J. Wood
Volume 8	Effects of the troposphere on radio communications Martin P. M. Hall
Volume 9	Schuman resonances in the earth-ionosphere cavity P. V. Bliokh, A. P. Nikolaenko and Y. F. Filippov
Volume 10	Aperture antennas and diffraction theory E. V. Jull
Volume 11	Adaptive array principles J. E. Hudson
Volume 12	Microstrip antenna theory and design J. R. James, P. S. Hall and C. Wood
Volume 13	Energy in Electromagnetism H. G. Booker
Volume 14	Leaky feeders and subsurface radio communications P. Delogne
Volume 15	The Handbook of Antenna Design Volume 1 Editors: A.W. Rudge, K. Milne, A.D. Olver, P. Knight
Volume 16	The Handbook of Antenna Design Volume 2 Editors: A.W. Rudge, K. Milne, A.D. Olver, P. Knight

Surveillance radar performance prediction

P. Rohan M.E. Ph.D
Department of Electrical and Electronic Engineering
The University of Adelaide.

Peter Peregrinus Ltd
On behalf of The Institution of Electrical Engineers

Published by: Peter Peregrinus Ltd., London, UK.

British Library Cataloguing in Publication Data

Rohan, P.
Surveillance radar performance prediction.
– (IEE electromagnetic waves series; 17)
1. Radar
I. Title II. Series
621.3848'5 TK6575

ISBN 0 906048 98 2

Printed in England by Short Run Press Ltd., Exeter

Memoriae meae matris
et uxori meae.

Contents

Acknowledgements

I am indebted to Prof. R. E. Bogner and Dr B. R. Davis of the Dept. of Electrical and Electronic Engineering, The University of Adelaide, for many stimulating discussions on various topics dealt with by this book.

My thanks are due to Mr P. Ramsey, Senior Navigational Aids Engineer, Dept. of Aviation, Adelaide, South Australia, for supplying me with graphs reproduced here in Figures 11.1, 11.2, 11.5 and 11.8 and to Mr D. Parr of the same department for the photograph of the Adelaide airport's radar station.

The book draws freely on publications by authors named in the text. I apologize to those whose name I have inadvertently omitted.

Acknowledgement is due to a number of authors, establishments and editors of periodicals for granting me permission to use illustrations and other selected material published by them; these are listed herewith in alphabetical order:

Australian Journal of Physics, Melbourne
Prof. K. G. Budden, Cambridge, England
Director, Comité Consultatif International des Radiocommunications, Genève
Dr J. V. DiFranco, Great Neck, N.Y.
Director, Electronic Research Laboratory, Defence Research Centre, Salisbury, South Australia
I.E.E.E. Transactions on Aerospace and Electronic Systems, New York
*L'Onde Electrique, Paris
*McGraw Hill Book Co. Inc., New York
Commanding Officer, Naval Research Laboratory, Washington, D.C.
Proceedings of the I.E.E.E., New York
Director General, Radar & Communications Technology, Communications Research Centre, Ottawa, Ontario, Canada
Radiotechnika i Elektronika, Moscow, USSR
Dr Wm L. Rubin, Great Neck, N.Y.
VEB Verlag Technik, Berlin, GDR.

Attempts, to obtain a response to several requests for permission to reproduce material published by those marked by an asterisk, have hitherto failed. Nevertheless, due acknowledgement is given to the above herewith as well as by reference in the appropriate passages or captions of the relevant illustrations of the book.

April, 1983 P. R.

Summary

A radar's performance depends not only on its and the target's parameters but also on various environmental factors and terrain features. It is, as a rule, ascertained by the time consuming and expensive method of flying targets of interest on radial paths and determining the detection ranges of the radar installed on the chosen site.

The book shows the use of mathematical models for surveying a proposed radar site and predicting the detection ranges of a surveillance radar installed there, thus obviating the necessity of costly methods for both of the above tasks. It gives complete listings of three models together with examples of their applications and the theoretical foundations required for their understanding.

The models are applicable to the performance predictions or assessment of land-based and ship-borne VHF, UHF and microwave radars. Model RDPRO has also been used for the performance prediction of air-borne radars.

Chapter 1

Introduction

The performance of a radar depends not only on the radar's design parameters but also on a number of other factors which have to be accounted for when assessing or predicting it. The simple radar equation based on free-space propagation yields, in most cases, far from accurate results and must be modified to account for multipath propagation and other effects.

The performance of radars is often assessed *a posteriori* by flying known targets on predetermined trajectories and measuring the detection ranges by the radar of interest installed on the chosen site. This method is time-consuming and costly.

An *a priori* prediction of the performance of a radar to be sited in a suitable position chosen on the basis of information gained from maps has been shown to be feasible and is the subject of the following chapters.

Radar performance assessment consists of a great number of repetitive calculations with only slight modifications of the data or computing processes; hence, it is particularly suited for processing by digital computers. This method is cheaper and faster than experimental methods and it permits the modelling, i.e. simulation, of conditions for the natural occurrence of which one might have to wait a very long time. Besides this, it permits the repetition of tests under conditions which may be difficult to obtain in practice.

Mathematical modelling is now used extensively, especially for the design and optimization of complex systems. It plays a prominent part in science, engineering, business and defence systems where it may be used for the solution of various scientific, economic, tactical and strategic situations.

A mathematical model is, in a mathematical sense, a representation of a concept, process or system. It is a collection of mathematical expressions, algorithms, arranged in a logical sequence to describe some properties of interest of the modelled prototype. It usually consists of a main computer program and a number of minor programs, subroutines, called by the main program whenever it requires data computed by the appropriate subroutine for a particular set of conditions.

In order to present a complete picture of the problem various aspects of radar theory have been included in the book. These rely to a large extent on the available literature and on relevant experience and studies in the field. The initial chapters present the background material in adequate depth for the understanding of the philosophy and techniques used in the assessment proper and in the mathematical models.

The book is concerned primarily with the performance prediction or assessment of land- and ship-based search or surveillance radars though model RDPRO has been used also for airborne radar performance assessment.

Chapter 2 assesses radar search qualitatively and proposes a search philosophy for radars equipped with phased array aerials.

Chapter 3 deals with the elements of decision and radar detection theory. It presents formulae used later in the book for the calculation of the signal-to-noise ratio and probability of detection.

A close look at the radar equation and an analytical examination of its parameters is presented in Chapter 4.

Radar wave propagation is dealt with in Chapter 5 which establishes the basic radar–ground–target geometry used in the models, i.e. the assessment tools, and in the performance assessment described later.

Radar radiation reaching the target consists partly of radiation propagating directly between radar and target and partly of radiation undergoing a reflection process between the two end-points. The magnitude of this latter component depends on the nature of the reflecting surface as shown later in Chapter 11. Chapter 6, a survey of the available literature on rough surface reflection, examines the phenomenon to permit the understanding of the process and evaluating its effect.

The radar target echoing area is the least defined parameter of the radar equation. Extensive literature on echoing area computation and modelling is available. However, for the sake of completeness of this treatise, a short survey of the most essential parts of the topic is included in Chapter 7.

Chapter 8 outlines the purpose of search or surveillance radars and examines the achievable accuracy of radar measurements and resolution.

Radar literature lacks guide-lines for site-selection. Chapter 9 considers environmental influences on radar performance and, based upon these, proposes steps to be followed in radar site selection.

The mathematical models used for the assessment are described in Chapter 10. The first described model which computes and plots the vertical coverage diagram may be used as a radar design tool since it permits the determination of the influence of the radar's parameters on the coverage and for performance assessment at a given site, as demonstrated in the following chapter. This model, named 'COVER', is applicable also to the derivation of various coverage diagram algorithms, as shown in this chapter.

The second model, named 'MAP', computes and plots contours of constant probability detection ranges of targets at specified heights all around the radar's site; it may be used for site selection and thus obviate the necessity of expensive surveys, while the third and most comprehensive of the three models, model 'RDPRO', determines the dependence of the probability of detection of a target moving on a specified trajectory on the distance from the radar. It also computes and plots the variation of the received signal power and clutter plus noise with target range.

The application and use of the above three models is shown on examples in Chapter 11, where the performance of Adelaide's West Beach Airport radar is assessed. Good agreement with observed performance figures, available by courtesy of the Australian Department of Aviation, is demonstrated.

Three appendices list the models used for the described assessments.

Appendix A contains the complete listing, in the FORTRAN IV computer language, of the COVER model together with all of its subroutines and algorithms, for the coverage diagram of the assessed radar's aerial. The listing for the computation of the first order Bessel function of the first kind, applicable for the computation of the coverage diagram of aerials with circular apertures, is also shown. Formulae for the coverage diagram of rectangular aperture aerials are given in Chapter 10.

Appendix B shows the complete listing of the MAP model which uses the same subroutines as model COVER.

The complete listing of model RDPRO is in Appendix C.

An extensive list of references provides titles of a large number of publications on the topics dealt with in the book.

Although parts of the book served as lecture notes in a course of lectures on radar theory and techniques taught to honours students in Electrical Engineering, it is basically addressed to the practising radar engineer confronted with the task of radar performance prediction, assessment or site selection. Consequently readers with adequate background in radar are referred directly to Chapters 10 and 11 describing the three models applicable for the task together with examples of their use. Readers with only a nodding acquaintance of radar will benefit from working through the initial chapters thus gaining the background information necessary for the understanding of the various statements of the programs.

Chapter 2

Radar search and its assessment

2.1 Radar search

The two basic operations that a surveillance radar carries out are searching for and detecting targets. Search and detection are intimately connected operations, especially since detection is the only concrete indication of a successful search. Judging by the great number of publications, the process of detection seems to have attracted much more attention than search. Yet, a judicious search program will lead to an earlier detection and hence possible economy in effort and power.

Search consists of scanning the space to be surveyed according to a predetermined program. Most radars perform this task by mechanically rotating their aerial in a horizontal plane at a uniform rate, thus performing a circular scan. This type of scan defines azimuth, one of the two coordinates on a plan position indicator (PPI) display. The rotating radiation cone surveys the space in a fan-like fashion. Its angular velocity must not exceed a value limited by the number of pulses which are to be intercepted by the target in order to achieve an acceptable probability of detection. If this number is designated by k and if f_r is the pulse repetition frequency and θ the horizontal half-power or 3 dB beamwidth of the aerial in radians, the dwell time of the beam on the target is $\Delta t = k/f_r$, from which the required angular velocity ω may be determined.

Since $\theta = \omega \Delta t = \omega k/f_r$, the maximum angular velocity for the given beamwidth and pulse repetition frequency becomes

$$\omega_{\mathbf{max}} = \theta f_r/k \quad \text{radians/s}$$

and the number of revolutions per second is

$$\frac{\theta f_r}{2\pi k}.$$

The angular displacement of the aerial per pulse is $\omega/f_r = \theta/k$, a value important for the rotating indicator beam.

The sector-scan is a special case of the circular scan; it is used when only part of the azimuthal plane is to be surveyed. It is evident, that a sector-scan may be performed in the vertical plane as in height-finder radars.

Combining the azimuthal scan with a vertical one provides a three-dimensional scan for which the surveyed space is in the shape of the section of a pyramid. This type of scan may be used in three-dimensional scanning radars which are capable of locating targets in range, azimuth and elevation.

A helical scan, in which a very narrow beam is shifted up or downwards by one beamwidth after each revolution of the aerial, is also capable of obtaining three-dimensional information.

Fire-control and guidance radars use spiral and conical scans in which a very narrow beam scans a conical sector having the aerial at its vertex.

The scanning motion of the aerial beams in some modern radars is controlled electronically. With these aerials the search procedure may be programmed in any required manner.

2.1.1 Theory of search assessment

The problem of formulating a search assessment theory lies in the derivation of mathematical expressions permitting the comparison of various search procedures on a qualitative basis.

In what follows, a search is considered to be successful if the target is in the searched element of space at the instant of searching. The notion of success is then purely abstract and another means, namely detection, is required to give a concrete proof of a search's success or failure.

Let there be, at random, a target in the space surveyed piecewise in small elemental volumes and let us assume that the probability of its being detected is $p = p(A)$. The probability of the search being successful k-times in m independent trials may be expressed, on the basis of the multiplication theorem for independent events, as

$$p^k (1 - p)^{m-k}.$$

Since the sequence in which successes occur is of no consequence and there are

$$C_k^m = \frac{m!}{k!(m-k)!}$$

possible combinations of k successes with $(m - k)$ failures, the probability of k successes in m events is given by the binomial distribution (Hogg and Craig, 1959)

$$P_m(k) = C_k^m p^k (1 - p)^{m-k}. \qquad (2.1)$$

The total probability

$$\sum_{k=0}^{m} P_m(k) = \sum_{k=0}^{m} C_k^m p^k (1 - p)^{m-k}$$

is the binomial expansion of $(p + (1 - p))^m$ and must be equal to unity.

The moment generating function of a binomial distribution is

$$M_k(t) = \sum_{k=0}^{m} f(k) \exp(kt) = \sum_{k=0}^{m} \frac{m!}{k!(m-k)!} (p \exp(t))^k (1-p)^{m-k}$$

$$= (p \exp(t) + (1-p))^m$$

and its mean value

$$\frac{dM_k(0)}{dt} = mp.$$

Let the whole searched volume of space be V, while an elementary volume searched at a given instant is v. The probability of the target being in v at the instant of search is

$$q = v/V = p(D).$$

The instantaneous probability of success when referred to a scanning radar becomes then the joint probability of events D and A, i.e.

$$p(A, D) = p(D)p(A|D),$$

where the conditional probability of the target being detected provided that it is in the volume v illuminated at the given instant must equal p. Then

$$p(A, D) = pv/V = p'$$

so that the total probability reads

$$P_m(k) = C_k^m p'^k (1-p')^{m-k} \tag{2.2}$$

and its mean value is

$$mp' = mpv/V = B \tag{2.3}$$

which when inserted into eqn. (2.2) results in

$$P_m(k) = C_k^m \left(\frac{B}{m}\right)^k \left(1 - \frac{B}{m}\right)^{m-k} \tag{2.4}$$

This is the probability of k successes in m independent trials. It expands into

$$P_m(k) = \frac{m!}{k!(m-k)!} \left(\frac{B}{m}\right)^k \left(1 - \frac{B}{m}\right)^m \left(1 - \frac{B}{m}\right)^{-k}$$

$$= \frac{m!}{m^k (m-k)! \left(1 - \dfrac{B}{m}\right)^k} \frac{B^k}{k!} \left(1 - \frac{B}{m}\right)^m \tag{2.5}$$

Letting m increase to infinity while keeping k and B constant, makes the first term on the right-hand side tend to unity and the third term to exp $(-B)$ so that the probability becomes

$$P_m(k) = \frac{B^k}{k!}\exp(-B). \tag{2.6}$$

Using eqn. (2.3), one then arrives at

$$P_m(k) = (pmv/V)^k \frac{\exp(-pmv/V)}{k!} \tag{2.7}$$

which is an approximation of expression (2.4).

The probability of no success in m independent trials is

$$P_m(0) = \exp(-pmv/V) \tag{2.8}$$

and that of at least one successful search in m independent trials is

$$Q_m(1) = 1 - P_m(0) = 1 - \exp(-pmv/V). \tag{2.9}$$

$P_m(k)$, eqn. (2.6), is a Poisson probability distribution. The maximum value of the probability given by eqn. (2.9) occurs when the second term on the right-hand side is zero. Since v/V cannot exceed unity, the maximum probability means an infinite number of trials or a continuous surveillance of the whole allocated space. This, however, presents technical difficulties.

For a circular scan by an aerial having a fan-like vertical coverage diagram the volume observed at any instant is

$$v = \theta\psi\frac{R^3}{3},$$

while the total volume is $V = 2\pi\psi R^3/3$ where R is the range and ψ and θ are the vertical resp. horizontal 3 dB beamwidth of the aerial in radians.

Assuming then that the vertical dimensions of the space allocated for surveillance and that scanned by the aerial are equal, the probability of the target being in the volume v becomes $p = v/V = \theta/2\pi$.

2.1.2 *Practical search assessment*

The use of the derived formulae is best demonstrated by examples. Assume that p in all trials is 0·5.

Example 1. Assuming a conventional radar the aerial of which has a 3 dB horizontal beamwidth of 1·3° and a revolution rate of 5 r.p.m., the ratio v/V becomes 0·003 61. There are 5 trials per minute and the probability of a success in five trials is

$$P_5(1) = 1 - \exp(-0{\cdot}009) = 0{\cdot}01,$$

a rather low probability which, however, may be improved by using a sector scan.

Example 2. Using the same aerial for the surveillance of a 90° sector yields a value of 0·012 78 for the ratio of the two volumes. Since the aerial may perform, in an equal time, more scans over the restricted volume, say twice as many over one quarter of the original volume, the probability of one success in 10 trials is

$$P_{10}(1) = 1 - \exp(-0{\cdot}0639) = 0{\cdot}06,$$

which represents a considerable improvement over the previous case.

In conclusion it may be said that while it is accepted that in certain situations, e.g. when the appearance of targets from any direction is possible, a circular search, though less efficient, is justified, the superiority of a sector scan, as demonstrated by the above examples, recommends its adoption wherever possible.

2.2 Proposal for a phased array search philosophy

Some of the most recent radars use phased array aerials which are basically groupings of many individual electronically steerable aerials with controlled phase-coherent radiation to produce a plane wave propagating in a desired direction. Phase or frequency variations are used for achieving coherence.

Since the direction of propagation may be chosen at random and need not follow a constant pattern, as in the case of mechanically steered or rotating aerials, the search philosophy used with phased array aerials may differ to a great extent from that described earlier.

The surveyed space may be divided into a number of cells. On the basis of the initial n scans of all cells, or on the basis of some other criterion, one establishes an *a posteriori* probability $P_i(n)$ of the target's presence in the individual cells and finds the cell j in which this probability is the largest.

Since the events observed on the individual scans of the same cell may be taken to be statistically independent, one may express the *a posteriori* probability of the event observed on the $(n+1)$th scan by

$$P_i(n+1) = \frac{P(x|i)P_i(n)}{\sum_i P(x|i)P_i(n)}, \quad i = 1, 2, \ldots, r,$$

when x is the event observed in the jth cell on this scan.

The conditional probabilities $P(x|i)$ and $P(x|0)$, i.e. the probabilities of presence resp. absence of a target in the scanned cell are

$$P(x|i) = P(x|0),$$

when $i \neq j$ and for $i = j$ it is

$$P(x|j) = P(x|y),$$

i.e. the probability of observing the event x on scanning the jth cell provided that

y, the target's presence in that cell, occurred. Then, the *a posteriori* probabilities for the ith and the jth cell become

$$P_i(n+1) = \frac{P(x|0)P_i(n)}{\sum_i P(x|i)P_i(n)}$$

and

$$P_j(n+1) = \frac{P(x|y)P_j(n)}{\sum_i P(x|i)P_i(n)}$$

respectively.

The choice of the cell to be scanned next is dictated by the largest of the $P_i(n+1)$ probabilities.

The search philosophy, in broad outlines, is then as follows:

At the beginning of the search, before the first scan, the probabilities $P_i(0)$ are allocated values obtained from the *a priori* probabilities of the target's presence in the corresponding cells. Cell j, for which the above probability is maximum, is scanned first and the value of the event so obtained is used for computing the probability $P_j(1)$ for this cell and similarly, from the scan of the other cells, the probabilities $P_i(1)$ for the rest of the cells are ascertained. The maximum of the $P_i(1)$ values determines the next cell to be scanned, and so forth.

Then, since the signal-to-noise ratio required for detection is proportional to the integration time, the illumination and sampling of cells with high probabilities of target appearance will last longer than that of those in areas of lower probabilities.

The time interval over which cell i will be illuminated is given by

$$t_i = \frac{P_i t}{\sum_{i=1}^{n} P_i} - t_s,$$

where P_i is the probability of a target appearing in cell i, t is the total search time and t_s is the time required for shifting the beam from cell i onto the next cell in the surveillance process. The total search time is a function of the dwell time yielding an adequate signal-to-noise ratio for detection with the specified probabilities of detection and false alarm on the lowest target approach probability cell.

Having determined the cell containing the most urgent target to be dealt with, a tracking mode assigned, in the first instance, to this target may be initiated and followed by the tracking of the other targets of lesser importance.

The information obtained on scanning the cells is stored in a computer memory. The reading of the stores is destructive, i.e. the originally stored information is erased and replaced by the information obtained from the last scan. For search purposes retention of the former is not required. However, for tracking the data will have to be transferred to another store.

The process of tracking appears likely, at this stage, to comprise the accurate establishment of the target's coordinates in range and azimuth at a given time, the storage of this information and its comparison in a prediction process with similar information measured at some later time in order to establish the next point on the track. The time interval between the compared memory contents is of the magnitude of the search scan period.

2.3 Sequential detection

The dwell-time of the radar beam on the individual cells of the surveyed space is a function of the required signal-to-noise ratio for the specified probability of detection and false alarm and this, in turn, depends on the target echoing area behaviour. An *a priori* knowledge of the echoing area variations is therefore required for the search procedure described above.

Sequential detection, as described by a number of authors, e.g. Bussgang and Middleton (1955), Blasbalg (1959), Preston (1960), Helstrom (1962), Marcus and Swerling (1962), Bussgang and Johnson (1965) and others, obviates the necessity of this knowledge.

A conventional detection system in which the sample size is fixed *a priori* uses a strategy, referred to as the Bayes' solution, for determining a decision boundary, or threshold, on the basis of the smallest average risk as it will be described in the chapter on detection theory. If the signal exceeds the boundary, i.e. the detection threshold is exceeded, it is decided that a target is present, while in the other case, when the signal does not reach the threshold, no target is indicated.

A search method which relaxes to some extent the constraint of a fixed sample size was suggested above. It is superior to Bayes' strategy in the sense that, on the average, a substantially smaller number of observations is required to achieve the same error probabilities.

In Bussgang's sequential detection, which is a further improvement of the suggested search strategy, the sample size required to terminate a particular test is a random variable.

Both the Bayes' test and the sequential probability ratio test are likelihood ratio tests. In both cases the operation of the receiver involves the computation of the likelihood ratio (as outlined in Section 3.2). The difference lies in the test durations and the threshold levels. The Bayes' test computes the likelihood ratio after a predetermined number of samples and compares it to a single threshold, while the sequential probability ratio test computes, after each sample, the likelihood ratio on the basis of information received so far and compares it with two thresholds. If the likelihood ratio reaches the upper or 'alarm' threshold, the presence of a target is indicated; the lower threshold is the 'dismissal' threshold. If the likelihood ratio does not reach the alarm threshold after a predetermined maximum admissible number of samples, the sampled cell is dismissed, the aerial beam is shifted and the next cell is sampled.

Since the detection time in sequential detection is not known *a priori*, a feedback from the receiver to the transmitter is required in order to inform it that a decision has been reached.

2.3.1 *Advantages of sequential detection*

Bussgang and Middleton (1955) have shown that a saving of 40 to 90 percent in the number of samples required to obtain the same error probabilities may be achieved with sequential detection when compared with fixed sample size Bayes' tests. This may lead to an appreciable reduction of search time since it automatically adjusts the dwell-time on the individual cells so as to permit the integration of only an adequate number of pulses to yield a specified detection and false alarm probability.

Chapter 3

Radar detection theory

3.1 Radar and communication

Communication theory avails itself freely of tools of mathematical statistics amongst others those of statistical decision theory. If then the radar problem can be reduced to that of communication, the techniques used for solving problems in communication may be applied to radar.

Although the majority of radars uses a common aerial for both the transmission and the reception, there are radars, the so-called bistatic radars, which use separate aerials for the above functions. If they are situated so as not to interfere with each other, the received signal is due only to reflections from targets within the volume of space illuminated by the radar. With this modification the radar problem becomes a problem of communication where, however, the method of imparting information to the carrier differs. While in communication the transmitted waveform contains information which may be modified by the medium of propagation, in radar the transmitted signal does not contain information on transmission but acquires it while propagating between the transmitter and the receiver.

Viewing the medium of propagation as a network, the received signal $f_2(t)$ may be expressed in the time domain in terms of the transmitted signal $f_1(t)$ and the network transfer function $h(t)$ by

$$f_2(t) = f_1(t) * h(t).$$

Using the Fourier transform of the above quantities, the received waveform in the frequency domain becomes (Papoulis, 1962)

$$F_2(\omega) = F_1(\omega) \,.\, H(\omega).$$

Since corresponding terms of the above two expressions occupy identical places it is permissible, for theoretical considerations, to interchange in both the positions of the right-hand side terms and consider the network transfer function to be the transmitted waveform containing the information. This reduces the radar problem to almost that of communication. The remaining difference is the *a priori* unknown amplitude, phase and time of arrival of radar signals as against the communication problem of an *a priori* unknown amplitude and phase only.

Statistical decision theory formulates a decision rule that uses the received data to decide what hypothesis of all available hypotheses is closest to the truth.

3.2 Decision theory

Radar detection is a binary decision problem in which the receiver makes a 'signal' or 'no signal' decision on the basis of input data about which there is an uncertainty of whether they consist of signal plus noise or noise alone.

Denoting the signal by $s(t)$ and the noise by $n(t)$ and supposing that these are additive, the received function is described by

$$x(t) = s(t) + n(t). \tag{3.1}$$

However, there is a possibility that no signal is present in which case the function $x(t)$ is noise alone, i.e.

$$x(t) = n(t). \tag{3.2}$$

Since noise is a random quantity, the received function will be random regardless of whether the signal $s(t)$ is random or not.

Basically, the receiver decides that a signal is present when the input to the decision device exceeds a given level x_0 which is usually referred to as 'threshold' or 'bias'.

Assuming that $x(t)$ may acquire any value within a given range of values, the threshold divides this range into two regions, one where hypothesis H_0 is valid and the other one for which hypothesis H_1 is valid. Since both of the above functions, eqns. (3.1) and (3.2), are random, they are defined by their probability density functions. These extend beyond the threshold x_0.

Accepting for the present for simplicity's sake a normal distribution, the probability distribution function of $x(t)$, defined by eqn. (3.2), is

$$p_0(x) = \frac{\exp(-x^2/2N)}{\sqrt{2\pi N}} \tag{3.3}$$

while that of eqn. (3.1) is

$$p_1(x) = \frac{\exp(-(x-S)^2/2N)}{\sqrt{2\pi N}}, \tag{3.4}$$

where N is the mean squared noise voltage and S is the amplitude of the signal (Fig. 3.1).

Obviously, there are two types of error which the receiver may make. It may decide that the function $x(t)$ consists of noise alone when, in reality, it consists of signal plus noise. In the radar case this would mean missing a target. Further, the receiver may decide that a signal is present in the received waveform when, in reality, there is no target return. This corresponds to a false alarm.

The areas representing these errors are those below the two probability distribution functions beyond the boundary at x_0.

Let the *a priori* probabilities in regions H_0 and H_1 be P_0 and P_1 respectively. The probability P_0 pertains to the correct decision of $x(t) = n(t)$ and to the error of missing a target, while P_1 is valid for those events which are decided upon by the application of hypothesis H_1, namely deciding correctly that $x(t) = n(t) + s(t)$ and the erroneous decision leading to a false alarm. The total probability, the sum of probabilities P_0 and P_1, must be equal to unity, therefore

$$P_1 = 1 - P_0$$

as it will be used below.

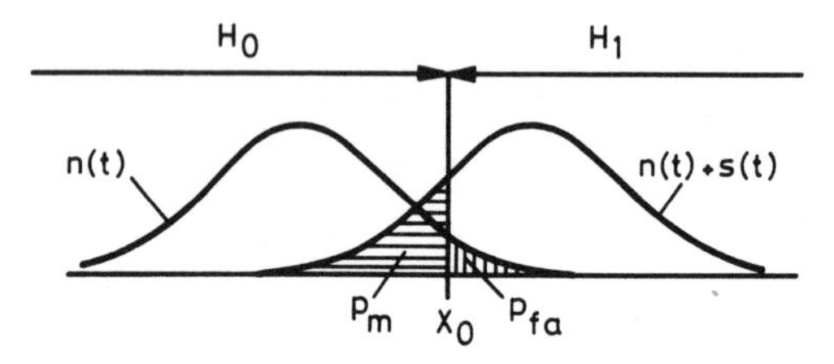

Fig. 3.1 *The probability distribution of noise and signal plus noise*

Decision theory is concerned then with finding a criterion for determining the threshold x_0. There are several approaches to the problem. Some of these assign costs and *a priori* probabilities to the four possible decisions.

One of the frequently used methods of decision theory for solving the problem of correctly defining the decision threshold is the Bayes criterion, the aim of which is to minimize the, so called, risk function (Helstrom, 1968). Each decision is assigned a cost and an *a priori* probability of occurrence. Calling C_{11} the cost of the correct decision about the received function consisting of signal plus noise, C_{00} the cost of the other correct decision, while C_m is the cost of missing the target and C_{fa} the cost of a false alarm, the average total cost may be expressed by

$$\bar{C} = (1 - P_0)\left[\int_{x_0}^{\infty} p_1(x)dx\, C_{11} + C_m\left(1 - \int_{-\infty}^{x_0} p_1(x)dx\right)\right] +$$
$$P_0\left[\left(1 - \int_{-\infty}^{x_0} p_0(x)dx\right)C_{00} + C_{fa}\int_{x_0}^{\infty} p_0(x)dx\right]. \tag{3.5}$$

The expressions in the square brackets are the risks connected with hypotheses H_1 resp. H_0. The minimum of eqn. (3.5) is obtained by differentiating it with respect to x and equating it to zero. This leads to

$$\frac{d\bar{C}}{dx} = (1 - P_0)[C_m - C_{11}]\,p_1(x) + P_0(C_{fa} - C_{00})p_0(x) = 0$$

from which the ratio

$$\Lambda(x) = \frac{p_1(x)}{p_0(x)} = \frac{P_0(C_{fa} - C_{00})}{(1 - P_0)(C_m - C_{11})}, \tag{3.6}$$

the, so-called, likelihood ratio may be obtained. The minimum value of the average risk is the Bayes risk. The ratio of the product of probabilities and costs, the right-hand side of eqn. (3.6), basically a cost ratio, is the threshold.

Since the above criterion depends only on the increase in costs due to erroneous decisions, the correct decisions may be arbitrarily assigned zero costs and one may minimize the risk function

$$R = C_m P_m + C_{fa} P_{fa}, \tag{3.7}$$

where P_m and P_{fa} are the *a priori* probabilities of miss and false alarm.

The application of Bayes' strategy requires a large number of observations and knowledge of the *a priori* probabilities of their occurrence.

If the probability of false alarm is known, the observer who determines the detection threshold by minimizing the total error given by eqn. (3.7) when $C_{fa} = C_m = 1$ is referred to as the 'ideal observer'.

In the case that the observer knows the four costs required for the application of Bayes' criterion but he is unaware of the *a priori* probabilities, the minimax criterion, which minimizes the risk for that value of the *a priori* probability P_0 for which the Bayes risk is maximum, may be used.

The Neyman–Pearson criterion guarantees that the expected frequency of occurrence of the more costly error will not exceed a preset quantity. It is the criterion most suited to radar applications where the *a priori* conditional probabilities of false alarm $P_{fa} = P(S|\sqrt{N})$, i.e. the probability that the observer decides erroneously that a target is present when there is only noise, and the probability of miss $P_m = P(\sqrt{N}|S)$, i.e. the probability of missing a target by attributing a target indication to noise only. The Neyman–Pearson criterion maximises the detection probability (or minimizes the probability of miss) while keeping the false alarm rate below a prescribed level.

On the basis of probability theory the above probabilities may be written in terms of conditional probabilities as

$$\begin{aligned} P_{fa} &= P(S|\sqrt{N}) = \int_s p(S|\sqrt{N})\,dS, \\ P_m &= P(\sqrt{N}|S) = \int_{\sqrt{N}} p(\sqrt{N}|S)\,d\sqrt{N}, \end{aligned} \tag{3.8}$$

where $p(S|\sqrt{N})$ and $p(\sqrt{N}|S)$ are the appropriate probability density functions.

Since the probability of false alarm and the probability of miss are dependent on each other, it is obvious that the larger the false alarm probability the lower the probability of miss so that the probability of detection increases with an increasing probability of false alarm.

Choosing the permissible probability of false alarm to be equal to a value ϵ and changing the variable of integration from S to Λ the conditions set by this criterion are

$$P_{fa} = \int_{\Lambda_0}^{\infty} p(\Lambda|\sqrt{N})\,d\Lambda = \epsilon$$

and

$$P_m = \int_0^{\Lambda_0} p(\Lambda|S)\,d\Lambda = \text{minimum.} \tag{3.9}$$

Λ_0 is the critical likelihood ratio from which the detection threshold may be determined.

3.3 Detection, detection threshold and the probability of false alarm

3.3.1 Detection and integration of pulses

In what follows, the term detector will signify any device the instantaneous output of which is a function of the envelope of the input signal only. Marcum (1948) found little difference between the performance of linear and square-law detectors. His findings will be given later. Radar receivers predominantly use square-law detectors.

It has been mentioned earlier that the pulse repetition frequency, i.e. the number of transmitted pulses per second, and the scanning rate of the aerial are chosen so as to permit more than one pulse intercepting and being returned by the target in order to improve detection. The received pulses are then integrated. Detection is achieved when the envelope of the integrated waveform exceeds a threshold.

Integration may be performed in different parts of the radar receiver. If the pulses passed through a matched filter in the intermediate frequency (IF) part of the receiver are integrated before the second detector, the integration is referred to as coherent or predetection integration.

For achieving the best possible performance of a predetection integrator the phase of the received signals must be preserved. Since signal returns arrive and are added at more or less regular intervals, the process of integration may be considered in terms of sampling of a waveform consisting of a periodic signal embedded in noise. The period of sampling is usually such that the noise samples are independent of each other. The signal has approximately the same value from sample to sample and does not contribute therefore to the variance of the sample mean. Only the variance of the random component, noise in the radar case, has to be considered. The sum of k samples of the variance N of noise becomes kN, i.e. it is equal to k-times the power of a single sample.

If S is the r.m.s. value of the signal voltage, then kS is the sum of k signal voltage samples so that the signal to noise power ratio becomes

$$\frac{(kS)^2}{kN} = k\,\frac{S^2}{N}.$$

Integrating k returns of equal signal to noise ratio in an ideal predetection integrator yields a k-times higher signal to noise ratio than that of a single return.

A low-pass filter, an $R-C$ network, having at least half the predetection matched filter bandwidth, constitutes the simplest form of postdetection or noncoherent integration. This type of integration is easier to implement and is therefore used in most of the current radars.

The improvement of postdetection integration compared with ideal predetection integration, for the case when all pulses are equal, has been determined by Marcum (1948) as

$$kE(k) = \frac{SR_1}{SR_k},$$

where SR_1 is the single pulse signal to noise ratio required for a specific probability of detection and SR_k is the value of signal to noise ratio per pulse for the same probability of detection when k pulses are integrated. The quantity $kE(k)$ is usually called the integration improvement factor. It indicates the improvement relative to that achievable by an ideal coherent integrator.

The most common method of integration takes advantage of the persistency of the cathode-ray tube screen's phosphor together with the integrating properties of the operator's eye–brain combination.

Experimental evidence gained during the Second World War shows that the improvement factor of this type of integration is proportional to $\sqrt{k}$ (Lawson and Uhlenbeck, 1950).

3.3.2 Detection threshold and the probability of false alarm

Inserting the probabiliy density functions (3.3) and (3.4) into eqn. (3.6) for the likelihood ratio, one obtains

$$\Lambda = \exp(xS/N - S^2/2N) \tag{3.10}$$

which when solved for x yields

$$x = \frac{N}{S} \ln \Lambda + \frac{S}{2}. \tag{3.11}$$

Choosing this to be the value $x = x_0$ and assuming that the costs of both errors are equal, the whole decision surface is divided by

$$x_0 = Y_b = \frac{N}{S} \ln \frac{P_0}{P_1} + \frac{S}{2} \tag{3.12}$$

into two mutually exclusive regions which may be described by $x \gtrless Y_b$ region where hypothesis $\begin{Bmatrix} H_1 \\ H_0 \end{Bmatrix}$ is valid.

Both error probabilities, i.e. the probability of false alarm and the probability of miss, are functions of the detector input signal to noise ratio. Curves of average risk as a function of this ratio are decision curves in terms of which specific system comparisons may be made (Middleton, 1960).

Only one sample was assumed so far but, as it was pointed out earlier, there are several pulses reaching the target and being reflected by it during each aerial sweep. Assuming that the pulse spacing is such that the noise is independent from pulse to pulse so that the probability distributions for a miss and false alarm of the sequence of pulses are given by the product of the distributions applicable to the individual pulses, the likelihood ratio becomes

$$\Lambda = \frac{p_1(x_1)p_1(x_2)\ldots p_1(x_k)}{p_0(x_1)p_0(x_2)\ldots p_0(x_k)} > \frac{P_0}{P_1}, \tag{3.13}$$

a ratio of joint probability functions of the kth order.

The previously accepted normally distributed densities for the voltage samples x_i in regions H_0 and H_1 are

$$p_0(x_i) = \frac{\exp(-(x_i - S_0)^2/2N)}{\sqrt{2\pi N}}, \quad i = 1, 2, \ldots, k \tag{3.14}$$

and

$$p_1(x_i) = \frac{\exp(-(x_i - S_1)^2/2N)}{\sqrt{2\pi N}},$$

where, for generality's sake, signals S_0 and S_1 pertaining to regions H_0 resp. H_1 were introduced (Schwartz, 1965).

The likelihood criterion, using the approach outlined earlier, becomes

$$\sum_{i=1}^{k} x_i = Y_b = k\,\frac{S_0 + S_1}{2} + \frac{N}{S_1 - S_0}\ln\frac{P_0}{P_1}, \tag{3.15}$$

where again Y_b is the threshold.

The decision is for region H_1 when the sum of the x_i samples exceeds the threshold and vice versa.

Since the threshold has been determined it is possible to define the limits of the integrals for the two error probabilities. Accepting the previous normal probability distributions, the error probabilities are

$$P_{fa} = \int_{Y}^{\infty} \frac{\exp(-x^2/2N)}{\sqrt{2\pi N}}\,dx \tag{3.16}$$

and

$$P_m = \int_{-\infty}^{Y} \frac{\exp(-(x-S)^2)/2N}{\sqrt{2\pi N}}\,dx, \tag{3.17}$$

which, with a change of variables, may be written as

$$P_{fa} = \frac{1}{\sqrt{\pi}} \int_{Y/\sqrt{2N}}^{\infty} \exp(-z^2)dz \qquad (3.18)$$

and

$$P_m = \frac{1}{\sqrt{\pi}} \int_{-\infty}^{(Y-S)/\sqrt{2N}} \exp(-z^2)dz. \qquad (3.19)$$

It is interesting to note that in case of k samples the total signal voltage is kS but the total noise voltage is only $\sqrt{\Sigma_{i=1}^{k} N_i} = \sqrt{kN}$ since, as it was pointed out before, the mean squared noise voltage was used in the above expressions. The voltage signal to noise ratio for k samples is then $\sqrt{k}\,S/\sqrt{N}$.

Since radar receivers have a limited bandwidth, the noise at the output of the i.f. amplifier is not white but band-limited; it has, according to Rice (1944, 1945), a Rayleigh distribution given by

$$q_k(x) = \frac{x}{N} \exp(-x^2/2N), \qquad (3.20)$$

where x is the noise voltage envelope and N, as before, is the mean squared average noise voltage.

The probability of false alarm, i.e. the probability that noise will exceed the preset detection threshold Y_b is

$$P_{fa} = \int_{Y_b}^{\infty} q_k(x)dx = \exp(-Y_b^2/2N) \qquad (3.21)$$

from which the bias level for specified values for the probability of false alarm may be determined.

In radar practice the signal returns during a sweep of the aerial are passed through a matched filter and sampled at the instant when they are expected to reach their maximum. They are then fed into an integrator the output of which feeds the decision device. The sum of a specified number of pulses has to exceed the threshold in order to effect a detection. It can happen, however, that noise pulses will add and by exceeding the threshold cause a false alarm.

The moments, i.e. the mean value, the mean square value and the variance of a random variable X, may be obtained by differentiating its characteristic function

$$C_x(\omega) = E(e^{j\omega x})$$

$$= \int_{-\infty}^{\infty} p(x)\, e^{j\omega x}\, dx$$

(Beckmann, 1967). The above integral expression is the Fourier transform of the probability density function $p(x)$. Similarly, the inverse Fourier transform, the anticharacteristic function, determines the probability density function as

$$p(x) = \frac{1}{2\pi} \int_{-\infty}^{\infty} C_x(\omega)\, e^{-j\omega x}\, d\omega.$$

The Fourier transform pair is uniquely related and therefore the characteristic function uniquely defines the probability density function and vice versa.

The characteristic and anticharacteristic function pairs of many probability distributions have been tabulated by Campbell and Foster (1948).

Marcum (1948) has shown that the characteristic function of k variates of signal plus noise for a square law detector is

$$C_k = \frac{\exp(-kx)}{(p+1)^k} \exp(kx/(p+1)) \tag{3.22}$$

from which, by using the Fourier transform pair 650·0 of Campbell and Foster (1948),

$$F(f) = \frac{1}{(p+p)^\alpha} \exp\left[\frac{1}{\lambda(\rho+p)}\right];$$

$$G(g) = (\lambda g)^{(\alpha-1)/2}\, e^{-\rho g}\, I_{\alpha-1}\left(2\frac{g}{\lambda}\right), \quad g > 0 \tag{3.23}$$

he obtained, for a zero power signal-to-noise ratio, the probability density function of noise alone

$$dP_{fa} = \frac{Y^{k-1}\, e^{-Y}}{(k-1)!} dY. \tag{3.24}$$

x in eqn. (3.22) is the signal-to-noise power ratio and $p = j\omega$ is a transformation variable. The integral of this expression

$$P_{fa} = \int_{Y_b}^{\infty} \frac{Y^{k-1}\, e^{-Y}}{(k-1)!} dY \tag{3.25}$$

is an incomplete Gamma-function defined by Pearson (1946) as

$$I(u,k) = \int_0^{u\sqrt{k+1}} \frac{v^k \exp(-v)}{k!} dv. \tag{3.26}$$

Using Pearsons notation, eqn. (3.25) becomes

$$P_{fa} = 1 - I\left(\frac{Y_b}{\sqrt{k}}, k-1\right). \tag{3.27}$$

In the above expressions Y_b is the bias level, Y the instantaneous noise amplitude at the output of a square law detector normalized to the r.m.s. video noise voltage and k is the number of integrated pulses.

For k up to 50 and false alarm numbers* up to 10^6 the bias is obtainable from Pearson's tables. It may be obtained from Pachares' tables (1958) for up to 150 pulses integrated and false alarm probabilities down to 10^{-12}.

Marcum solved eqn. (3.25) by successive integration by parts and by treating the so obtained finite series as an infinite one. His solution is

$$P_{fa} = \frac{k Y_b^k \exp(-Y_b)}{k!(Y_b - k + 1)}, \tag{3.29}$$

which, with Stirling's approximation for the factorial term (Merritt, 1962)

$$k! \approx k^k \sqrt{2\pi k} \exp(-k) \tag{3.30}$$

becomes

$$P_{fa} \cong \sqrt{\frac{k}{2\pi}} \frac{\exp[-Y_b + k(1 + \ln(Y_b/k))]}{Y_b - k + 1}. \tag{3.31}$$

Substituting for P_{fa} Marcum's eqn. (21)

$$P_{fa} \cong \frac{k}{n_{fa}} \ln \frac{1}{P_0}, \tag{3.32}$$

where n_{fa} is the previously mentioned false alarm number and P_0 the probability that noise alone will not exceed the bias level, one arrives, with $P_0 = 0{\cdot}5$, at

$$\begin{aligned} \log n_{fa} &= 0{\cdot}24 + 0{\cdot}5 \log k + \log(Y_b - k + 1) \\ &\quad + 0{\cdot}434(Y_b - k) - k \log(Y_b/k) \end{aligned} \tag{3.33}$$

from which the bias Y_b may be determined.

Now that the theoretical aspects of bias level determination have been dealt with a few words on the practical way of threshold setting are in place.

While threshold setting can be made by an operator viewing the PPI (plan position indicator) display, it is obvious that it will differ not only from operator to operator but, depending on the operator's alertness and motivation, also from one instant to another for the same operator. The capability of the human operator as part of the radar detection process was studied experimentally (Lawson and Uhlenbeck, 1950; Baker, 1962). It was found that an operator's threshold is

*The false alarm time τ_{fa} is the average time interval between false alarms. In a pass-band of B-Hz there may be B independent noise samples applied to the threshold per second. The number of opportunities for the occurrence of a false alarm in the false alarm time is the false alarm number

$$n_{fa} = B\tau_{fa} \tag{3.28}$$

Its reciprocal is the probability of false alarm.

difficult to predict since it does not remain fixed. For this reason more modern radars use electronic threshold setting.

It was mentioned that the output of a matched filter is fed into a square law detector which is usually followed by a linear noncoherent integrator the output of which is taken to the threshold device. The reference input of this device comes from a r.m.s. noise measuring circuit fed by the matched filter and followed by a multiplier in which the measured noise is multiplied by a factor to obtain the bias level. The threshold device compares the integrated output of the detector with the bias level and if the detector output is larger than the bias a probable presence of a target is indicated. The probability of detection will be dealt with later.

In order to implement the Neyman–Pearson criterion which, as was mentioned earlier, requires a specified fixed false alarm probability, one has to measure the received noise power.

The probability of false alarm for a noncoherent receiver, i.e. one where integration occurs after the second detector, in an additive noise environment when the noise is assumed to have a Rayleigh distribution was given by eqn. (3.21). If the mean squared noise voltage N were known the threshold would be set at

$$Y = \sqrt{2N \ln \frac{1}{P_{fa}}}. \tag{3.34}$$

However, since N is not known with adequate accuracy, an estimate N', which is a maximum likelihood estimate of N obtained from k independent observations x_i, has to be used in the above equation (Rappaport, 1969). This estimate is defined by

$$N' = \frac{1}{2k} \sum_{i=1}^{k} x_i^2, \tag{3.35}$$

where the x_i terms are peak values, each with a Rayleigh distribution. Further, since, due to statistical variations of N', the specified probability of false alarm will not be achieved, the new threshold has to contain a factor u to be determined below.

The probability of false alarm is then

$$P_{fa} = E \int_{\sqrt{2N'u \ln(1/P_{fa})}}^{\infty} q_k(x)\, dx \tag{3.36}$$

with E the mathematical expectations with respect to the ramdon variable N'. Inserting eqn. (3.20) into eqn. (3.36), the probability of false alarm becomes

$$P_{fa} = E \exp\left[-\frac{N'}{N} u \ln \frac{1}{P_{fa}}\right]. \tag{3.37}$$

Remembering that N and N' denote noise power, setting their ratio N'/N equal to $X/2k$ shows that X is a chi-square variate with $2k$ degrees of freedom so that the probability density function of X is given by (Kapur and Saxena, 1972)

$$p_{2k}(X) = \frac{1}{2^k(k-1)!} X^{k-1} \exp(-X/2), \quad X \geqslant 0. \tag{3.38}$$

Using this expression for the expectation in eqn. (3.37) yields

$$P_{fa} = \frac{1}{2^k(k-1)!} \int_0^\infty X^{k-1} \exp\left[-\left(\frac{X}{2}\left(1 + \frac{u}{k}\ln\frac{1}{P_{fa}}\right)\right)\right] dX \tag{3.39}$$

where the integral, with the substitution

$$\xi = \frac{X}{2}\left(1 + \frac{u}{k}\ln\frac{1}{P_{fa}}\right),$$

is a Gamma-function $\Gamma(k)$ equal to $(k-1)!$ causing

$$P_{fa} = \frac{1}{\left(1 + \frac{u}{k}\ln\frac{1}{P_{fa}}\right)^k}$$

from where

$$u = \frac{k(1 - P_{fa}^{1/k})}{P_{fa}^{1/k}\ln\frac{1}{P_{fa}}}. \tag{3.40}$$

Dividing both, the numerator and denominator, by $P_{fa}^{1/k}$ one obtains

$$u = \frac{k(1 - P_{fa}^{-1/k})}{\ln P_{fa}}. \tag{3.41}$$

In order to achieve the specified false alarm probability the threshold given by eqn. (3.34) with N replaced by uN' is to be used for the electronic setting of the bias.

3.4 False alarm time

Some authors prefer using false alarm time instead of false alarm probability. The definition of false alarm time was given in the footnote following eqn. (3.27).

Hollis (1954) pointed out that two different definitions of false alarm time can be found in the relevant literature. Kaplan and McFall (1951) define false alarm time as the average time interval between false target indications, while

Marcum (1960) calls false alarm time the time in which the probability of noise not exceeding the bias level is 0·5.

Using the first definition and determining the probability that one false alarm will occur within the false alarm time, by writing the Poisson distribution (Lee, 1964)

$$P(k;\tau) = \frac{(m\tau)^k \exp(-m\tau)}{k!}, \tag{3.42}$$

where $P(k;\ \tau)$ is the probability of k successes and m the average or expected number of successes in the interval τ, one arrives, with $m\tau = 1$ in accordance with the definition, at a probability $\exp(-1) = 0{\cdot}368$ for one false alarm occurring within the false alarm time.

Using Marcum's definition, the probability $P(k;\ \tau) = \frac{1}{2} = \exp(-m\tau)$, from where the average number of false alarms occurring within the considered time interval is $m\tau = 0{\cdot}693$ and the probability of one false alarm within this interval is 0·346.

The relation of the probabilities of false alarm corresponding to the two definitions, considering that Kaplan–McFall's mean is unity while Marcum's is 0·693, may be expressed as

$$\frac{\text{Kaplan–McFall}}{\text{Marcum}} \approx \frac{0{\cdot}736}{0{\cdot}693^k}. \tag{3.43}$$

The false alarm time defined by the first of the two definitions is 44% longer than that defined by the second one although the probabilities of one false alarm within their respective alarm times differ by 6% only.

It should be mentioned that Marcum's definition of the false alarm time seems to have been more widely accepted.

Since the sum of probabilities of the noise not exceeding the bias and that of its exceeding it is unity, the latter is also 0·5 and Marcum's determination of the bias level is identical with that used in a binary communication system except for considering noise alone instead of signal plus noise.

If more than one target return is received during one sweep of the aerial, as is usually the case, the definition of false alarm probability has to take the increased number of threshold crossings into consideration as shown above.

Out of n_{fa} possible decisions in the false alarm time τ_{fa} there will be, on average, one false decision (Skolnik, 1962). The average number of possible decisions n_{fa} between false alarms is usually referred to as the false alarm number and, in the time τ_{fa}, it is equal to the number of range intervals per pulse period $1/f_r\Delta t = B/f_r$ multiplied by the pulse repetition frequency f_r and the false alarm time τ_{fa} or $n_{fa} = B\tau_{fa}$ as in eqn. (3.28).

Uncorrelated or only partly correlated signals are required for range determination. Two signals separated by the pulse width are uncorrelated. Since there is a fixed relation between pulse width and bandwidth in a radar system, the range interval, i.e. the interval over which data may be accepted, may be taken to be inversely proportional to the bandwidth. The probability of false alarm is the reciprocal of the false alarm number and hence $P_{fa} = 1/(B\tau_{fa})$, as before. If k pulses are integrated in order to improve detection, the number of independent decisions in the false alarm time is reduced k-times and the probability of false alarm increased k-times so that it becomes $P'_{fa} = k/n_{fa}$.

3.5 Probability of detection

The other important probability, namely that appertaining to the correct decision that a signal is present when it is present for which therefore $x > x_0 = Y_b$ and decision hypothesis H_1 applies, is the probability of detection. It depends obviously on the signal to noise ratio which, in the radar case, depends to a great extent on the nature of the target and on its behaviour.

A steady, non-fluctuating, target echoing area was generally accepted for radar detection range calculations in the early days of radar. In fact, the now classical paper of Marcum's (1948–1960) is based on this assumption. This type of target would yield target echos of constant amplitude which, as was shown in practice, is valid only for a very limited range of targets. Since real targets, e.g. aircraft and ships, consist of a large complex structure with smaller features which, however, are large when compared with the wavelength, multiple reflections of the impinging electromagnetic radiation are possible. These combine on reradiation with different phases due to which the reradiated energy may be larger or smaller than that which could be assumed for a non-fluctuating target of comparable size.

Swerling (1954) introduced two groups of targets by using two different probability distribution functions with two different degrees of correlation. They were proved to resemble the behaviour of radar cross-sections of real targets.

Swerling's first group, cases I and II, consider target cross-sections to have a Rayleigh probability density function which, using Swerling's notation, is given by

$$w(x, \bar{x}) = \frac{1}{\bar{x}} \exp(-x/\bar{x}), \quad x \geqslant 0, \tag{3.44}$$

where x is the input signal to noise ratio, and $\bar{x}$ is the average of x over all target fluctuations.

In case I the returned signal power per pulse is assumed to have a constant amplitude during a single scan but to be fluctuating independently from scan to scan. In other words, the normalized autocorrelation function of the target cross-section is approximately unity during a single scan and it is approximately zero during the interval between scans. This case neglects the effect of the aerial's beam shape on the amplitude of the received signal during one sweep of the aerial.

The target echo in Swerling's case II fluctuates from pulse to pulse.

The above two types of target echo fluctuations are applicable to complex targets which are large compared with the wave length of the used radiation and may be represented by several independently fluctuating reflectors of approximately equal cross-sections. This type of fluctuation would describe aircrafts and ships.

The chi-square probability density

$$w(x, \bar{x}) = \frac{4x}{\bar{x}^2} \exp(-2x/\bar{x}), \quad x \geqslant 0, \tag{3.45}$$

with an appropriately selected number of degrees of freedom, is applicable to target cross-sections exhibiting scan to scan fluctuations of Swerling's case III and pulse to pulse fluctuations of case IV. These latter cases are used for targets represented by one large reflector and a number of small ones or by one large reflector subject to fairly small changes of orientation.

Cases I and III with scan-to-scan fluctuations are best suited to such targets as jet aircraft and missiles intercepted by radars using high pulse repetition frequencies and aerial scan rates.

Cases II and IV exhibiting pulse-to-pulse fluctuations are appropriate to propeller driven aircraft if the propeller contributes a large part of the echoing area or to such targets for which small changes in aspect angle cause large changes in the radar cross-section. These targets would include long thin objects observed by high frequency radars having a low pulse repetition frequency.

Most actual targets would be intermediate between the above cases.

The mathematical modelling of target cross-sections will be described later; the modifications to the signal to noise ratio caused by the above target echoing area fluctuations are of interest at the present.

3.5.1 The non-fluctuating target

Although single hit operation is found rarely in radar practice, the single hit probability of detection is the most often used criterion of system assessment. It considers the signal during one correlation interval, i.e. the duration of a single pulse only.

The expression for the appropriate probability density was obtained by Rice (1945) by considering the distribution of the envelope of the sum of two phasors, the signal in this case, and a rotating phasor representing the noise (Fig. 3.2), at the output of a linear filter having a bandpass which is narrow compared with its centre frequency. The instantaneous amplitude of the sum is

$$R = S \cos \omega t + N'. \tag{3.46}$$

The initial phase of the signal S is immaterial since it may be absorbed into the orientation of the coordinate system.

The noise phasor may be represented by its components.

Accepting that the orthogonal components $n' \cos \omega t$ and $n' \sin \omega t$ of the noise phasor are normally distributed about zero and have a variance N, then applying the transformation

$$p(r, \theta) = r p_{xy}(r \cos \theta, r \sin \theta) \tag{3.47}$$

(Wozencraft and Jacobs, 1965), the probability density of the envelope of a sine wave and random noise becomes

$$dP = \frac{R}{N} \exp\left(-\frac{R^2 + S^2}{2N}\right) I_0\left(\frac{RS}{N}\right) dR, \tag{3.48}$$

where $I_0(\ )$ is a modified zero order Bessel function of the first kind (Watson, 1922). Integrating eqn. (3.48) with respect to R within the limits Y_b and infinity yields the sought probability of detection.

The analytic solution of the integral of the probability distribution function is difficult; however, the Bessel function may be expanded into a series by

$$I_0(z) = \sum_{k=0}^{\infty} \frac{z^{2k}}{2^{2k} k!\, k!} (-1)^k.$$

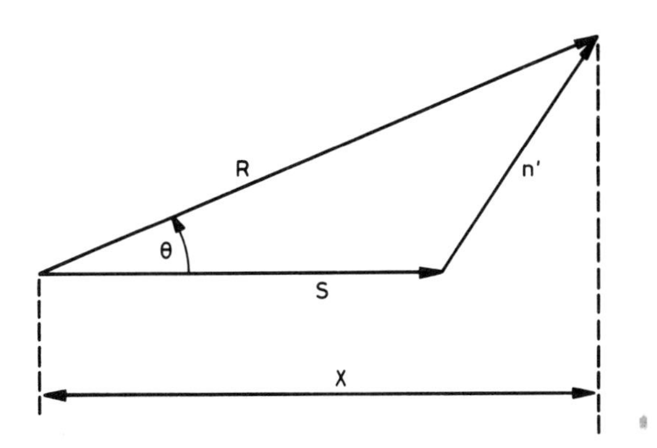

Fig. 3.2 *Signal and noise phasors*

Its values are tabulated (Jahnke and Emde, 1945). Rice computed the probability that the ratio $R/\sqrt{N}$ will not exceed a specified value of this ratio. His calculations were used to construct curves (Fig. 3.3) of the probability of detection as a function of the ratio of the bias level to noise with signal to noise as parameter (Hancock, 1961). Since the latter ratio determines the probability of false alarm (eqn. (3.34)), the probability of false alarm scale was added thus permitting the determination of the mean square signal-to-noise ratio for specified values of the probability of detection and false alarm. It may be used then for the calculation of the single hit detection range as will be demonstrated later.

Marcum (1960) derived the characteristic function for the sum of k equal amplitude pulses embedded in noise at the output of a square law detector to be

$$C_k = \frac{\exp(-kx)}{(p+1)^k} \exp(kx/(p+1)) \tag{3.49}$$

as given by eqn. (3.22) above. From this he obtained the probability density function

$$\begin{aligned} dP_k &= (Y/kx)^{(k-1)/2} \exp(-Y-kx) I_{k-1}(2\sqrt{kxY})dY, \quad & Y > 0 \\ dP_k &= 0, & Y < 0 \end{aligned} \tag{3.50}$$

by using pair 650·0 (see eqn. (3.23)) of Campbell and Foster (1947).

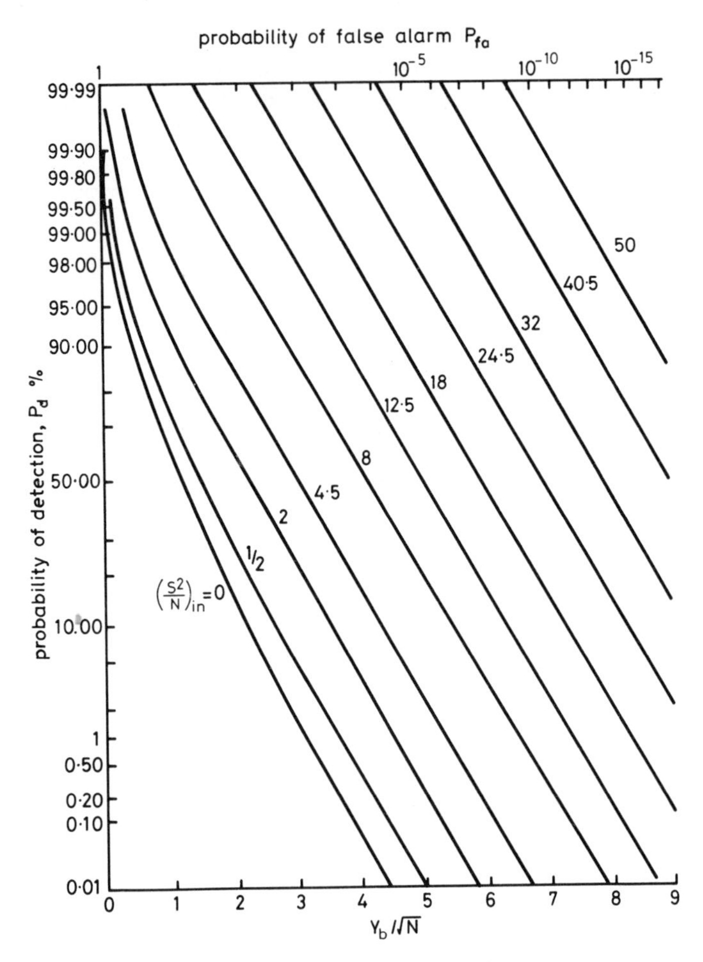

Fig. 3.3 *Probability of detection as a function of Y_b/N for S^2/N as parameter (After Hancock, 1961, modified by Rohan)*

Having established the bias level for the specified probability of false alarm, the probability of detection may be obtained by integrating eqn. (3.50) within the limits Y_b, i.e. the said bias level, and infinity. However, since this expression is not directly amenable to solution in terms of known functions, Marcum used

Campbell and Foster's pairs 210

$$F(f) = F/p, \qquad G(g) = \int_{-\infty}^{g} G\, dg = D_g^{-1} G,$$

where D_g^{-1} denotes integration with respect to g, and 655·1

$$F(f) = \frac{1}{p+\rho} \exp \frac{1}{\lambda(p+\rho)}, \qquad G(g) = e^{-\rho g} I_0 \left(\frac{2\sqrt{g}}{\sqrt{\lambda}} \right),$$

$$0 < g$$

together with the pair 650·0, given earlier by eqn. (3.23), to find the probability of detection as

$$P_k = 1 - \exp(-Y_b - kx) \sum_{r=k+1}^{r=\infty} (Y_b/kx)^{(r-1)/2} I_{r-1}(2\sqrt{kxY_b}). \tag{3.51}$$

Since, according to the central limit theorem of mathematical statistics, the sum of a great number of random variables approaches a normal distribution, Marcum overcame the difficulty of solving the above probability function numerically for large values of k by expanding it into a Gram–Charlier series (Fry, 1928) which is an orthogonal polynomial expansion of the form

$$f(y) = \sum_{i=0}^{\infty} c_i \phi^i(y),$$

where

$$\phi^0(y) = \phi(y) = \frac{1}{\sqrt{2\pi}} \exp(-y^2/2),$$

the first term of the expansion is a Gaussian density function and generally

$$\phi^i(y) = \frac{d^i \phi(y)}{dy^i}.$$

The coefficients

$$c_i = \frac{(-1)^i}{i!} \int_{-\infty}^{\infty} H_i(y) f(y)\, dy$$

are evaluated in terms of the Hermite polynomials $H_i(y)$ and $[H_0(y) = 1]$.

It should be noted here that according to Marcum's investigations there is very little difference between the performance of a linear and square law detector. The two are identical for $k = 1$. At $k = 10$ the linear detector exceeds the square law one by 0·11 dB to become equal again at $k = 70$ after which the square law detector becomes better by asymptotically exceeding the linear one by 0·19 dB as k tends to infinity having reached 0·16 dB at $k = 1000$.

3.5.2 Swerling's fluctuating targets

The nature of the four targets with fluctuating radar cross-sections has been described earlier, the following few lines will be devoted to the expressions for their probabilities of detection.

Assuming x, the input signal to noise power ratio, to be a random variable having the probability density function $w(x, \bar{x})$ but being constant for each group of k pulses and fluctuating independently from group to group, the characteristic function for the sum of k pulses becomes

$$\bar{C}(p) = \int_{-\infty}^{\infty} w(x, \bar{x})\, C(p)\, dx. \tag{3.52}$$

However, since $w(x, \bar{x}) = 0$ for $x < 0$, the lower limit of the above integral may be taken to be zero. $\bar{x}$, to use again Swerling's definition, is the average of x over all target fluctuations and $p = j\omega$ is a transformation variable.

3.5.2.1 Case I: For this case Swerling considered the amplitude of the whole pulse train returned by the target during one aerial sweep to be a single random variable with a distribution function given by eqn. (3.44).

Using then this Rayleigh probability density Swerling found the characteristic function of eqn. (3.52) to be

$$\bar{C}_k(p) = \frac{1}{(p+1)^{k-1}\,[1 + p(1 + k\bar{x})]}, \tag{3.53}$$

which, by using Campbell and Foster's pair 438

$$F(f) = \frac{1}{p + \beta}, \qquad G(g) = e^{-\beta g} \tag{3.54}$$

yields, for $k = 1$, the probability density function

$$p(v) = \frac{\exp\,(-v/(1 + \bar{x}))}{1 + \bar{x}}, \qquad v \geqslant 0, \tag{3.55}$$

where v is the sum of k normalized pulses of signal plus noise.

The probability of detection for this case is

$$P_d = \exp\,(-Y_b/(1 + \bar{x})). \tag{3.56}$$

For $k > 1$ the probability density function

$$p(v) = \left(1 + \frac{1}{k\bar{x}}\right)^{k-2} \frac{1}{k\bar{x}}\, I\left[\frac{v}{\left(1 + \frac{1}{k\bar{x}}\right)\sqrt{k-1}},\ k-2\right] \exp\,(-v/(1 + k\bar{x})) \tag{3.57}$$

was found by using Campbell and Foster's pair 581·7

$$F(f) = \frac{1}{(p+\beta)(p+\rho)^{\alpha-1}}$$

$$G(g) = \frac{\exp(-\beta g)\gamma[\alpha-1,(\rho-\beta)g]}{\Gamma(\alpha-1)(\rho-\beta)^{\alpha-1}}, \quad g > 0, \tag{3.58}$$

where

$$\gamma(\nu, z) = \Gamma(\nu) I\left[\frac{z}{\sqrt{\nu}}, \nu - 1\right]$$

and $I(\mu, k)$ is the Pearson (1946) notation for the incomplete gamma function.

The probability of detection is given by

$$P_d = 1 - I\left[\frac{Y_b}{\sqrt{k-1}}, k-2\right] + \left(1 + \frac{1}{k\bar{x}}\right)^{k-1}$$

$$I\left[\frac{Y_b}{\left(1+\frac{1}{k\bar{x}}\right)\sqrt{k-1}}, k-2\right] \exp(-Y_b/(1+k\bar{x})) \tag{3.59}$$

obtained from eqn. (3.52) by using Campbell and Foster's pair 101

$$F(f)\ G(g) = \int_{-\infty}^{\infty} F(f) \operatorname{cis}(2\pi fg)\, df, \tag{3.60}$$

with

$$\operatorname{cis}(Z) = \exp(jZ),$$

and pair 208

$$F(f) = pF, \qquad G(g) = D_g G, \tag{3.61}$$

where D_g means differentiation with respect to g, and pair 431

$$F(f) = \frac{1}{(p \pm \beta)^k}, \qquad G(g) = \frac{\pm 1}{(k-1)!} g^{k-1} e^{\mp\beta\rho}, \quad 0 < \pm g. \tag{3.62}$$

3.5.2.2 Case II: Swerling assumed for his case II target fluctuations the amplitude of each individual pulse during an aerial scan to be a statistically independent random variable having the probability density function (3.44).

For $k = 1$ the characteristic function (3.53) is

$$\bar{C}_1(p) = \frac{1}{1 + p(1+\bar{x})}, \tag{3.63}$$

which for $k > 1$, due to the multiplication theorem applicable to independent random variables, becomes

$$\bar{C}_k(p) = [\bar{C}_1(p)]^k = \frac{1}{[1 + p(1 - \bar{x})]^k}. \tag{3.64}$$

With the aid of Campbell and Foster's pair 431, given by eqn. (3.62), Swerling obtained the probability density function

$$f(v) = \frac{v^{k-1} \exp(-v/(1 + \bar{x}))}{(1 + \bar{x})^k (k-1)!} \tag{3.65}$$

from which the probability of detection

$$P_d = 1 - I\left[\frac{Y_b}{(1 + \bar{x})\sqrt{k}}, k - 1\right] \tag{3.66}$$

may be computed. Here again $I(\mu, k)$ is an incomplete gamma function.

3.5.2.3 Case III: The probability density function which Swerling attributed to his cases III and IV was given by eqn. (3.45). Inserting it into eqn. (3.52) yields the characteristic function

$$\bar{C}(p) = \frac{1}{(1+p)^{k-2}\left[1 + p\left(1 + \dfrac{k\bar{x}}{2}\right)\right]^2}, \tag{3.67}$$

from which Campbell and Foster's pair 201

$$F(f) = F_1 \pm F_2, \qquad G(g) = G_1 \pm G_2, \tag{3.68}$$

pair 442

$$F(f) = \frac{1}{(p \pm \beta)^2}, \qquad G(g) = \pm g\, e^{\mp \beta g} \tag{3.69}$$

and pair 449·5

$$F(f) = \frac{p}{(p \pm \beta)^2}, \qquad G(g) = (\pm 1 - \beta g)^{\mp \beta g}, \quad 0 < \pm g \tag{3.70}$$

permits to determine the probability density function for $k = 1$ as

$$f(v) = \frac{1}{\left(1 + \dfrac{\bar{x}}{2}\right)^2}\left[1 + \frac{v}{1 + \dfrac{2}{\bar{x}}}\right] \exp\left(-v\Big/\left(1 + \frac{\bar{x}}{2}\right)\right), \tag{3.71}$$

while for $k = 2$

$$f(v) = \frac{v}{(1+\bar{x})^2} \exp(-v/(1+\bar{x})). \tag{3.72}$$

With pair 581·1

$$F(f) = \frac{1}{(p+\rho)^{\alpha+\nu}(p+\sigma)^{\alpha-\nu}},$$

$$G(g) = \frac{g^{\alpha-1}\exp\left(-\dfrac{g(\rho+\sigma)}{2}\right)}{\Gamma(2\alpha)(\rho-\sigma)^{\alpha}} M_{\nu,\alpha-1/2}[(\rho-\sigma)g], \quad 0 < g. \tag{3.73}$$

after appropriate transformation Swerling arrived at

$$f(v) = \frac{v^{k-1}e^{-v}}{(k-1)!\left(1+\dfrac{k\bar{x}}{2}\right)^2} {}_1F_1\left[2, k, \frac{v}{1+\dfrac{2}{k\bar{x}}}\right] \tag{3.74}$$

as the probability density function for $k > 2$. The function

$${}_1F_1[a;b;z] = 1 + \frac{az}{b1!} + \frac{a(a+1)}{b(b+1)}\frac{z^2}{2!}\cdots$$

is a confluent hypergeometric function (Irving and Mullineux, 1959) and

$$M_{\mu,\nu}(z) = z^{1/2+\nu} e^{-1/2z} {}_1F_1(\tfrac{1}{2}+\nu-\mu; 2\nu+1; z)$$

is a Whittaker function (Slater, 1960).

With the simplifying assumptions that the false alarm probability in radar practice is, as a rule, much less than one and for reasonable detection probabilities $k\bar{x}/2 > 1$, for $v > Y_b$ the probability of detection for case III targets, for all values of k, is given by Swerling as

$$P_d = \left(1+\frac{2}{k\bar{x}}\right)^{k-2}\left[1+\frac{Y_b}{1+\dfrac{k\bar{x}}{2}} - \frac{2(k-2)}{k\bar{x}}\right]\exp\left(\frac{-Y_b}{1+\dfrac{k\bar{x}}{2}}\right) \tag{3.75}$$

3.5.2.4 Case IV: The difference between Swerling's cases III and IV is the same as that between cases I and II, i.e. while in case III the whole pulse train is accepted to be a single random variable, in case IV individual pulses in the train are taken to be independent random variables. The probability density function and therefore also the characteristic function is common to both latter cases.

For $k = 1$ eqn. (3.67) becomes

$$\bar{C}_1(p) = \frac{1+p}{\left[1+p\left(1+\frac{\bar{x}}{2}\right)\right]^2} \tag{3.76}$$

and for $k > 1$, since in this case the pulses are independent of each other, the characteristic function is the kth power of eqn. (3.76), or

$$\bar{C}_k(p) = \frac{(1+p)^k}{\left[1+p\left(1+\frac{\bar{x}}{2}\right)\right]^{2k}}. \tag{3.77}$$

The exact probability density function contains a confluent hypergeometric function

$${}_1F_1\left(-k, k; \frac{-\bar{x}/3}{1+\bar{x}/3} Y\right)$$

the higher terms of the series expansion of which are identically zero so that the expression for the above probability density becomes

$$p(Y) = \frac{Y^{k-1} k! \exp\left(-\frac{Y}{1+\bar{x}/4}\right)}{(1+\bar{x}/4)^{2k}} \sum_{m=0}^{k} \left(\frac{\bar{x}/4}{1+\bar{x}/4}\right)^m \frac{Y^m}{m!(k+m-1)!(k-m)!}.$$

DiFranco and Rubin (1968) give

$$P_d = 1 - \frac{k!}{\left(1+\frac{\bar{x}}{2}\right)^k} \sum_{m=0}^{k} \left(\frac{\bar{x}}{2}\right)^m \frac{I\left[\frac{Y_b}{(1+\bar{x}/2)\sqrt{k+m}}, k+m-1\right]}{m!(k-m)!} \tag{3.78}$$

for the probability of detection. The expression $I(u, s)$ is an incomplete gamma function defined as

$$I(u, s) = \int_0^{u\sqrt{1+s}} \frac{e^{-v} v^s}{s!} dv.$$

3.6 Computational methods

3.6.1 Numerical computations

Fehlner (1962) used Marcum's and Swerling's results to compute the probability of detection as a function of signal to noise ratio with false alarm number as parameter for various values of the number of integrated pulses k and present them in practically applicable graphs.

Since some of the expressions for the probability of detection and bias level given by the above two authors are not suitable for machine computation in their original form, Roll and Trotter (see Fehlner, 1962) used a recursion relation to express the incomplete gamma functions as series which may be evaluated numerically to any desired precision depending on the allocated time.

Integrating the incomplete gamma function by parts

$$f(a,p) = \int_0^a \frac{e^{-u}\, u^p}{p!}\, du = 1 - \int_a^\infty \frac{e^{-u}\, u^p}{p!}\, du \tag{3.79}$$

yields

$$f(a,p) = -\frac{e^{-a}\, a^p}{p!} + \int_0^a \frac{e^{-u}\, u^{p-1}}{(p-1)!}\, du,$$

$$f(a,p) = -\frac{e^{-a}\, a^p}{p!} - \frac{e^{-a}\, a^{p-1}}{(p-1)!} - \ldots + 1 - e^{-a},$$

which may be expressed as

$$f(a,p) = 1 - \sum_{k=0}^{p} \frac{e^{-a}\, a^k}{k!}, \quad a > p$$

$$= \sum_{k=p+1}^{\infty} \frac{e^{-a}\, a^k}{k!} \qquad a < p. \tag{3.80}$$

The probability density function for the non-fluctuating target may be obtained by contour integration of the characteristic function (3.49) in the form

$$f(v) = \frac{1}{2\pi j} \int_{-j\infty}^{j\infty} \frac{e^{-kx}\, e^{kx/(p+1)}\, e^{(p+1-1)v}}{(p+1)^k}\, dp$$

$$= e^{-kx} \sum_{m=0}^{\infty} \frac{(kx)^m}{m!} \frac{v^{k-1+m}}{(k-1+m)!}\,. \tag{3.81}$$

The probability distribution then becomes

$$p(x, Y_b) = e^{-kx} \sum_{m=0}^{\infty} \frac{(kx)^m}{m!} \sum_{n=0}^{k-1+m} \frac{e^{-Y_b} Y_b^n}{n!},$$

$$Y_b > k(x+1) \tag{3.82}$$

$$= 1 - e^{-kx} \sum_{m=0}^{\infty} \frac{(kx)^m}{m!} \sum_{n=k+m}^{\infty} \frac{e^{-Y_b} Y_b^n}{n!},$$

$$Y_b < k(x+1).$$

The maximum of the above expressions is identified by the value of m which makes the products of the expressions under the summing symbols a maximum. Summation is carried out around the so identified points.

Applying the second part of eqns. (3.80) to eqns. (3.57) for the probability density function of Swerling's case I when

$$a = \frac{v}{1 + \dfrac{1}{k\bar{x}}} \quad \text{and} \quad p = k - 2,$$

one arrives at

$$p(v) = \left(1 + \frac{1}{k\bar{x}}\right)^{k-2} \frac{1}{k\bar{x}} \exp\left(-\frac{v}{1 + k\bar{x}}\right)$$

$$\sum_{m=k-1}^{\infty} \frac{\exp\left(-\dfrac{v}{1 + 1/k\bar{x}}\right)\left(\dfrac{v}{1 + 1/k\bar{x}}\right)^m}{m!} \tag{3.83}$$

as an alternative expression for the above density function.

Similarly, when in the second expression of eqn. (3.80) one uses $a = Y_b$ for the series expansion of the first, $a = Y_b/[1 + (1/k\bar{x})]$ for that of the second incomplete gamma function in eqn. (3.59) and $p = k - 2$ in both of them, one arrives at the probability of detection given by

$$P_d = 1 - \sum_{m=k-1}^{\infty} \frac{Y_b^m \exp(-Y_b)}{m} + \left(1 + \frac{1}{k\bar{x}}\right)^{k-1} \exp\left(\frac{-Y_b}{1 + k\bar{x}}\right)$$

$$\sum_{m=k-1}^{\infty} \left(\frac{Y_b}{1 + 1/k\bar{x}}\right)^m \frac{\exp\left(-\dfrac{Y_b}{1 + 1/k\bar{x}}\right)}{m!}. \tag{3.84}$$

The probability of detection of Swerling's case II targets given by eqn. (3.66) may be expanded in the above manner into the series

$$P_d = 1 - \sum_{m=k}^{\infty} \left(\frac{Y_b}{1+\bar{x}}\right)^m \frac{\exp\left(-\frac{Y_b}{1+\bar{x}}\right)}{m!}, \tag{3.85}$$

when $a = Y_b/(1+\bar{x})$ and $p = k-1$ is adopted in the second expression of eqn. (3.80).

Expressions suitable for numerical computations of the probability densities and detection probabilities of the other two Swerling cases may be obtained by a similar approach.

Using eqn. (3.67) one may obtain for case III the probability density function

$$f(v) = -\frac{(k-2)c^2 e^{-cv}}{(1-c)^{k-1}} + \frac{c^2 v e^{-cv}}{(1-c)^{k-2}}$$
$$+ c^2 e^{-v} \sum_{m=0}^{k-3} \frac{(k+1)v^{k-3-m}}{(k-3-m)!\,(1-c)^{m+2}}, \tag{3.86}$$

where

$$c = \frac{1}{1+\frac{k\bar{x}}{2}}.$$

The expression for the probability distribution then becomes

$$p(c, Y_b) = \frac{Y_b^{k-1} e^{-Y_b} c}{(k-2)!} \sum_{m=0}^{k-2} \frac{e^{-Y_b} Y_b^m}{m!} \tag{3.87}$$
$$+ \frac{e^{-cY_b}}{(1-c)^{k-2}} \left[1 - \frac{(k-2)c}{1-c} + cY_b\right]$$
$$\left[1 - \sum_{m=0}^{k-2} \frac{e^{-Y_b(1-c)}[Y_b(1-c)]^m}{m!}\right],$$

where the summations are incomplete gamma functions evaluated by using eqn. (3.80).

The expression for the probability of detection for Swerling's case IV is obtained from the characteristic function (3.77) by deriving first the probability density function

$$f(v) = e^{-cv} c^{2k} \sum_{m=0}^{2k-1} \frac{k!}{m!\,(k-m)!} (1-c)^{k-m} \frac{v^{2k-1-m}}{(2k-1-m)!} \tag{3.88}$$

using in this case

$$c = \frac{1}{1+\frac{\bar{x}}{2}}.$$

The probability distribution then becomes

$$\begin{aligned} p(c, Y_b) &= c^k \sum_{m=0}^{k} \frac{k!}{m!(k-m)!} \left(\frac{1-c}{c}\right)^{k-m} \sum_{n=0}^{2k-1-m} \frac{e^{-cY_b} (cY_b)^n}{n!} \\ &= 1 - c^k \sum_{m=0}^{k} \frac{k!}{m!(k-m)!} \left(\frac{1-c}{c}\right)^{k-m} \sum_{n=2k-m}^{\infty} \frac{e^{-cY_b} (cY_b)^n}{n!} \end{aligned} \tag{3.89}$$

The first one of these equations converges faster for $Y_b > k(2-c)$ while the second one for $Y_b < k(2-c)$. The summations are performed around the maximum identified in the first case by that value for m which makes

$$\frac{k!}{m!(k-m)!} \left(\frac{1-c}{c}\right)^{k-m} \frac{(cY_b)^{2k-1-m}}{(2k-1-m)!}$$

a maximum while in the second case the summation is around the value of m making

$$\frac{k!}{m!(k-m)!} \left(\frac{1-c}{c}\right)^{k-m} \frac{(cY_b)^{2k-m}}{(2k-m)!}$$

a maximum.

3.6.2 Analytic approximations

Since the signal-to-noise ratio in the above exact expressions is stated implicitly and is rather difficult to express explicitly, one has to resort to analytic approximations proposed by DiFranco and Rubin (1968) whenever it is to be computed for specified probabilities of detection and false alarm. And, although the accuracy of the approximations is rather poor for values of $k < 100$, it may be improved by applying corrections based on analytic formulae given below.

3.6.2.1 The non-fluctuating target: According to the central limit theorem of mathematical statistics the probability density function of the sum random variable of a great number of random variables approaches a normal distribution. Marcum (1948) retained only the first term of the Gram–Charlier series, an orthogonal polynomial expansion the first term of which is the normal density function, for representing the probability density function of the sum random variable Y

$$p(Y;\bar{x})_{\text{approx}} = \frac{1}{\sqrt{2\pi}\sigma} \exp - \frac{(Y-\bar{Y})^2}{2\sigma^2}, \tag{3.90}$$

where σ is the r.m.s. value, σ^2 the variance and $\bar{Y}$ the mean value of Y.

The first order Gaussian approximation of the above probability density function is described then by the true mean and variance of Y. The variance may be computed from the first two moments.

The moments of a random variable, as mentioned earlier, may be obtained from its characteristic function. The first moment then becomes

$$m_1 = \bar{Y} = k(1+\bar{x}),$$

while the second one becomes

$$m_2 = k^2(1+\bar{x})^2 + k(1+2\bar{x})$$

and the variance

$$\sigma^2 = m_2 - m_1^2 = k(1+2\bar{x}).$$

Inserting these into eqn. (3.90) yields the probability density function

$$p(Y;x)_{\text{approx}} = \frac{1}{\sqrt{2\pi k(1+2\bar{x})}} \exp\{-[(Y-k(1+\bar{x})]^2/2k(1+2\bar{x})\}, \tag{3.91}$$

which for noise alone, i.e. for $\bar{x} = 0$, becomes

$$p(Y;0)_{\text{approx}} = \frac{1}{\sqrt{2\pi k}} \exp[-(Y-k)^2/2k]. \tag{3.92}$$

The probability of false alarm may be obtained by integrating this expression within the appropriate limits as

$$P_{fa} \cong \int_{Y_b}^{\infty} \frac{1}{\sqrt{2\pi k}} \exp[-(Y-k)^2/2k]\, dY. \tag{3.93}$$

Substituting

$$z = \frac{Y-k}{\sqrt{k}}$$

the probability of false alarm becomes

$$P_{fa} \cong \frac{1}{\sqrt{2\pi}} \int_{(Y_b-k)/\sqrt{k}}^{\infty} e^{-z^2/2}\, dz, \tag{3.94}$$

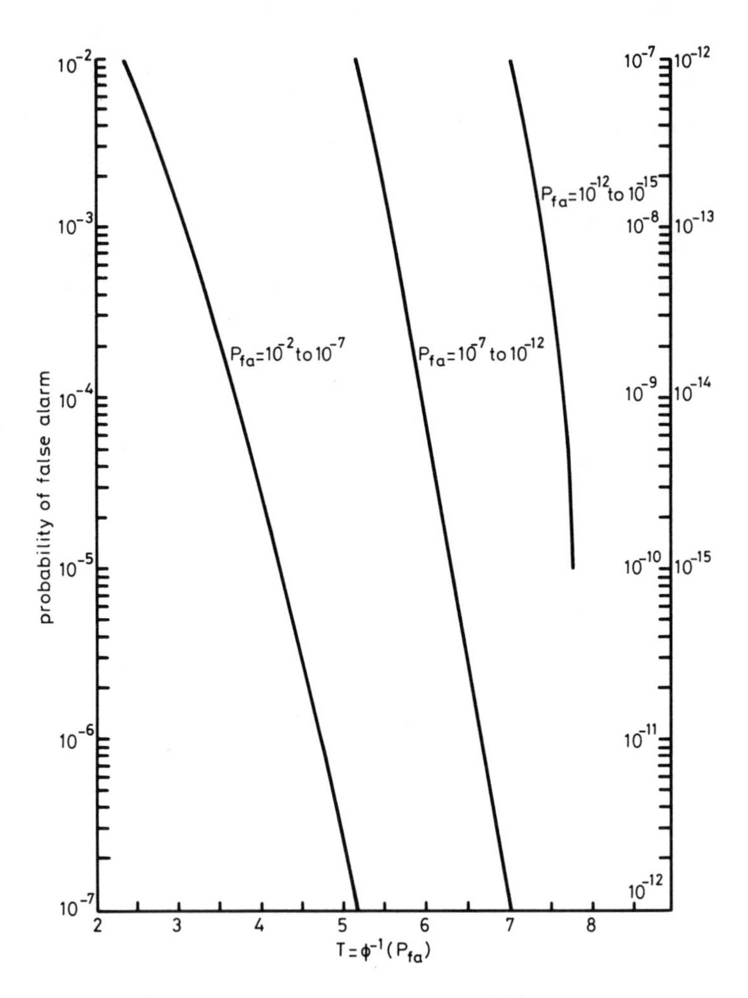

Fig. 3.4 *P_{fa} vs. ϕ^{-1} (P_{fa}) (After DiFranco and Rubin, 1968)*

$$\Phi(t) = \frac{1}{\sqrt{2\pi}} \int_T^{\infty} e^{-x^2/2}\, dx = P_{fa}$$

which written concisely is

$$P_{fa} \cong \phi\left(\frac{Y_b - k}{\sqrt{k}}\right)$$

and its inverse

$$\phi^{-1}(P_{fa}) \cong \frac{Y_b - k}{\sqrt{k}}$$

The approximate value of the bias is then

$$Y_b \cong \sqrt{k}\,\phi^{-1}(P_{fa}) + k. \tag{3.95}$$

DiFranco and Rubin (1968) plot the probability of false alarm for values of P_{fa} between 10^{-2} and 10^{-15} as a function of $\phi^{-1}(P_{fa})$ in their Fig. 9.6-1 reproduced here in Fig. 3.4.

Values read from this graph may be used for the determination of the required value of $\phi^{-1}(P_{fa})$ for the specified probability of false alarm.

The approximate probability of detection may be obtained by integrating the probability density function (3.91) within the limits Y_b and infinity which with the substitution

$$w = \frac{Y - k(1+\bar{x})}{\sqrt{k(1+2\bar{x})}}$$

reads

$$P_d \cong \frac{1}{\sqrt{2\pi}} \int_{Y_b - k(1+\bar{x})/\sqrt{k(1+2\bar{x})}}^{\infty} e^{-w^2/2}\,dw, \tag{3.96}$$

or in concise form

$$P_d \cong \phi\left(\frac{Y_b - k(1+\bar{x})}{\sqrt{k(1+2\bar{x})}}\right). \tag{3.97}$$

Inserting eqn. (3.95) and using the approximation $\sqrt{1+2\bar{x}} \cong 1+\bar{x}$, applicable for $\bar{x} < 1$, yields the signal-to-noise ratio

$$\bar{x} \cong \frac{\phi^{-1}(P_{fa}) - \phi^{-1}(P_d)}{\phi^{-1}(P_d) + \sqrt{k}}. \tag{3.98}$$

For $k \gg 1$, due to $\sqrt{k} \gg \phi^{-1}(P_d)$, this expression reduces further to

$$\bar{x} \cong \frac{\phi^{-1}(P_{fa}) - \phi^{-1}(P_d)}{\sqrt{k}}. \tag{3.99}$$

However, since in radar practice one uses the ratio of peak instantaneous signal power to the average noise power for probability of detection calculations, which ratio is equal to $2\bar{x}$, the above two expressions, namely eqns. (3.98) and (3.99), must be multiplied by 2.

Values of $P_d = f(\phi^{-1}(P_d))$ were graphed by DiFranco and Rubin (1968) in their Fig. 9.6-2 reproduced here in Fig. 3.5. Expressed analytically, by redrawing the above function on linear coordinates permits obtaining a curve symmetrical about $P_d = 0{\cdot}5$. Fitting then a quadratic curve to its positive half for values of P_d from 0·01 to 0·5 and taking its negative for P_d between 0·5 and 0·99 one obtains

$$YM = 0{\cdot}008\,478\,6/P_d - 3{\cdot}1881\,P_d + 1{\cdot}534\,021, \quad 0{\cdot}01 < P_d < 0{\cdot}5 \tag{3.100}$$

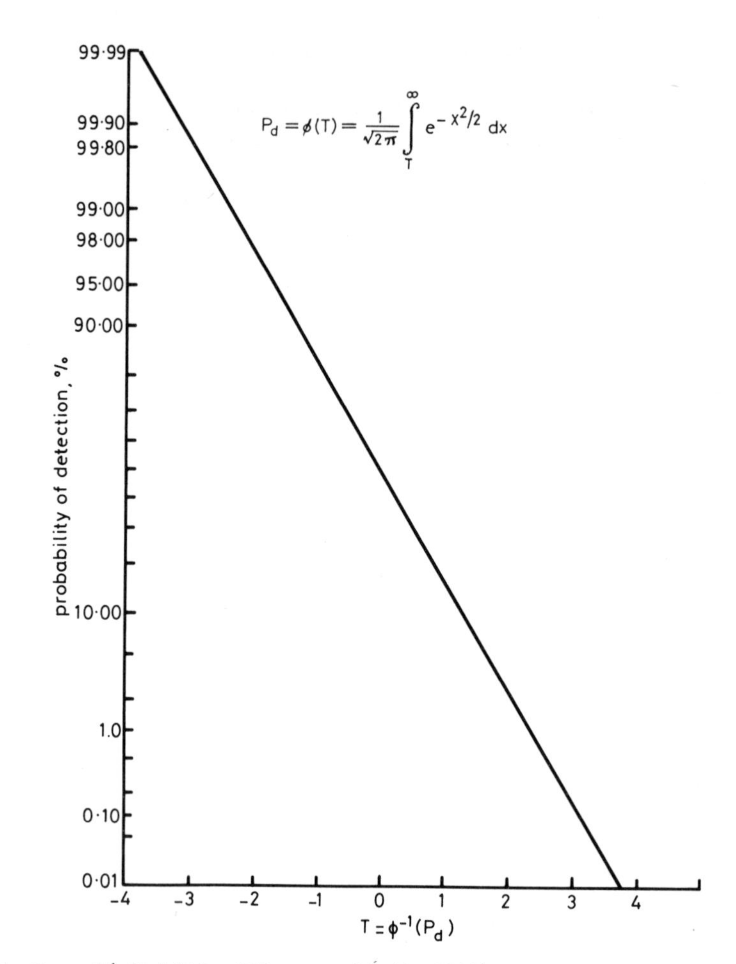

Fig. 3.5 *P_d vs. ϕ^{-1} (P_d) (After DiFranco and Rubin, 1968)*

and

$$YM = 3{\cdot}1881(1 - P_d) - 0{\cdot}008\,478\,6/(1 - P_d) - 1{\cdot}534\,021, \quad 0{\cdot}5 < P_d < 0{\cdot}99 \tag{3.101}$$

with

$$YM = 3{\cdot}73 \text{ for } P_d < 0{\cdot}01 \quad \text{and} \quad YM = -3{\cdot}73 \text{ for } P_d > 0{\cdot}99.$$

The signal to noise ratios obtained from eqn. (3.99) are claimed to be accurate to within 1 dB for values of $k > 100$. The deviation from the exact values of the signal-to-noise ratio for $k < 100$ is rather large. It may be reduced, however, to acceptable values by using a correction

$$C_{NF} = 0{\cdot}055\,51 \ln^2 k - 0{\cdot}564\,85 \ln k + 2{\cdot}623\,88 \qquad (3.102)$$

derived by dividing the accurate signal-to-noise ratios obtained from graphs published in DiFranco and Rubin's book, by those computed from eqn. (3.99) and solving simultaneous quadratic equations of the type

$$R = a \ln^2 k + b \ln k + c$$

written for values of k below 100. R in these equations are the mentioned ratios of the exact and approximate signal-to-noise ratios.

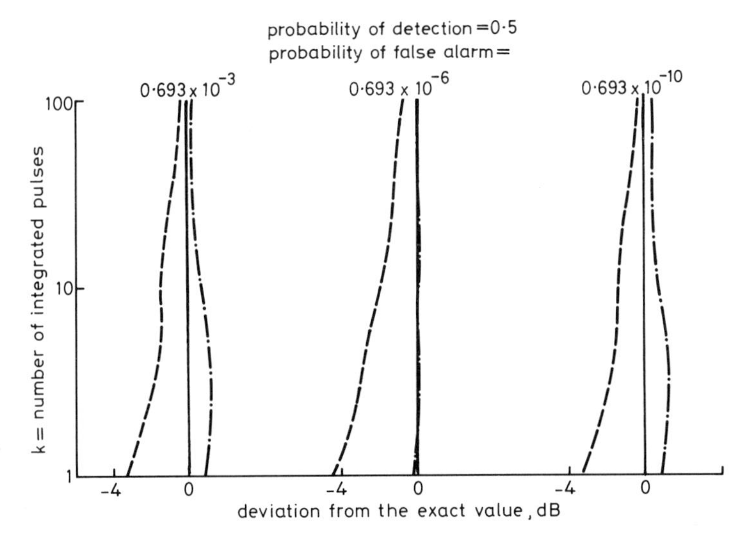

Fig. 3.6 *Error in the computed signal-to-noise ratio. Non-fluctuating target*

The improvement obtained by using this correction is shown by Fig. 3.6 where curves of the signal-to-noise ratio for $k < 100$ computed from eqn. (3.99) for a probability of detection of 0·5 and probabilities of false alarm of $0{\cdot}693 \times 10^{-3}$, $0{\cdot}693 \times 10^{-6}$ and $0{\cdot}693 \times 10^{-10}$ are graphed together with those corrected by the factor given by eqn. (3.102). The full straight line drawn vertically above the 0 dB mark is the accurate value, the dashed line is computed from eqn. (3.99), while the dot–dashed one shows the corrected values of the signal-to-noise ratio.

The dependence of the signal-to-noise ratio on the number of integrated pulses is not linear. However, since the graph shows the deviation of the computed values of the signal-to-noise ratio from the exact ones and the exact values have obviously a 0 dB deviation, these are drawn as a straight vertical line above the 0 dB mark. The meandering of the lines computed from the analytical approximation given by eqn. (3.99) and those corrected by the application of factors obtained from eqn. (3.102) is due to the above chosen linear representation of the exact values of the signal-to-noise ratio.

3.6.2.2. Swerling's fluctuating targets. (*a*) *Case I*: An expression for the approximate value of the probability of detection for this target return fluctuation type may be obtained from the exact expression (3.59). Both incomplete gamma functions of this equation may be shown (DiFranco and Rubin, 1968) to be close to unity for $k\bar{x} \gg 1$ and $P_{fa} \ll 1$ so that the approximate probability of detection becomes

$$P_d \cong \left(1 + \frac{1}{k\bar{x}}\right)^{k-1} \exp\left[-Y_b \Big/ \left(k\bar{x}\left(1 + \frac{1}{k\bar{x}}\right)\right)\right], \tag{3.103}$$

which yields the exact value of the probability of detection for $k = 1$.

Taking the natural logarithm of both sides, expanding the expressions in the brackets into series and retaining only the first term results in the expression

$$\ln P_d \cong \frac{-1}{k\bar{x}}(Y_b - k + 1). \tag{3.104}$$

Inserting Y_b determined from eqn. (3.95) leads to

$$\ln P_d \cong -\frac{1}{k\bar{x}}(\sqrt{k}\,\phi^{-1}(P_{fa}) + 1) \tag{3.105}$$

from which the approximate value of the signal-to-noise ratio, for $k \gg 1$,

$$\bar{x} \cong \frac{\phi^{-1}(P_{fa})}{\sqrt{k}\,\ln(1/P_d)} \tag{3.106}$$

may be determined.

This expression must again be multiplied by 2 in order to obtain the customary peak signal to average noise power ratio used in radar practice.

In Fig. 3.7 the dashed line shows deviations of the signal-to-noise ratio obtained from this formula from the exact value. Similarly to the non-fluctuating target case, a correction factor

$$C_{Sw1} = 0.062\,45 \ln^2 k - 0{\cdot}572 \ln k + 2{\cdot}435 \tag{3.107}$$

was determined as explained above. The dot–dashed lines of Fig. 3.7 show deviations of the corrected values from the accurate ones (full lines) for a probability of detection of 0·5 and the three probabilities of false alarm used in the previous figure.

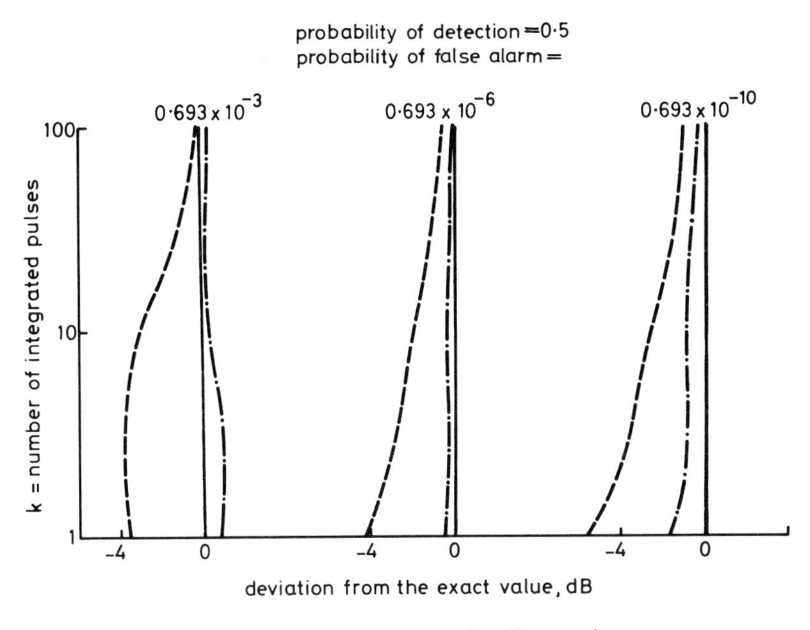

Fig. 3.7 *Error in the computed signal-to-noise ratio. Swerling case I*

(*b*) *Case II*: Using the Gaussian approximation for the probability density function

$$p(Y) \cong \frac{1}{\sqrt{2\pi k}\,(1+\bar{x})} \exp\left\{-\frac{[Y-k(1+\bar{x})]^2}{2k(1+\bar{x})^2}\right\} \tag{3.108}$$

with the variance determined from the moments of Y

$$m_1 = k(1+\bar{x}),$$

$$m_2 = k(k+1)(1+\bar{x})^2,$$

the probability of detection in its concise form is

$$P_d \cong \phi\left(\frac{Y_b - k(1+\bar{x})}{\sqrt{k}\,(1+\bar{x})}\right) \tag{3.109}$$

and its inverse

$$\phi^{-1}(P_d) \cong \frac{Y_b - k(1+\bar{x})}{\sqrt{k}\,(1+\bar{x})} \tag{3.110}$$

from which, using eqn. (3.95) for the approximate value of the bias, the signal-to-noise ratio may be expressed as

$$\bar{x} \cong \frac{\phi^{-1}(P_{fa}) - \phi^{-1}(P_d)}{\sqrt{k} + \phi^{-1}(P_d)} . \tag{3.111}$$

Since for $k \gg 1$ the inverse of the probability of detection is negligible when compared with $\sqrt{k}$, the peak signal to average noise ratio becomes

$$SN = 2\bar{x} \cong \frac{2[\phi^{-1}(P_{fa}) - \phi^{-1}(P_d)]}{\sqrt{k}} . \tag{3.112}$$

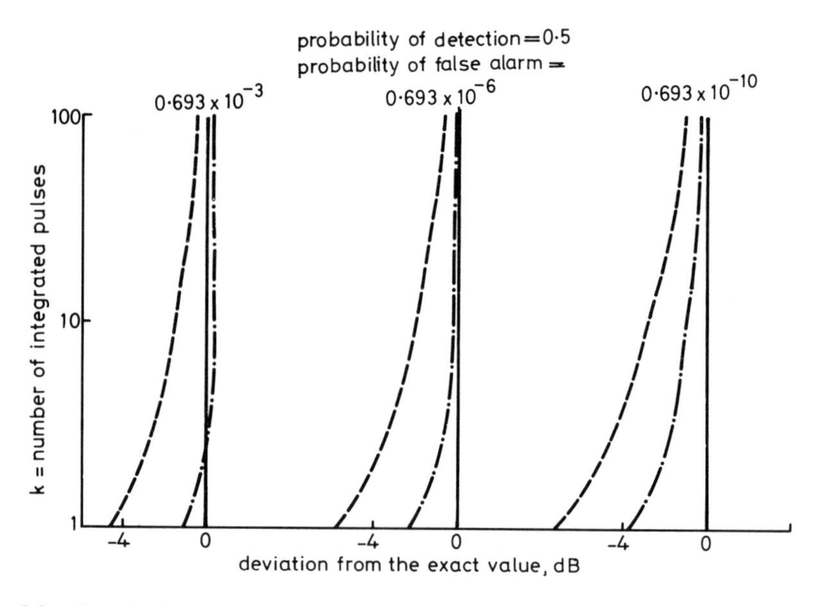

Fig. 3.8 *Error in the computed signal-to-noise ratio. Swerling case II*

The accuracy of SN given by this approximation is acceptable for $k > 100$. For cases where the number of integrated pulses is less than 100 the deviation from the accurate value is excessive and should be corrected by a factor computed from

$$C_{Sw2} = 0{\cdot}036\,81 \ln^2 k - 0{\cdot}403\,62 \ln k + 2{\cdot}257\,12, \tag{3.113}$$

an expression determined in the manner described earlier.

Figure 3.8 shows the deviation of the signal-to-noise ratios obtained from eqn. (3.112) (dashed lines) and those corrected by eqn. (3.113) (dot–dashed lines) from the accurate values shown by the full line. The previously used values for P_d and P_{fa} were used for the comparison.

(*c*) *Case III*: Accepting that for large values of k the fraction $k\bar{x}/2$ in eqn. (3.75) for the probability of detection of Swerling's case III target fluctuations is much larger than unity and that

$$\ln(1+z) \cong z, \quad z \ll 1$$

and

$$1 + k\bar{x}/2 \cong k\bar{x}/2,$$

taking logarithms on both sides and combining terms leads, for values of $k > 100$, to the acceptable approximation

$$\ln P_d \cong -\frac{Y_b - k + 2}{k\bar{x}/2} + \ln\left(1 + \frac{Y_b - k + 2}{k\bar{x}/2}\right), \tag{3.114}$$

which, with the value of Y_b given by eqn. (3.95), may be expressed as

$$\ln(1/P_d) \cong \frac{\sqrt{k}\,\phi^{-1}(P_{fa}) + 2}{k\bar{x}/2} - \ln\left(1 + \frac{\sqrt{k}\,\phi^{-1}(P_{fa}) + 2}{k\bar{x}/2}\right). \tag{3.115}$$

Since for the customary values of P_{fa}, $\sqrt{k}\,\phi^{-1}(P_{fa}) \gg 2$, the above expression reduces to

$$\ln(1/P_d) \cong M - \ln(1 + M), \tag{3.116}$$

where

$$M = \frac{2\phi^{-1}(P_{fa})}{\sqrt{k}\,\bar{x}}. \tag{3.117}$$

The peak signal to average noise ratio from this expression becomes

$$SN = 2\bar{x} \cong \frac{4\phi^{-1}(P_{fa})}{\sqrt{k}\,M}. \tag{3.118}$$

The probability of detection as a function of M is shown by DiFranco and Rubin as their Fig. 11.4–12 reproduced here in Fig. 3.9 with the expression for M modified to suit this treatise. This dependence was expressed analytically, by fitting a curve to the representation of the curve on linear coordinates, in order to make it suitable for numerical computation. It reads

$$M = 3{\cdot}115\,73 - 3{\cdot}487\,72 P_d + 0{\cdot}265\,22 P_d^2 + \frac{0{\cdot}114\,77}{P_d}. \tag{3.119}$$

The deviation of the signal-to-noise ratio as computed from eqn. (3.118), from the accurate value is shown, for a probability of detection of 0·5 and the three earlier used values of the probability of false alarm in Fig. 3.10. Corrections derived in the same way as those for other target types and computable from

$$C_{Sw3} = 0{\cdot}000\,52 \ln^2 k - 0{\cdot}164\,29 \ln k + 1{\cdot}931\,31 \qquad (3.120)$$

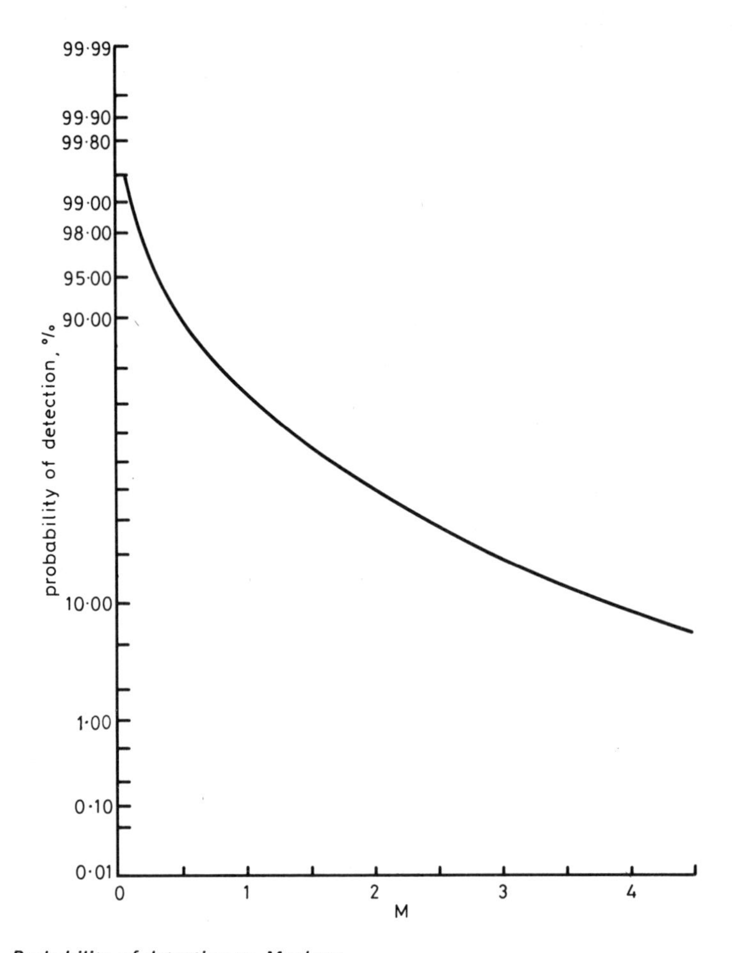

Fig. 3.9 *Probability of detection vs. M where*

$$M = 2\phi^{-1}(P_{fa})/\sqrt{k}\,\bar{x}$$

(After DiFranco and Rubin, 1968)

reduce the deviations to those shown by the dot–dashed curves of the same figure.

(*d*) *Case IV*: Retaining again only the first term of the Gram–Charlier expansion of the probability density function $p(Y)$ with the true mean and variance of Y leads to the approximation

$$p(Y) \cong \frac{\exp\left\{-\dfrac{[Y-k(1+\bar{x}/2)]^2}{2k\,[2(1+\bar{x}/4)^2-1]}\right\}}{\sqrt{2\pi k\,[2(1+\bar{x}/4)^2-1]}}$$

from which, due to $\bar{x}/4 \ll 1$, the concise form of the probability of detection

$$P_d \approx \phi\left(\frac{Y_b - k(1+\bar{x}/2)}{\sqrt{k}}\right) \tag{3.121}$$

may be obtained.

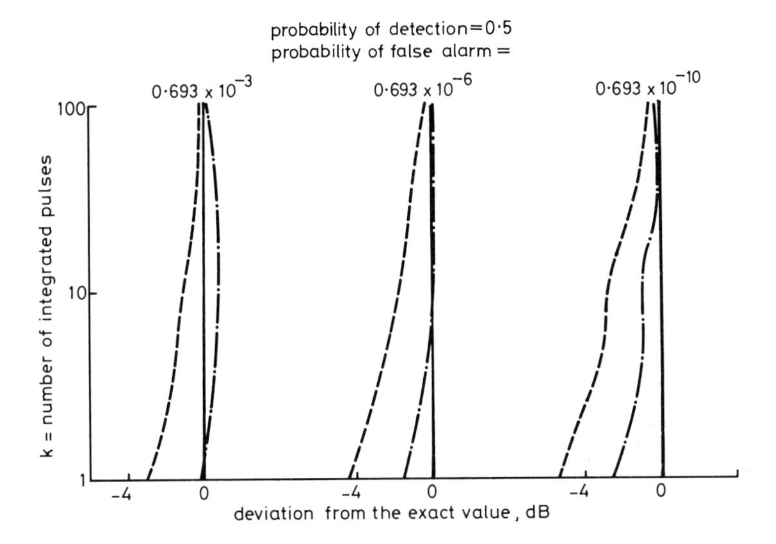

Fig. 3.10 *Error in the computed signal-to-noise ratio. Swerling case III*

Inserting the approximate value of the bias from eqn. (3.95) yields the inverse function of eqn. (3.121)

$$\phi^{-1}(P_d) \cong \frac{\sqrt{k}\,\phi^{-1}(P_{fa}) - k\bar{x}/2}{\sqrt{k}} \tag{3.122}$$

from which, with some simplifying assumptions, one arrives at

$$SN \cong \frac{2\,[\phi^{-1}(P_{fa}) - \phi^{-1}(P_d)]}{\sqrt{k}}, \tag{3.123}$$

the approximate value of the peak signal to average noise ratio.

The values of SN computed from this expression are claimed to be accurate within 1 dB for $k > 100$. Those computed for lower values of k may be corrected by the correction factor

$$C_{Sw4} = 0{\cdot}059\,81 \ln^2 k - 0{\cdot}632\,53 \ln k + 2{\cdot}809\,34. \qquad (3.124)$$

This was obtained by the earlier indicated method. Deviations of the approximate and corrected values of the signal-to-noise ratios for a probability of detection of 0·5 and probabilities of false alarm of $0{\cdot}693 \times 10^{-3}$, $0{\cdot}693 \times 10^{-6}$ and $0{\cdot}693 \times 10^{-10}$ are shown by dashed resp. dot–dashed curves in Fig. 3.11. The accurate values are shown by the vertical straight full lines.

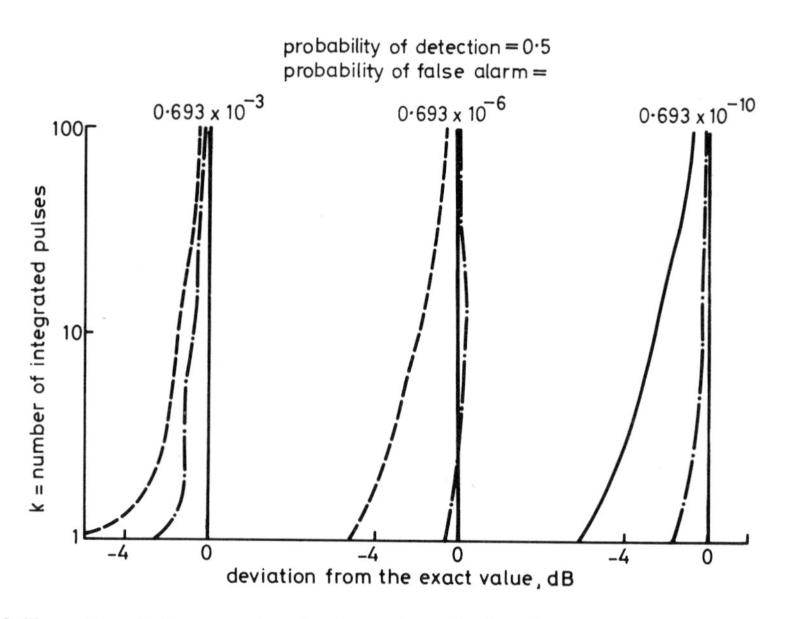

Fig. 3.11 *Error in the computed signal-to-noise ratio. Swerling case IV*

3.7 The required number of hits per scan

From what has been said so far it transpires that the signal-to-noise ratio for a specified probability of detection and false alarm depends on the type of the considered target return fluctuations. A non-fluctuating target requires the lowest, while a Swerling type I target the highest signal-to-noise ratio.

Plotting then the limits of the signal-to-noise ratio, using the plots of DiFranco and Rubin (1968), e.g. for a probability of detection of 0·5 and a false alarm probability of $0{\cdot}693 \times 10^{-6}$ against the number of hits per scan, as in Fig. 3.12, permits the quick determination of the required range of hits per scan for a specified signal-to-noise ratio. If the signal-to-noise ratio of, e.g. 5 dB is accepted as the standard for detection with the above probabilities, then the number of hits per scan will have to be in the range of about 20 to 32 and the dwell-time of the aerial beam on the target and the pulse repetition frequency will have to be chosen accordingly.

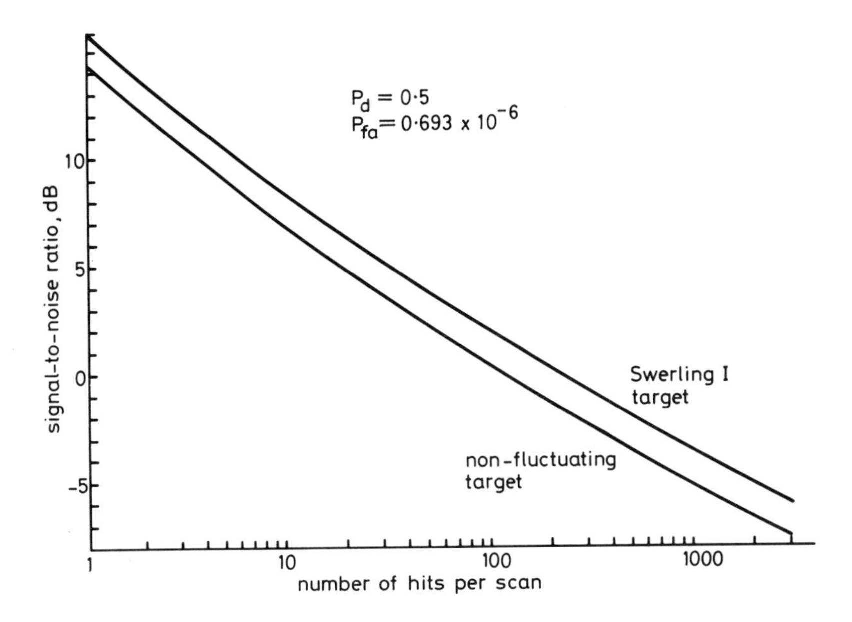

Fig. 3.12 *Required number of hits/scan for specified signal-to-noise ratios*

3.8 Further thoughts on radar detection

Costs mentioned in Section 3.2 which are ascribed to errors in deciding whether a target is present when there isn't one and vice versa become real when one considers the costs connected with alerting the services at an airport or, in the military case, the defences of an area, or failing to do so. It was decided therefore to investigate the probability of detection on successive aerial sweeps.

3.8.1 The probability of detection on consecutive aerial sweeps

Due to the uncertainty of whether a PPI (plan position indicator) paint is caused by a target or noise, the radar operator usually waits for the appearance of the echo in the same or adjacent position on subsequent aerial sweeps before raising alarm. The first echo could be termed the detection while those following the first one the confirmation of the presence of a target.

3.8.1.1 Probabilities of detection and confirmation: With the above in mind, a general formula for the determination of the probabilities of detection and confirmation has been derived.

It can be shown that the probability that at least one echo out of k integrated echoes will exceed the detection threshold is

$$P_k(1) = 1-(1-P_d)^k.$$

If successive echoes are considered to be dependent on one another, the joint probability of observing an echo on two successive sweeps of the aerial is

$$P(1,2) = (P(2|1)P_d,$$

i.e. the product of the probability P_d of detecting the first echo and the conditional probability $P(2|1)$ that the second echo will occur after the first one has been observed.

Similarly, the joint probability of observing echoes on three consecutive sweeps of the aerial is

$$P(1,2,3) = P_d\,P(2|1)\,P(3|2,1).$$

It is again the product of the probabilities that the first echo was observed, that the second occurred after the observation of the first one and that the third occurred after the second and first echo.

The probabilities of interest for the determination of the required signal-to-noise ratios for specified false alarm probabilities are the conditional probabilities $P(2|1)$ and $P(3|2,1)$. Since, however, the joint probabilities $P(1,2)$ and $P(1,2,3)$ are difficult to assess, the above conditional probabilities are rather difficult to determine.

On making the simplifying assumptions that the range between the radar aerial and the target does not change considerably from aerial sweep to aerial sweep, and since the sweep to sweep correlations of echoes from the same target are negligible, the second and successive echoes have the same chance of being observed as the first one, i.e. the observations are independent, the expression for the single echo detection may be modified to yield the probability of at least one echo exceeding the detection threshold on each of m sweeps.

The joint probability of m observations is then

$$P(1,2,3,\ldots,m) = P_{md} = [1-(1-P_d)^k]^m$$

from where the probability of detection

$$P_d = 1-[1-P_{md}^{1/m}]^{1/k}. \tag{3.125}$$

The signal-to-noise ratio for the specified false alarm probability is read from the graph (Fig. 3.3) for this P_d.

The probability of detection P_d increases with the number of aerial sweeps m.

The radar operator sets his detection threshold such that the PPI display should not be obliterated by noise, thus permitting the observation of weak echoes.

By setting the threshold, i.e. selecting the ratio $Y_b/\sqrt{N}$, the probability of false alarm is also determined.

An increase in the probability of detection at a constant value of the probability of false alarm is accompanied by an increase in the signal-to-noise ratio. The increase in signal-to-noise ratio is due to the subconscious visual integration of the PPI paints by the observer.

The observation of an echo on a number of consecutive aerial sweeps results then, due to a constant probability of false alarm, in an increase of the signal-to-noise ratio and hence an increase of the probability of detection.

3.8.1.2 The Probability of detection and signal-to-noise ratio: The single sweep probabilities of detection for one, two and three aerial sweeps, a 10^{-6} probability of false alarm and joint probability $P_{md} = 1$, 10 and 30% have been calculated and the appropriate peak signal to average noise $(2S^2/N)$ ratios determined, using Fig. 3.3. The number k of integrated pulses per aerial sweep was assumed to be ten and the observer's visual integration was taken into consideration by using $\sqrt{k}$ in the calculations. The results are tabulated below:

	Number of consecutive aerial sweeps					
	1		2		3	
P_md % (see text)	P_d %	SN_{dB}	P_d %	SN_{dB}	P_d %	SN_{dB}
1	0·3173	7·6	3·277	10·7	7·386	11·7
10	3·277	10·7	11·326	12·2	17·906	12·8
30	10·66	12·0	22·19	13·1	29·54	13·4

Theoretically, the false alarm probability for a constant signal-to-noise ratio increases with an increasing probability of detection since both are functions of $Y_b/\sqrt{N}$, the ratio of detection threshold to noise. This, however, applies strictly to detection on a single aerial sweep only. Since the probability of the occurrence of noise pulses strong enough to be mistaken for target returns in approximately the same position of a PPI display is very low, detection and confirmation of a target, by observing it on a number of consecutive sweeps, has the same effect as an increase of the signal-to-noise ratio or an increase of the probability of detection at a constant probability of false alarm.

3.8.2 The probability of confirmation dependent on an immediately preceding detection

The preceding section considered the probability of detection when detection on a particular sweep of the aerial is independent of any of the possible detections on previous sweeps. The following section considers the probability of detection when detection on one sweep is dependent on detection on the immediately preceding sweep.

3.8.2.1 The probability of two consecutive detections: A simple Markov chain is a process which assumes randomly discrete states with the probability of assuming a certain state depending on the last, and only the last, previous state (Bechenbach, 1956). Its behaviour for the studied problem is best described by the 'stochastic' matrix

		2nd sweep	
		miss X	hit Y
1st sweep	miss X	P_{xx}	P_{xy}
	hit Y	P_{yx}	P_{yy}

where, adopting the substript 'X' for a miss and 'Y' for a hit or detection, the probabilities P_{xx}, P_{xy}, P_{yx} and P_{yy} are:

P_{xx} the probability of miss on both sweeps,

P_{xy} the probability of miss on the first, hit on the second sweep,

P_{yx} the probability of hit on the first, miss on the second sweep,

P_{yy} the probability of hit on both sweeps.

Since, near threshold, the probabilities of two successive misses and two successive hits are equal and also the probability of a hit following a miss and a miss following a hit are equal, i.e. there is an equal chance that a very weak signal pulse added to noise or a noise pulse alone will or will not exceed the threshold, the above matrix becomes:

$$\begin{vmatrix} P & Q \\ Q & P \end{vmatrix}$$

where $P = P_{xx} = P_{yy}$ and $Q = P_{xy} = P_{yx}$.

Further, since the sum of the probabilities of missing and hitting a target on a single sweep is unity, the sum of the terms in each row of the above matrix is unity and the matrix may be written in the form

$$\begin{vmatrix} P & 1-P \\ 1-P & P \end{vmatrix}$$

It remains then to determine the conditional probability P in order to be able to calculate the above four probabilities of interest.

The general expression for a conditional probability is

$$P(X \mid Y) = \frac{P(X, Y)}{P(Y)}$$

with X denoting a miss and Y a hit, as before.

The probabilities of miss and hit will be assumed to have normal distributions, hence, the joint probability $P(X, Y)$ and also the conditional probability $P(X \mid Y)$ will be normally distributed. This is a characteristic of the normal distribution.

Since misses or detections on two successive sweeps are taken to be dependent on one another their dependence may be expressed by a correlation coefficient (Hogg and Craig, 1959).

The mean product of the deviation of two random variables Y and Y_1, or their covariance, is

$$\operatorname{cov}(Y, Y_1) = \overline{(Y - \bar{Y})(Y_1 - \bar{Y}_1)} = \overline{YY_1} - \bar{Y}\,\bar{Y}_1.$$

If the standard deviations of Y and Y_1, σ_y and σ_{y1}, are positive, the ratio

$$\rho = \frac{\operatorname{cov}(Y, Y_1)}{\sigma_y\,\sigma_{y1}}$$

is the correlation coefficient of Y and Y_1.

The covariance and hence also the correlation of independent random variables is zero. It is unity and positive for fully correlated quantities varying in harmony, and unity and negative for quantities varying in opposition. The correlation coefficient then lies within the limits

$$-1 \leqslant \rho \leqslant 1.$$

Assuming, for simplicity's sake, the probability density function of the two dependent events of interest, namely X and Y, with mean values $\bar{X}$ and $\bar{Y}$, to be normal, their joint distribution may be expressed (Davenport and Root, 1958) by

$$p(X, Y) = \frac{\exp\left\{-\dfrac{1}{2(1-\rho^2)}\left[\dfrac{(X-\bar{X})^2}{\sigma_x^2} - \dfrac{2(X-\bar{X})(Y-\bar{Y})}{\sigma_x\,\sigma_y} + \dfrac{(Y-\bar{Y})^2}{\sigma_y^2}\right]\right\}}{2\pi\sigma_x\,\sigma_y\sqrt{1-\rho^2}}$$

from which the probability may be determined by integration between the limits of variation.

Considering now the observation of a paint after detection on the immediately preceding aerial sweep, one has to examine the conditional probability density function $p(Y_1 \mid Y)$ where Y is written for both events to be in agreement with the hitherto used notation, i.e. Y for detection and X for miss. Y denotes the detection while Y_1 the confirmation of the detection. This probability density function is the ratio of the joint probability density function $p(Y, Y_1)$ and the probability density function of the first paint $p(Y)$. Hence, taking for simplicity's sake $\sigma_y = \sigma_{y1}$ and $\bar{Y} = \bar{Y}_1 = 0$, the probability density function of interest becomes

$$p(Y_1 | Y) = \frac{\exp\left[-\dfrac{Y^2 - 2\rho Y Y_1 + Y_1^2}{2\sigma^2(1-\rho^2)}\right] \Big/ 2\pi\sigma^2\sqrt{1-\rho^2}}{\exp\left[-\dfrac{Y^2}{2\sigma^2}\right] \Big/ \sigma\sqrt{2\pi}}. \qquad (3.126)$$

The probability of observing paints on two consecutive sweeps is the integral of this expression. However, a better approach, from the heuristic point of view, is to examine eqn. (3.126) for the extreme values of the correlation coefficient.

When ρ is zero, i.e. the case when there is no correlation between the occurrence of the first and second paint, the conditional probability density function reduces to

$$p(Y_1 | Y) = \frac{\dfrac{1}{2\pi\sigma^2}\exp\left[-\dfrac{Y^2 + Y_1^2}{2\sigma^2}\right]}{\dfrac{1}{\sigma\sqrt{2\pi}}\exp\left[-\dfrac{Y^2}{2\sigma^2}\right]}$$

$$= \frac{\exp\left[-\dfrac{Y_1^2}{2\sigma^2}\right]}{\sigma\sqrt{2\pi}}$$

yielding the probability of Y_1, or the probability of the second paint alone.

This is obvious, since for completely uncorrelated or independent events, the occurrence of the second event is given by its probability alone and does not depend on the occurrence of the first event.

When ρ is unity and positive, i.e. the events are completely correlated, the conditional probability density function is

$$p(Y_1 | Y) = \frac{1}{\exp\left[-\dfrac{Y^2}{2\sigma^2}\right] \Big/ \sigma\sqrt{2\pi}} = 1,$$

since the detection of the first paint has been assumed.

For $\rho = -1$, the result is unity again. Here, however, the probability of a miss following a detection is arrived at since a negative correlation coefficient relates events varying in opposition.

Considering again the $\rho = 0$ case, i.e. no correlation between events, it is obvious that there is a 0·5 probability of observing a paint on the second sweep. Then the empirical expression

$$P(Y_1 \mid Y) = 0{\cdot}5(1+\rho)$$

suggests an acceptable solution for the case of two consecutive paint observations with correlation ρ between them.

Since the probability of a miss is $P(X) = 1 - P(Y)$, for cases when $\rho < 0$, i.e. a miss following a detection, the above expression changes to

$$P(X_1 \mid Y) = 1 - P(Y_1 \mid Y) = 0{\cdot}5(1-\rho).$$

The above study confirms theoretically that the probability of a correct decision about the presence of a target increases when the sweep following a detection is observed. While the probability of observing a target on the sweep following detection, when there is no correlation between echoes, is only 0·5, it rises, e.g. to 0·75 for a 0·5 correlation.

3.9 Cumulative probability of detection

The generally accepted meaning of the term cumulative probability of detection in radar practice is the probability of detecting an approaching target by a scanning radar at least once by the time the target reaches a range of interest. It has been dealt with by several authors, e.g. Mallet and Brennan (1963), Johnson (1966) and DiFranco and Rubin (1968).

It will be shown later that the range R at which a radar can detect a target is

$$R^4 = \text{const.}\,\frac{1}{SN}, \tag{3.127}$$

where SN is the signal-to-noise ratio required for detection with a given probability at a specified probability of false alarm and const. is a 'figure of merit' containing the radar's and target's parameters.

Denoting the range at unity signal-to-noise ratio by R_0, one may express a normalized range

$$\left(\frac{R}{R_0}\right)^4 = \frac{\text{const.}\,\dfrac{1}{SN}}{\text{const.}} = \frac{1}{SN} \tag{3.128}$$

in terms of the signal-to-noise ratio.

Assuming that the probabilities of detecting a target on individual scans are independent of each other and that the probability of detection on the ith scan is P_i, the probability of not detecting the target is $1 - P_i$ and that of missing the target on n consecutive scans is

$$P_m = \prod_{i=1}^{n} (1 - P_i). \tag{3.129}$$

The probability of detecting the target at least once in n scans, the cumulative probability of detection, is

$$P_c = 1 - P_m = 1 - \prod_{i=1}^{n} (1 - P_i). \tag{3.130}$$

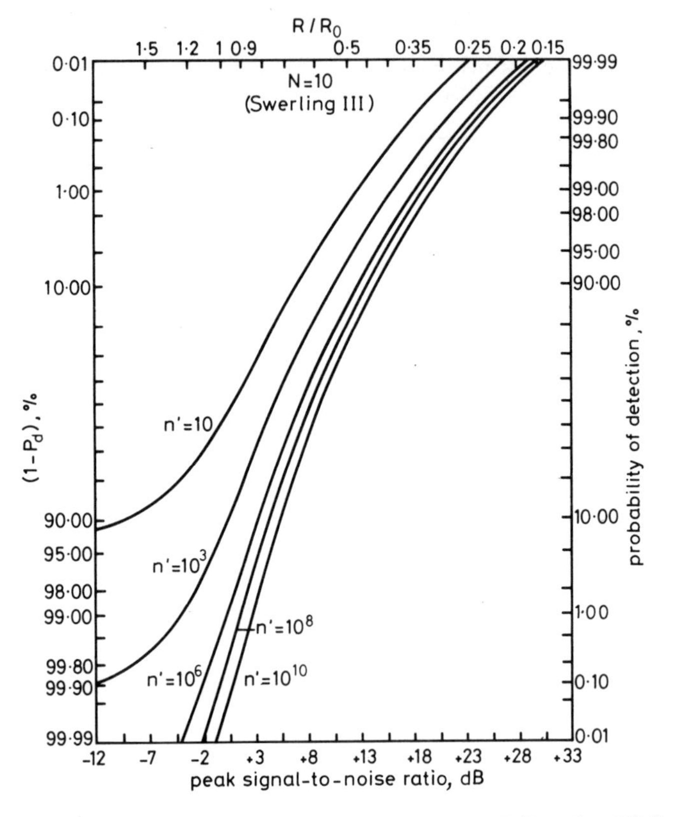

Fig. 3.13 *Probability of detecting a fluctuating target, $k = 10$ (Swerling III) (k = number of pulses incoherently integrated; $P_{fa} = 0{\cdot}693/n'$) (After DiFranco and Rubin, modified by Rohan)*

A target will travel a distance $\Delta = v/r$ between scans when moving with a velocity v and the aerial's scanning rate is r. The distance between the radar and the target on individual scans may be expressed by $R - i\Delta$, where R is the initial range and i is the number of scans of the radar's aerial since the target was at that range.

The cumulative probability of detection then becomes

$$P_c = 1 - \prod_{i=1}^{n} [1 - P_i(R - i\Delta)]. \tag{3.131}$$

Rohan (1971) added the scales for

$$1 - P = f\left(\frac{R}{R_0}\right)$$

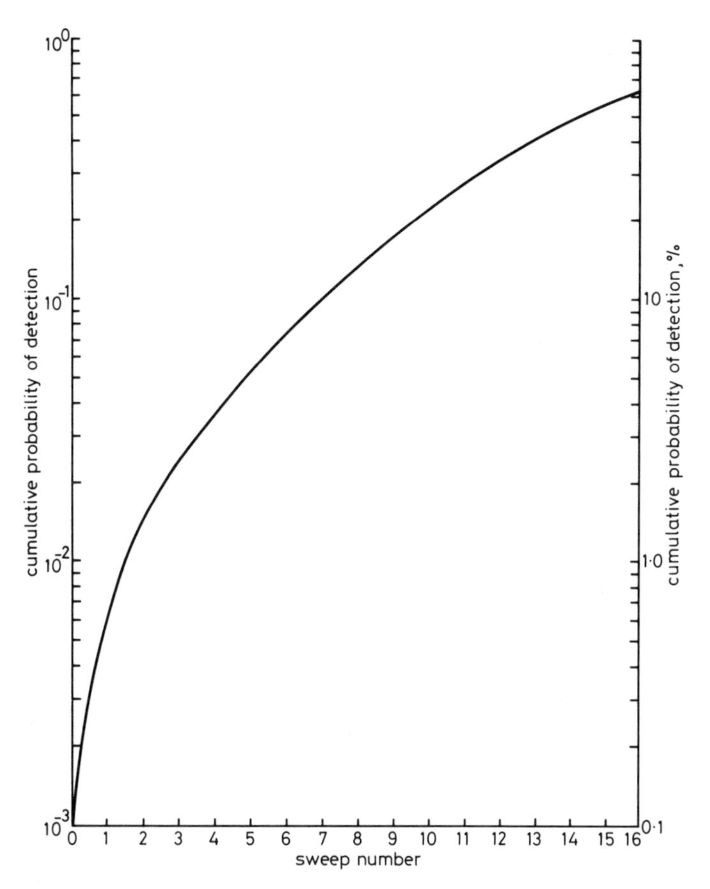

Fig. 3.14 *Growth of the cumulative probability of detection (After Rohan, 1971)*

to Fehlner's curves as given by DiFranco and Rubin (1968), one of which is shown by Fig. 3.13 (top and left scales). They permit the direct reading of the probability of misses for normalized ranges and use them in eqn. (3.131).

3.9.1 Example of cumulative probability of detection calculation

Assuming that for a given radar integrating ten pulses and using an aerial scanning rate of 10 r.p.m. the range R_0 at unity signal-to-noise ratio, at a probability of false alarm of 10^{-6}, is 250 km when looking at a Swerling III target approaching the radar at a velocity of approximately Mach 2, the distance travelled by the target between successive scans is about 4 km. Then, if the radar's horizon for the target's altitude is 260 km, the normalized ranges and the corresponding probabilities of miss are

$$\frac{R}{R_0} = \frac{260}{250} = 1 \cdot 04 \qquad 1 - P_i = 0 \cdot 994$$

$$\frac{256}{250} = 1 \cdot 024 \qquad 0 \cdot 992$$

$$\frac{252}{250} = 1 \cdot 008 \qquad 0 \cdot 991$$

$$\frac{248}{250} = 0 \cdot 992 \qquad 0 \cdot 988$$

Inserting these into eqn. (3.128), one obtains

$$P_c = 1 - 0 \cdot 994 \times 0 \cdot 992 \times 0 \cdot 991 \times 0 \cdot 988 = 0 \cdot 0347 = 3 \cdot 5\%.$$

The probability of detecting the target by the time it reaches the specified range of 250 km is about 3·5%.

Figure 3.14 shows the growth of the cumulative probability of detection for sixteen sweeps of the considered radar aerial.

Chapter 4

The radar equation

4.1 The simplest form of the radar equation

Although the radar equation in its simplest form yields far from accurate figures of radar performance since it does not consider, amongst others, the effect of the environment and site effects, it is usually the starting point of most radar performance investigations. Besides that, it provides an easily evaluated figure of merit applicable to the comparison of radars.

Its derivation is based on the evaluation of the power density at the radar's receiving aerial due to the reradiation by a target in space illuminated by a source of electromagnetic energy.

The power density on the surface of a sphere of radius R due to an isotropic source of power P at its centre, is $P/4\pi R^2$, where a target having an echoing area σ (to be dealt with later) intercepts a portion $\sigma P/4\pi R^2$ of the total power and reradiates it towards the source. If the source is equipped with an aerial of gain G in the direction of the target, the reradiated power will be $G\sigma P/4\pi R^2$; an aerial of aperture, or effective area A, at the source will, in turn, intercept the power

$$P_R = \frac{PGA\sigma}{(4\pi R^2)^2}. \tag{4.1}$$

Most surveillance radars use a common aerial for transmission and reception.

The relation between aerial aperture and gain may be derived as follows (Starr, 1953):

Let the aerial's radiation in a given direction ϵ, relative to that in the direction of maximum radiation, be $F(\epsilon)$, while its gain in the direction of maximum radiation is G_0. The gain in the direction ϵ then becomes $G_0F(\epsilon)$ and the relative power density at a distance R is $G_0F(\epsilon)/4\pi R^2$, so that the energy flowing through an annular zone of height $R \sin\epsilon\, d\epsilon$ of the sphere's surface between angles ϵ and $\epsilon + d\epsilon$ is proportional to

$$\frac{G_0F(\epsilon)}{4\pi R^2}\, 2\pi R^2 \sin\epsilon\, d\epsilon = \tfrac{1}{2} G_0F(\epsilon) \sin\epsilon\, d\epsilon.$$

The total radiated relative energy is

$$\int_0^{\pi} \tfrac{1}{2} G_0 F(\epsilon) \sin \epsilon \, d\epsilon = 1$$

from where

$$G_0 = \frac{2}{\int_0^{\pi} F(\epsilon) \sin \epsilon \, d\epsilon}, \tag{4.2}$$

i.e. the maximum gain as a function of the distribution of radiation may be computed.

A small dipole's radiation distribution function is $F(\epsilon) = \sin^2 \epsilon$ so that the maximum gain becomes

$$G_0 = \frac{2}{\int_0^{\pi} \sin^3 \epsilon \, d\epsilon} = \frac{2}{4/3} = 1{\cdot}5.$$

The electromagnetic force intercepted by a dipole of length ds is $E\,e^{j\omega t}\,ds$ and if its radiation resistance, as shown by Starr (1953), is $80\pi^2\,ds^2/\lambda^2$, the maximum received power is

$$\frac{\tfrac{1}{2} E^2\,ds^2}{4(80\pi^2\,ds^2/\lambda^2)} = \frac{E^2\lambda^2}{640\pi^2} \quad \text{Watt.}$$

The power flow in a wave is $E^2/(2 \times 120\pi)$ where 120π is the impedance of free space. Dividing the maximum received power by the power flow yields the power intercepted by an aperture

$$A = \frac{E^2\lambda^2/640\pi^2}{E^2/240\pi} = \frac{3\lambda^2}{8\pi}.$$

Then, since the gain was found to be $G = 3/2$, one arrives at

$$G/A = \frac{3/2}{3\lambda^2/8\pi} = 4\pi/\lambda^2 \tag{4.3}$$

as the relation between aerial gain and aperture applicable to any aerial (Starr, 1953).

Returning now to eqn. (4.1) and remembering that the same aerial is used for both transmission and reception of radar signals, the basic radar equation becomes

$$P_R = \frac{PG^2\lambda^2\sigma}{(4\pi)^3 R^4}, \tag{4.4}$$

thus yielding an expression for the computation of the received power for known transmitter power, frequency, aerial gain, target echoing area and range, or the computation of the range when the above radar and target parameters and the received power are known.

In the latter case one requires the knowledge of the 'minimum detectable signal' if the maximum range at which a given target may be detected is to be computed. This signal depends on the required signal-to-noise ratio defined in Chapter 3.

It is a known fact that electrical networks and circuits generate noise. This

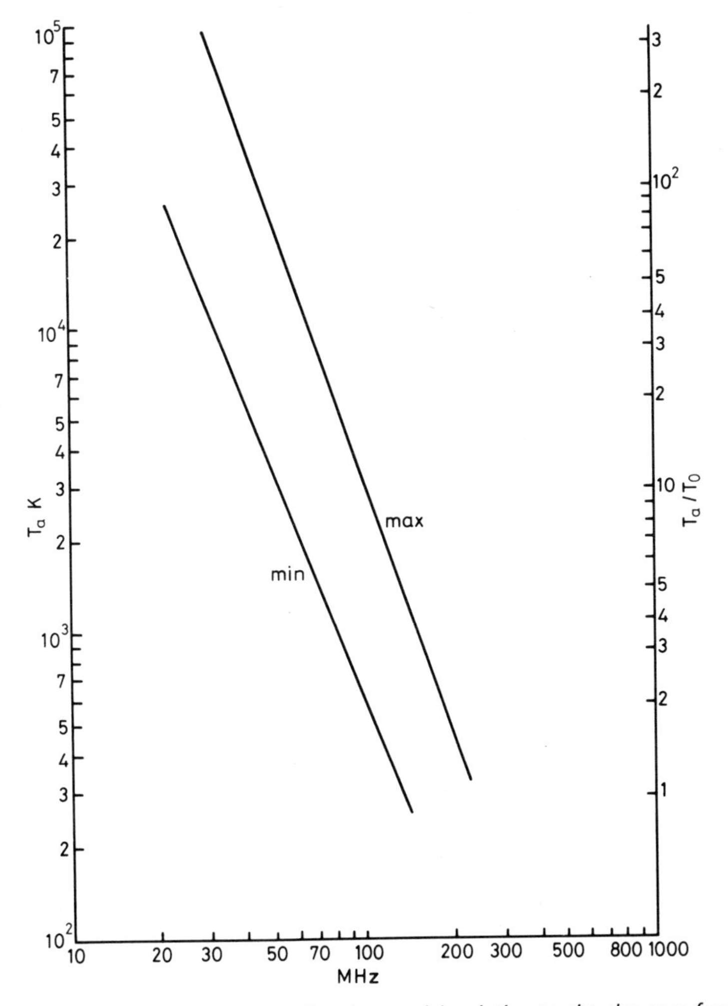

Fig. 4.1 *Range of noise temperatures T_a of an aerial pointing to the sky as a function of frequency (T_0 = 290 K) (After Rohan, 1964)*

leads to the specification of a receiver's sensitivity in terms of its noise figure (Friis, 1944):

$$F = \frac{\text{available input signal power}}{\text{available input noise power}} \Big/ \frac{\text{available output signal power}}{\text{available output noise power}}, \quad (4.5)$$

where the term 'available' means maximum power delivered to a matched load.

It has been shown (Rohan, 1964) that the minimum signal power which an aerial has to supply to the input of a receiver for a required signal-to-noise ratio is

$$S_{\min} = kT_0B\,[T_a/T_0 + F - 1]\,SN, \tag{4.6}$$

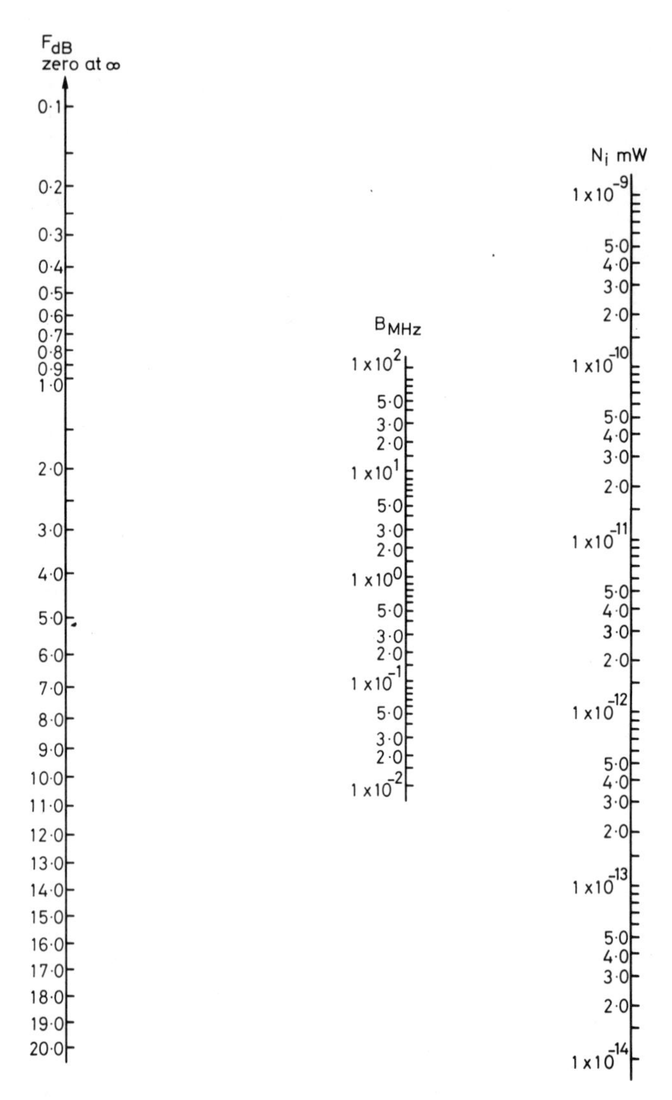

Fig. 4.2 *Equivalent internal noise power contribution N_i mW in terms of the noise figure F_{dB} and bandwidth B_{MHz} (After Rohan, 1964)*

where k is Boltzmann's constant ($1{\cdot}38 \times 10^{-23}$ Ws/K), T_0 is the absolute temperature of the receiver, usually taken to be 290 K, B is the pre-second detection bandwidth of the receiver which will be examined closer in the following, T_a is

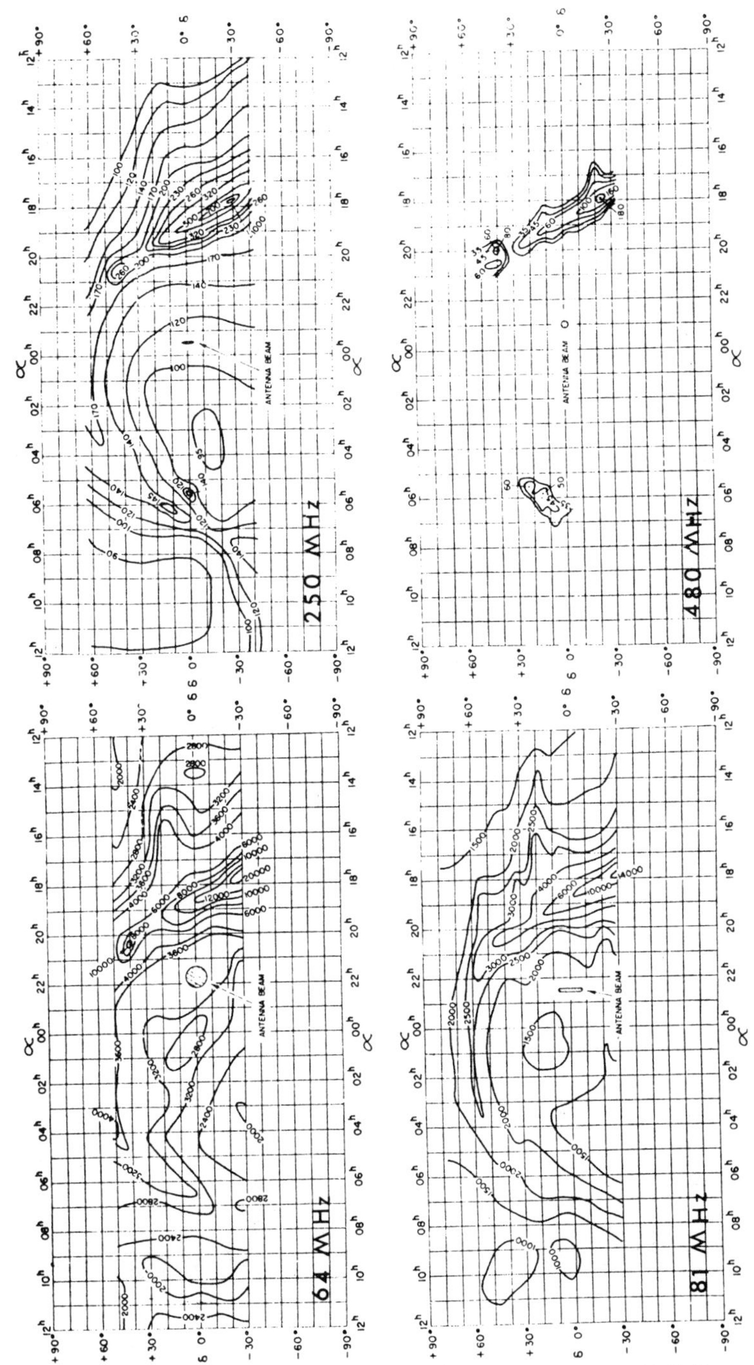

Fig. 4.3 *(a) Maps of the radio sky background (After Ko, P.I.R.E., 1958) (Copyright, 1958 IRE, now IEEE)*

δ = declination
α = right ascension

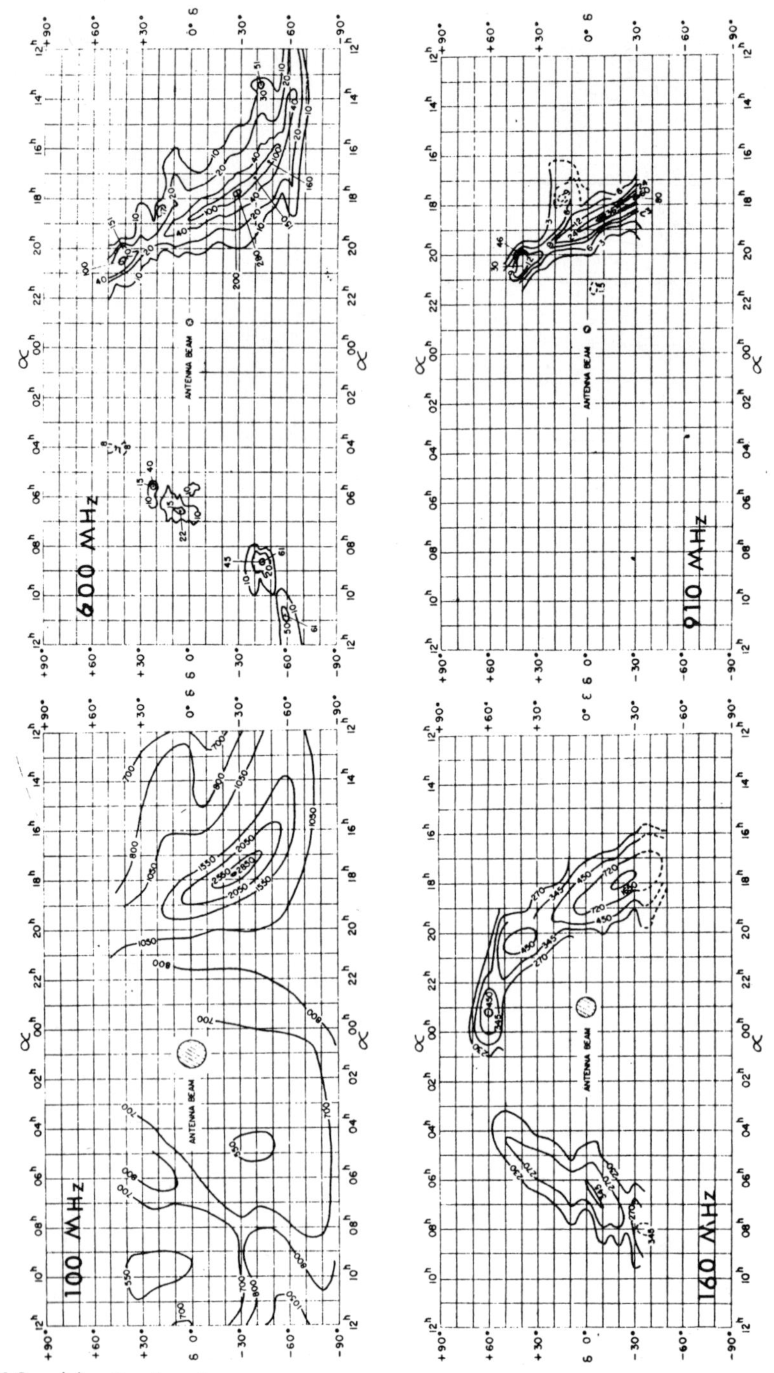

Fig. 4.3 *(a) – Continued*

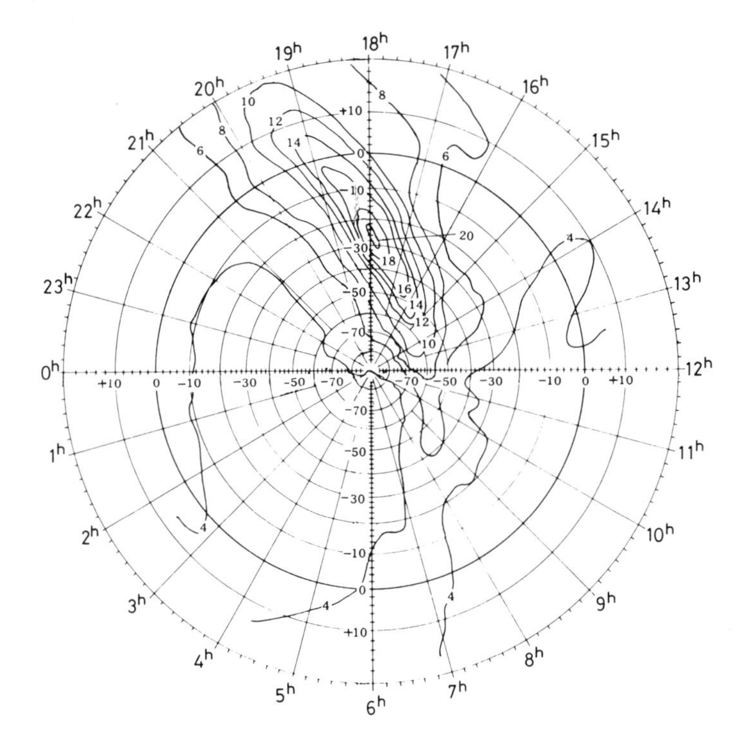

Fig. 4.3 *(b) Isophotal map of the southern sky at 55 MHz (After Rohan and Soden, 1970) contours labelled in 1000 K units.*

the aerial's noise temperature, F is the noise figure of the receiver, and SN is the required signal-to-noise ratio for a specified probability of detection and false alarm, as described in Chapter 3.

The limits of the noise temperature T_a of an aerial pointing at the sky as a function of frequency are shown in Fig. 4.1 (Rohan, 1964), while the nomograph of Fig. 4.2 (Rohan, 1964) permits the determination of the noise contribution of a network of bandwidth B MHz and noise figure F dB referred to its input terminals.

External noise, e.g. jamming, will be considered in Chapter 11.

Noise due to sky temperatures may be significant in the detection of small high flying objects. Maps of the sky temperatures for the northern hemisphere have been published by Ko (1958) and are reproduced in Fig. 4.3(*a*). A map of the southern sky temperatures has been determined by Rohan and Soden (1970) and it is shown in Fig. 4.3(*b*). The conversion to temperatures at other than the given frequency may be made by the application of the formula

$$\mathrm{dB}_f = \mathrm{dB}_{f_1} - k_1 \log(200/f_1) + k \log(200/f) \tag{4.7}$$

with $dB_f = 10 \log T_f$; f_1 and k_1 refer to the frequency of the sky background measurements given in the relevant illustrations, and f and k to the frequency of interest. The coefficients k and k_1 are given by

$$k, k_1 = 26, \qquad f < 200\,\text{MHz},$$

$$k, k_1 = 2{\cdot}7(35 - dB_{f_1} + k_1 \log (200/f_1)), \quad f > 200\,\text{MHz}.$$

The above formulae are based on Menzel (1960).

At frequencies above 0·1 GHz the ratio T_a/T_0 equals approximately unity (Livingston, 1961), so that eqn. (4.6) reduces to

$$S_{\min} = kT_0BF\,SN, \tag{4.8}$$

which, when inserted into eqn. (4.4) for the received signal power, yields the expression

$$R^4 = \frac{PG^2\lambda^2\sigma}{(4\pi)^3 kT_0BF\,SN} \tag{4.9}$$

for the computation of the free space range at which a target of echoing area σ may be detected with the probability of detection and false alarm specified by the used signal-to-noise ratio SN.

This is the simplest form of the radar equation which, however, does not apply to real situations since it neglects not only imperfections of the radar system, which may be expressed as losses, but also the influence of the radar's environment which modifies the performance to a great extent.

Before proceeding with the derivation of more complete forms of the radar equation, it is proposed to examine the parameters encountered so far.

4.2 Parameters of the radar equation's simplest form

4.2.1 The transmitter power

Since the signal-to-noise ratio used in detection theory, as described in Chapter 3, is the ratio of the peak instantaneous signal power to average noise power, it is obvious that P, as used in the above radar equation, must be peak transmitter power as distinct from peak transmitted power which must consider losses between the transmitter output terminals and the aerial's input terminals and also the efficiency of the aerial.

It is appreciated that the bandwidth B used in the denominator of the radar equation is not the customary 3 dB bandwidth but the 'equivalent noise bandwidth', the difference between which in most radar receivers is negligible, as is shown in Section 4.2.4. Optimum output signal-to-noise ratio, according to Lawson and Uhlenbeck (1950), is obtained for a bandwidth which is approximately equal to the reciprocal of the pulse duration τ. If one substitutes this for B in the radar equation, the numerator will contain the product of the peak power P and the pulse duration which is the pulse energy. The pulse duration is defined in these considerations as the interval between half-power points of the pulse power envelope.

The simplest way of determining pulse power is to compute it by dividing the measured average power by the duty cycle, i.e. the product of pulse duration and pulse repetition frequency. The average power is usually measured by the calorimetric method (Barlow and Cullen, 1950).

High peak powers and long pulses are required for high average powers. Achievable peak powers depend on the state of the art of microwave oscillator or power amplifier design, while usable peak powers are limited by the break-down voltage of the used microwave components. The required range resolution limits the applicable length of the pulses; the lower limit is dictated by the necessity of the pulse's extending over the dimensions of targets of interest in the direction of the measured range, while the upper limit determines the extent of the range cells. A 1 μs pulse is equivalent to a 150 m radar range as given by $R = c\tau/2$, where c is the velocity of propagation of electromagnetic waves.

Pulse compression techniques used in some modern radars achieve high average powers by transmitting relatively long pulses compressed by appropriate signal processing in the receiver in order to realize high range resolution.

4.2.2 Aerial gain, directivity, pattern factor, polarization and frequency

According to the *IEEE Standard Dictionary of Electrical and Electronics Terms* (1972), aerial gain in a given direction is 4π times the ratio of the radiation intensity in that direction to the net power accepted by the aerial from the connected transmitter, while the directive gain is 4π times the ratio of the radiation intensity in that direction to the total power radiated by the aerial.

Directivity is defined as the value of directive gain in the direction of its maximum value.

The relation between gain and directivity is

$$G = \eta D,$$

with G the gain, D the directivity and η the efficiency of the aerial.

The gain function describes the variation of the radiated power with direction,

$$G(\theta, \phi) = 4\pi \frac{P(\theta, \phi)}{P_T}.$$

Here $P(\theta, \phi)$ is the power radiated per unit solid angle in the direction θ, ϕ and P_T is the total radiated power.

The beamwidth of an aerial is the angular width of an aerial's radiation pattern between points at which the power level has fallen to one half of its maximum value. It is often referred to as the half-power or 3 dB beamwidth.

The aerial pattern factor defines the field strength in a given direction relative to that in the direction of maximum radiation.

Since the gain of an aerial is the function of its aperture and frequency, as shown by eqn. (4.3), a high gain aerial must have a large aperture to wavelength ratio. Aperture is limited by physical realizability, weight and wind load, while the choice of frequency depends on the intended application of the radar. Search radars usually operate at frequencies of 1 to 10 GHz where the influence of meteorological factors, i.e. oxygen and water vapour absorption does not impair their performance to a great extent.

Returning to eqn. (4.1) for the power intercepted by the radar's aerial and accepting that the product PG is the radiated power, one observes that the received power at a range R is a function of the product of the radiated power and receiving aerial aperture and its dependence on the frequency is implicit only through the dependence of the receiving aerial's gain on the wavelength, as given by eqn. (4.3). Search range is then indirectly dependent on the radar's frequency of operation.

It will be shown later that the elevation angle of the first maximum of the interference pattern due to multipath propagation, viewed from the base of the transmitting aerial is approximately $\lambda/4h$, where h is the aerial height; this intimates that higher frequencies or higher aerials are to be used if low flying targets are to be intercepted. Besides that, a high frequency permits the narrowing of the vertical beamwidth of the aerial thus reducing radiation towards the ground and diminishing the effects of multipath propagation.

On the other hand, the effects of clutter, both sea and ground, atmospheric attenuation and receiver noise are lower at lower frequencies.

It is obvious then that many aspects have to be considered for the judicious choice of the operating frequency if high performance is to be realized. However, the choice of the various radar parameters will, in the end, be a compromise.

Polarization of the used radiation must be also carefully considered since it may affect the radar's performance. Clutter, i.e. reradiation from unwanted targets, e.g. ground features, sea waves, rain, etc., depends on polarization. Generally, sea clutter is larger for vertical polarization than it is for horizontal one. Some target cross-sections are also polarization dependent. Circular polarization reduces rain clutter since, e.g. a right-hand circularly polarized wave becomes left-hand polarized when reflected by spherically-shaped rain drops. The complex shapes of aircrafts scatter both circular polarizations, so that transmitting and receiving circularly polarized waves of the same sense permits the detection of aircraft while reducing rain clutter. However, it was found that the echoing area of aircraft is, in general, less with circularly polarized radiation than with a linearly polarizaed one. The difference can be as much as 6 to 8 dB.

Some radars use frequency agility, i.e. changing their frequency of operation from pulse to pulse, in order to reduce the effect of frequency dependent clutter, to fill-in gaps between lobes in the vertical coverage diagram and to increase immunity to jamming. Others may change polarization of their radiation to facilitate detection or eliminate polarization dependent targets.

4.2.3 Target cross-section

Radar cross-section is defined by the *IEEE Standard Dictionary of Electrical and*

Electronics Terms (1972) as the portion of the back-scattering cross-section of a target associated with a specified polarization component of the scattered wave, while effective echoing area is the area of a fictitious perfect electromagnetic reflector that would reflect the same amount of energy back to the radar as the target. The terms radar cross-section and target echoing area are used in radar practice without distinction to denote the same concept, namely the ability to reflect incident electromagnetic radiation, and is generally defined as 4π times the ratio of the power reflected in the direction of the origin of radiation per unit solid angle to the incident power per unit area.

The echoing area of real targets varies greatly with the radar's aspect angle and the practice of using an equivalent point target has been generally adopted for radar calculations.

A sphere is the only target with an aspect independent radar cross-section.

Fluctuations of the received signal-to-noise ratio due to target cross-section fluctuations have been considered by Swerling (1962) as mentioned in Chapter 3.

A survey of the mathematical modelling of radar targets is given in Chapter 7.

4.2.4 Bandwidth

The effect of bandwidth on the performance of radar has been thoroughly studied in the early days of radar during the Second World War. The findings of the studies are extensively reported by Lawson and Uhlenbeck (1950).

It was mentioned earlier that it is the equivalent noise bandwidth and not the 3 dB bandwidth that is to be used in the radar equation. The equivalent noise bandwidth of an amplifier may be defined by

$$B_N = \int_0^\infty \frac{P(\omega)}{P(\omega_r)} d\omega,$$

where $P(\omega)$ is the power gain as a function of frequency and $P(\omega_r)$ the power gain at a reference frequency.

Then, since the general expression for the selectivity function of a unit of m stages, each consisting of a network of n tuned circuits, is

$$(1 + X^{2n})^{-m},$$

where X, as given by Valley and Wallman (1948), is the normalized detuning from the reference frequency ω_r, the noise bandwidth normalized to the 3 dB bandwidth of a single unit becomes

$$B_N/B_n = \int_0^\infty \frac{dX}{(1 + X^{2n})^m}$$

from where the noise bandwidth B_N for any synchronous or flat-staggered amplifier may be computed. B_n is the 3 dB bandwidth of a single unit.

Table 4.1 (Rohan, 1981) showing the bandwidth of m stages in terms of B, the

Table 4.1. *Cascade of identical single-tuned circuits*

Number of stages m	Relative bandwidth B_m/B	Noise bandwidth to 3 dB bandwidth of cascade B_N/B_m
1	1	1·57
2	0·64	1·22
3	0·51	1·16
4	0·44	1·13
5	0·39	1·11
6	0·35	1·1
7	0·32	1·1
8	0·3	1·09
9	0·28	1·08
10	0·27	1·07

bandwidth of a single stage, permits the determination of the noise bandwidth of m cascaded stages.

The 3 dB bandwidth of two synchronous cascaded stages ($n = 1, m = 2$) is then

$$B_2 = 0{\cdot}64\,B$$

and the equivalent noise bandwidth is

$$B_N = 1{\cdot}22\,B_2 = 0{\cdot}78\,B.$$

The error due to using the 3 dB bandwidth instead of the equivalent noise bandwidth diminishes with an increasing number of cascaded units in the intermediate frequency (i.f.) amplifier of a radar receiver. It may be shown (Rohan, 1981) that it is only 0·5 dB in the case of a single double-tuned stage and 0·2 dB for a triple-tuned one. The error introduced by using the 3 dB bandwidth of two single-tuned stages is only 0·9 dB.

Since the i.f. amplifier of most radar receivers consists of more than two stages, the 3 dB bandwidth may be used for radar performance calculations in most practical cases. Besides that, the 3 dB bandwidth is readily available since it is usually quoted in radar specifications and easily determined by measurement.

The expanded bandwidth of pulse-compression radars is accompanied by an appropriate change in peak power. Since the radar equation uses the ratio of peak power and bandwidth, it is immaterial which bandwidth is used in performance calculations provided that one uses the corresponding values of peak power and pulse duration.

The mathematical models to be described later require the insertion of the expanded bandwidth, compressed pulse duration and uncompressed peak power for the determination of pulse-compressed radar performance.

4.3 Extensions of the radar equation

An ideal radar in free space has been used for the derivation of the simplest form of the radar equation. It was assumed that all the energy generated by the transmitter is transferred, without losses, to the ideal aerial from whence it is radiated in a single direction towards the target, again without losses, reradiated by the target towards the radar and that the received power is such as to yield the specified signal-to-noise ratio to effect detection with the required probabilities.

This ideal case is never achieved in practice since losses even in the most carefully designed and maintained system are unavoidable; further, all aerials radiate also in other than the desired direction thus causing multipath propagation not accounted for by the radar equation as encountered so far.

Ranges computed from eqn. (4.9) will therefore be inaccurate and in order to yield more realistic results the radar equation must be expanded to account for losses and effects of propagation. Nevertheless, as pointed out earlier, the simple form of the radar equation is often used in radar practice for comparing the performance of radars.

4.3.1 System losses

All losses encountered in a radar system may be divided into three categories. Losses associated with the equipment, due to possibly a compromise in design or inadequate maintenance will, in this treatise, be termed 'equipment losses', while those caused by operational shortcomings will be referred to as 'operating losses', and losses connected with electromagnetic wave propagation will be called 'propagation losses'. A short survey of losses in these groups, not necessarily in order of their magnitude or importance, follows.

4.3.1.1 Equipment losses. (*a*) *Aerial losses*: The aerial gain, as used hitherto, was considered to be proportional to the aerial's aperture where this was taken to be equal to the physical aperture. In reality, not all the energy reaching the input terminals of the aerial is radiated in the desired direction; some is lost in side-lobes and in spill-over. To account for this loss, the effective aperture of the aerial, i.e. the aerial's projected area in the direction of the target is to be reduced by 30 to 60 percent.

Aerial gain, as given in manufacturers' catalogues, is computed from measured vertical and horizontal 3 dB beamwidths by using the expression for the power gain (Skolnik, 1962)

$$G(\theta, \phi) = \frac{\text{power gain per unit solid angle in azimuth } \theta \text{ and elevation } \phi}{\text{power delivered to the aerial}/4\pi},$$

which for θ and ϕ in radians becomes

$$G(\theta, \phi) = \frac{4\pi}{\theta\phi}. \tag{4.10}$$

Integrating the measured power density over 4π steradians of solid angle around the aerial, thus considering all the power radiated by the aerial, i.e. that in the main lobe, side lobes, spill-over, etc., leads to a more accurate gain appreciation.

(*b*) *Beam-shape loss*: Since the aerial beam is not rectangular but has a quasi elliptical cross-section, pulses transmitted towards a target and received from it by a scanning aerial will vary in amplitude according to the aerial's radiation pattern. The total signal power due to the sum of a given number of pulses radiated by a real aerial will be less than that radiated by an idealized aerial having a beam of rectangular cross-section. To account for this discrepancy, one introduces a loss factor which Marcum (1948–1960) called the beam shape loss.

Marcum assumed that the aerial beam within its 3 dB beamwidth may be approximated by a Gaussian curve

$$\exp\left(-a^2(\theta/B_\theta)^2\right),$$

where θ is an angle measured from the centre of the beam, B_θ is the 3 dB beamwidth in the direction of θ and the constant a is determined by the requirement that the value of the above function should be equal to $\frac{1}{2}$, when $\theta = B_\theta/2$, i.e.

$$\exp(-a^2/4) = \tfrac{1}{2}$$

from where $a^2 = 2{\cdot}773$.

A Gaussian beam approximation is acceptable within the 3 dB beamwidth. However, since the Gaussian pattern does not simulate side-band radiation, it is not usable beyond the above beamwidth.

If a target is off the beam centre in both the vertical and horizontal directions then the above normal curve approximation becomes

$$\exp\{-2a^2[(\theta/B_\theta)^2 + (\phi/B_\phi)^2]\}$$

and the aerial's gain dependence on the angles θ and ϕ is

$$G = G_0 \exp\{-2a^2[(\theta/B_\theta)^2 + (\phi/B_\phi)^2]\}.$$

For the derivation of the total signal power of k pulses received by the Gaussian aerial pattern and integrated without any additional losses, one assumes, for the sake of convenience, that in the series of the k pulses one pulse coincides with the centre of the beam (Skolnik, 1962) and that a transmitted pulse and its echo are subjected to the same modification in amplitude by the aerial's pattern. If the received signal power due to a pulse transmitted and received at the centre of the beam is S_1, then the total power of k pulses received by the Gaussian aerial pattern and losslessly integrated is, according to Skolnik (1962),

$$S_k = S_1\left[1 + 2\sum_{j=1}^{(k-1)/2} \exp\{-2a^2 j^2[(\Delta\theta/B_\theta)^2 + (\Delta\phi/B_\phi)^2]\}\right]$$

with $\Delta\theta = B_\theta/k$ and $\Delta\phi = B_\phi/k$, the vertical resp. horizontal angular separation between pulses.

The beam shape loss of a radar integrating k pulses is $L = kS_1/S_k$; denoting then the reciprocal of the sum in the inner square brackets of the above expression, i.e. the sum of pulses within the 3 dB beamwidth in the two considered and mutually perpendicular directions, by k_B^2, the beam shape loss becomes

$$L = \frac{k}{1 + 2\,\Sigma_{j=1}^{(k-1)/2} \exp\,[-(2a^2 j^2)/k_B^2]}, \qquad (4.11)$$

where $k < k_B$.

If there is no pulse transmitted and received at the centre of the beam, i.e. if the target is not intercepted by the beam's centre, the value of S_k will be lower and the increase in the beam shape loss will be the ratio of the square of the maximum achieved aerial gain to the square of the aerial gain at the beam's centre.

(*c*) *Scanning loss*: If the aerial's scanning rate is such that the assumption of equal gain on transmission and reception is no longer valid, an additional loss, the so-called scanning loss, has to be introduced into the radar equation. This loss is but rarely considered in currently used search radars. However, it may be of significance when very fast aerial scanning rates are used, e.g. in some special naval radars where the high scanning rate helps sea clutter suppression (Croney, 1966) and in very long range radars used for the observation of satellites or celestial bodies.

(*d*) *Feeder losses*: Aerials, except in the case of low-power and/or portable equipment, are seldom connected directly to the output of transmitters or input of receivers. A length of transmission line, be it an open line, coaxial cable or waveguide, is usually interposed between the aerial and the terminals of the equipment. Losses in the line decrease the signals fed from the transmitter to the aerial and also those received by the aerial to the receiver.

While these losses may have a tolerable effect on the high transmitted powers, they are objectionable when very weak signals are fed into receivers since the deterioration of the input signal reduces the overall signal-to-noise ratio which is tantamount to increasing the aerial's noise temperature.

Denoting the attenuation of the transmission line by α and expressing the output power as $P_{out} = (1-\alpha)P_{in}$, the loss factor becomes $L = P_{in}/P_{out} = 1/(1-\alpha)$.

The noise contribution of the line at a temperature T_{amb} is $\alpha k T_{amb} B$ and that of the aerial at a temperature T, at the output end of the transmission line, is $(1-\alpha)kTB$ with k denoting Boltzmann's constant. The quantity B is the equivalent noise bandwidth of the system.

The total noise power, $kT_{out}B$, must equal the sum of the above two noise powers so that the noise temperature at the input to the receiver becomes

$$T_{out} = \alpha T_{amb} + (1-\alpha)\,T,$$

which, when expressed in terms of the loss factor, reads

$$T_{out} = T_{amb}\frac{L-1}{L} + T/L. \tag{4.12}$$

Figure 4.4 (Rohan, 1964) plots this expression as a function of line losses for an ambient temperature of 290 K and a number of temperatures T as parameters. It may be seen that the noise temperature of a 50 K aerial rises to over 100 K at the output terminals of a line having a loss of 1 dB.

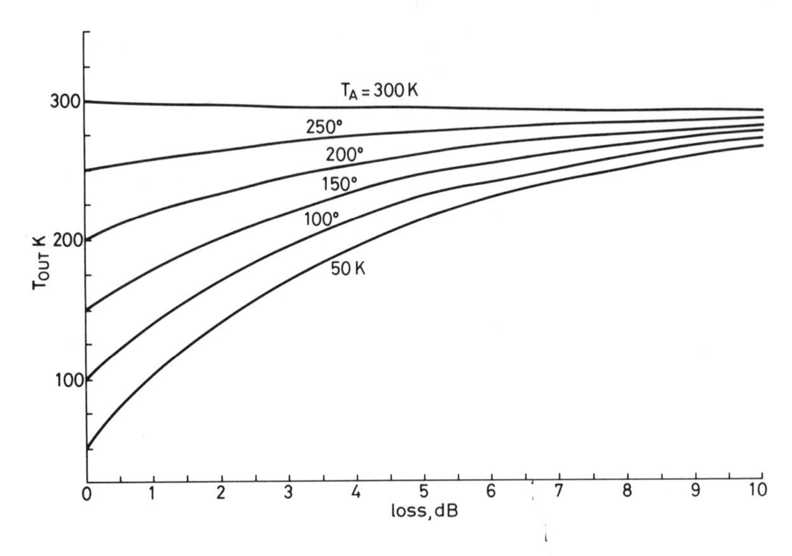

Fig. 4.4 *Line output noise temperature as a function of line losses in dB for an ambient temperature of 290 K (After Rohan, 1964)*

(*e*) *Integration loss*: It has been mentioned earlier (Section 3.3.1) that most radars integrate received signal pulses in order to increase the probability of detection and reduce the probability of false alarm.

The integration improvement factor was defined as

$$I(k) = kE(k),$$

where

$$E(K) = \frac{SN_1}{k\,SN_k} \tag{4.13}$$

is the integration efficiency (Skolnik, 1962) with SN_1 the single pulse ($k = 1$) signal-to-noise ratio required for a specified probability of detection, and SN_k the signal-to-noise ratio per pulse necessary to yield the same probability of detection when k pulses are integrated.

Since the signal-to-noise ratio per pulse SN_k of an ideal coherent integrator, when k equal amplitude pulses are integrated, is SN_1/k, the integration efficiency is unity or 100%.

The improvement achieved by other modes of integration is always less than that achievable by coherent integration. Marcum (1948) introduced the concept of integration loss relative to coherent integration with all pulses of equal amplitude and expressed it by the reciprocal of the integration efficiency. This, in dB, reads

$$L_i = 10 \log \frac{1}{E(k)} = 10 \log \frac{k\, SN_k}{SN_1}. \tag{4.14}$$

The integration loss depends on the probability of detection and false alarm. However, this dependence is rather weak.

(*f*) *Collapsing loss*: Collapsing loss denotes the difference between the signal-to-noise ratio, expressed in dB, required for a given probability of detection when additional noise pulses are integrated together with the received signal pulses and the signal-to-noise ratio required for the same probability of detection when only signal pulses are integrated.

It accompanies any loss of resolution and may be due to, e.g. improper receiver bandwidth, improper gate width, excessive cathode-ray tube spot size, collapsing of three-dimensional data into two dimensions, etc. Marcum (1948) defines the general form of the collapsing ratio, used for the determination of the above effects, as

$$\rho = \frac{M+N}{N},$$

where M is the number of integrated additional effective pulses and N is the number of integrated signal plus noise pulses.

He defines the effective collapsing ratio for the case when the loss is caused by an improper video bandwidth B_v as

$$\rho_{\mathrm{eff}} = \frac{B+B_v}{B_v}, \tag{4.16}$$

where B is the overall receiver 3 dB bandwidth, while for the case of loss caused by low writing speed of the cathode-ray beam, the effective collapsing ratio, according to Marcum, is

$$\rho_{\mathrm{eff}} = \frac{d+s\tau}{s\tau} \tag{4.17}$$

with d denoting the spot diameter, s the writing speed and τ the pulse length.

The collapsing loss, based on the above definition, may be computed from the ratio of the integration losses due to the integration of $M+N$ pulses to that of N pulses alone (Skolnik, 1962). These, as shown earlier, are proportional to the reciprocals of the corresponding integration improvement factors $I(M+N)$ resp.

$I(N)$ so that the expression for the collapsing loss becomes

$$L_c = \frac{M+N}{N} \frac{I(N)}{I(M+N)} = \rho \frac{I(N)}{I(M+N)}. \tag{4.18}$$

Since the number of pulses, when the additional noise pulses are integrated, is larger than that considered for the determination of the detection threshold, it is obvious that a new, higher threshold has to be used if the false alarm probability is to remain the same as that specified for the case of integration without the additional noise pulses.

It should be pointed out that since noise is a random quantity, it cannot be integrated coherently so that there will be a collapsing loss even in systems using coherent integration. The integration improvement factor $I(N)$ due to coherently integrating N equal amplitude signal plus noise pulses is N which, when inserted into the above expression, yields the collapsing loss for a coherently integrating receiver as

$$L_c = \frac{M+N}{I(M+N)}. \tag{4.19}$$

(*g*) *Limiting loss*: Since limiting of the amplitude of the received signal reduces the signal-to-noise ratio, limiting occurring in any part of the receiver will reduce the probability of detection if everything else is held constant.

While a properly designed and constructed receiver does not limit the received signals, intensity modulated cathode-ray displays, e.g. PPI displays, have a limited dynamic range and may limit.

Marcum (1948) defines the limiting ratio as the ratio of the limit level to the r.m.s. noise level and concludes that for a large number of integrated pulses the limiting loss is only a fraction of a dB if the limiting ratio is as large as 2 or 3, but if only one or two pulses are integrated the limiting ratio must be in the neighbourhood of 10 in order to prevent a serious loss.

Limiting in the output of an integrator may also cause a loss which, however, in most practical cases, is small compared with the loss due to limiting individual pulses.

Davenport (1953) has shown, for small signal-to-noise ratios in a band-pass limiter, that the reduction of the signal-to-noise ratio of a sine-wave embedded in narrowband Gaussian noise is about 1 dB. The output signal-to-noise ratio is proportional only to the input signal-to-noise ratio because frequencies remote from the centre frequency are filtered out before they reach the limiter's input.

(*h*) *CFAR and MTI loss*: Some more sophisticated radar receivers are equipped with signal processing features aimed at reducing the effect of clutter either by automatically adjusting the detection threshold to a sampled noise, jamming or clutter background in order to maintain a constant false alarm rate (CFAR), or by discriminating between stationary and moving targets in a moving target indicator (MTI) circuit.

A second reference channel, in the CFAR case, samples the noise background in cells adjacent to the cell of interest and determines the noise to be expected in the signal channel. The detection threshold of the signal channel is usually set above the actually required level in order to avoid excessive false alarm rates due to erroneous estimation. A loss in detection sensitivity caused by the raised threshold is called CFAR-loss. It depends on the number of noise samples used for the estimation and is shown by Fig. 4.5, based on data given by Hall (1962–63) for a probability of detection of 0·5. Hansen and Ward (1972) provide an empirical expression

$$m_{\log} = 1{\cdot}65\, m_{\text{lin}} - 0{\cdot}65$$

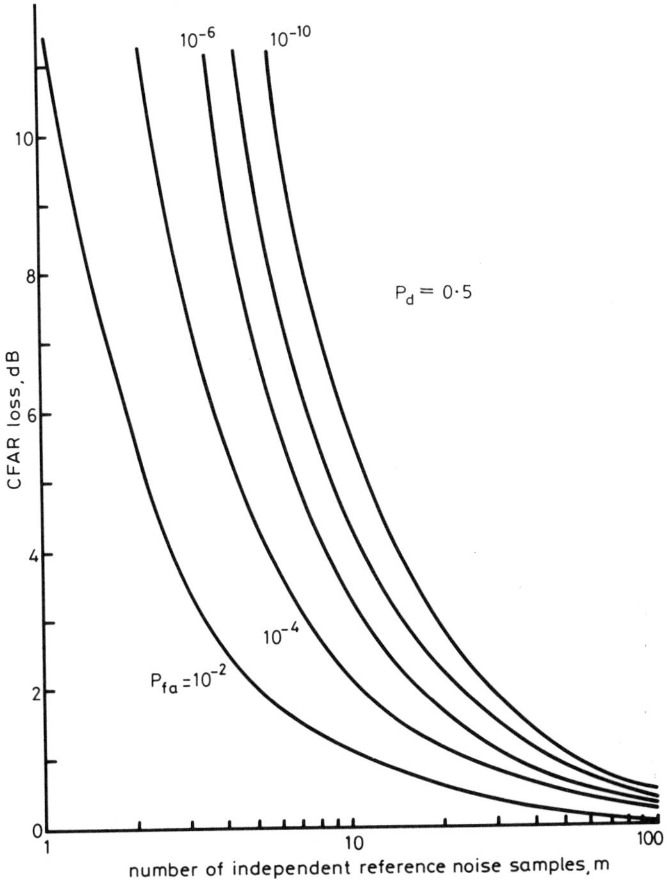

Fig. 4.5 *CFAR loss for $P_d = 0{\cdot}5$ as a function of the number of reference noise samples (After Hansen and Ward, 1972) (Copyright, 1972, I.E.E.E.)*

for the computation of the required number of reference samples $m_{\log}$ in a logarithmic detector receiver for its CFAR losses to equal those of a linear detector receiver.

The CFAR loss falls rapidly with an increasing number of reference noise samples.

In the moving target indicator (MTI) the target echo is compared with a delayed phase-coherent sample of the transmitted pulse. The amplitude of the video signal at the output of a phase discriminator is constant for echoes of fixed targets but fluctuates for those originating from moving ones. Subtracting the coherent video signals of successive pulses from one another in a canceller cancels echoes of constant phase and hence amplitude, while echoes of moving targets having variable phase relations with respect to the transmitted pulse produce video signals.

All components of the MTI system must exhibit high phase stability if good fixed echo cancellation is to be achieved. The system's inherent phase fluctuations determine the minimum phase difference $\phi_{\mathbf{min}}$ which produces a still observable target echo.

While MTI reduces clutter, it introduces losses due to a number of inherent characteristics.

The MTI canceller causes pulse-to-pulse correlation of the output noise (Hall and Ward, 1968); hence, if only a single output pulse is used in a two-pulse canceller, this pulse will have to be produced by two output pulses. For targets uniformly distributed over all velocities, the average canceller output signal power is twice the power of the individual pulses. However, the noise power is also the sum of the noise power of the two pulses, so that the output and input signal-to-noise ratios will remain equal.

Integrating only alternate pulses in a two-pulse canceller halves the number of integrated pulses so that the increase in input signal-to-noise ratio required to yield the same probability of detection and false alarm, as achievable in a non-MTI radar, may be expressed as a loss

$$L_{\mathbf{MTI}a} = \frac{E(k_e)}{E(k)},$$

where the measure of the integrator's effectiveness is

$$k_e = \frac{k^2}{k + 2\,\Sigma_{r=1}^{k-1}\,(k-r)R^2(r)}$$

with k the total number of independent signal samples and

$$R(r) = \frac{\overline{e_2(i)\,e_2(i+r)}}{\overline{e_2^2}}$$

the output correlation coefficient of the interference. Expressions for the $e_2(\,)$ terms are given by the above two authors. $R(r) = 2/3$ for a two-pulse canceller and 18/35 for a three-pulse one.

An MTI canceller operating only on the in-phase component of the receiver output exhibits a fluctuation loss (Vannicola, 1973) of

$$L_{\mathrm{MTI}b} = \frac{L_f(Kk/2)}{L_f(Kk)}$$

given by the ratio of the fluctuation loss L_f due to halving the number of degrees of freedom to the fluctuation loss for MTI-less operation.

Besides that, the reduction in the number of target samples introduces an additional loss

$$L_{\mathrm{MTI}c} \approx \frac{E(k/2)}{E(k)}$$

proportional to the required signal-to-noise ratios in the MTI and non-MTI cases for equal values of the probability of detection and false alarm.

A further type of MTI loss in systems operating at a low pulse repetition frequency occurs for targets moving at a velocity near the 'blind speed' of

$$v_b = \frac{\lambda}{2} f_r$$

or its integral multiples. The magnitude of this loss depends on the phase of the sampled phase detector output and the fraction of the observed cycle.

Kuhrdt (1961) gives

$$L_{\mathrm{MTI}d} = \frac{R_{\mathrm{MTI}}}{R_{\mathrm{normal}}} = \sqrt{\frac{4}{\pi}\left[\left|\sin\frac{\phi}{2}\right| - \left|\cos\frac{\phi}{2}\right| \frac{\phi_{\min}}{2}\right]}$$

for $2\pi - \phi_{\min} > \phi > \phi_{\min}$ and $R_{\mathrm{MTI}} = 0$ for $\phi < \phi_{\min}$ and $\phi > (2\pi - \phi_{\min})$ as the MTI radar range referred to that of a radar without MTI. The angle ϕ, for target velocities v, is

$$\phi = 2\pi \frac{v}{v_b}.$$

The above range ratio is periodical in ϕ with a period of 2π. No video output is obtained, even at short ranges, for values of the angle $\phi < \phi_{\min}$.

(*i*) *Equipment degradation*: Since some components used in the production of electronic equipment have a limited life, their performance and consequently the equipments' performance deteriorates with time. Electronic valves are the best example of such components.

It is obvious then that transmitter power, as quoted by manufacturers, measured under laboratory conditions and using possibly selected valves, will differ from the transmitter power of the same or same type of radar used and maintained often by semi-skilled personnel.

Other possible sources of losses due to deterioration are waveguide and cable joints, detuned circuits, power dividers, transmit–receive switches, rotary joints, etc.

The evaluation of losses caused by the above and possibly other factors is arbitrary and although many radar users monitor the performance of their equipment continuously and statistical data for individual radars are available, it is not possible to extrapolate the findings for the prediction of the future performance of the same radar let alone that of other radars of the same or other types.

One is therefore restricted to making a considered estimate of these losses and correct the ranges calculated on the basis of manufacturers' data to account for the inevitable deterioration of the equipment.

Besides direct equipment degradation, there are other causes of performance deterioration, namely instabilities and drifts.

The probabilities of detection and false alarm which determine the signal-to-noise ratio used in the radar equation are functions of the detection threshold. If the threshold drifts upwards, the above probabilities decrease and vice versa. Some tolerance in the threshold level is usually allowed for in practice. The loss due to the above instabilities depends to a certain extent on the frequency and thoroughness of adjustments during maintenance and the quality of the used circuits and components. A skilled operator watching a PPI display and knowing the detection ranges of known targets is usually capable of detecting a deterioration of performance.

4.3.1.2 Operating losses: As it was indicated earlier, operating losses are due to improper operation of the radar equipment. Radar users adapt operating procedures to suit their particular requirements; besides that, individual operators might overlook or change, perhaps inadvertently, some procedures and so affect the overall efficiency of the system. Although some of the most modern radars use automatic detection, the human operator observing a PPI display still remains the prevalent type of detector for surveillance radars and his performance, the last link in the process of detection, is therefore of considerable importance.

Extensive experimental work aimed at assessing the radar operator's performance was undertaken during the Second World War and its findings are reported on by Lawson and Uhlenbeck (1950). Some of these influenced the design of radar equipment. Baker (1962) describes the findings of additional investigations performed by a large group of investigators over about fifteen years immediately after the war.

(*a*) *Operator loss*: An operator's capacity to concentrate adequately to detect on a CRT screen the appearance of paint due to a target approaching from an unknown direction depends on his state of mind. It is obvious that a rested operator is likely to perform better than one who is tired, but even there, threat, e.g. by an enemy, or penalty, will spur a tired operator to perform better than he would perform under untroubled conditions. Further, the operator's senses, his eyes and ears, and brain are subject to certain limitations which may restrict the assimilation of the amount of useful information. The information bandwidth of an alert human operator is, according to Skolnik (1962), of the order of 10 Hz (20 bits/s) but the rate at which information can be displayed on a PPI may be many times above the operator's capacity of its absorption. Typically, a bandwidth of about 150 Hz would be required for the rate at which information is displayed by conventional search radar PPI's. The bandwidth mismatch causes losses in the operator's performance.

In the experiments described by Lawson and Uhlenbeck (1950), operators, aware of the azimuth and six possible ranges at which target returns could appear, were to determine on an A-scope and PPI display, in sets of trials in which various parameters were varied, the range at which they thought that they have seen targets.

The influence of the following parameters on the detectability of radar signals by human operators was studied:

Trace brightness, average noise deflection and sweep parameters.
The product of the intermediate frequency bandwidth and pulse length $B\tau$.
The product of video bandwidth and pulse length $b\tau$, the product of sweep speed and pulse length $s\tau$ and focus.
Pulse repetition frequency PRF, sweep repetition frequency SRF.
Signal presentation time and screen material.
Attention interval, number and spacing of possible signal positions.
Video mixing.

While it is not intended to dwell on the experiments themselves, their results are of adequate interest to warrant their brief description.

A method of objective threshold setting had to be found first. This essentially correlates the observer's answers with some desired signal parameter which, in case of an A-scope display (signal amplitude vs. range) is the range position of the signal, while in case of a PPI display it could be the azimuth angle. The operator's guesses are correlated with the known signal positions in order to determine whether the signal is perceptible, thus minimizing the judgement required from the observer. A great number of tests with a random variation of range positions and involving many operators is required if the result is to be meaningful.

Before quoting the results of the investigations of signal threshold level dependence on trace brightness, it is important to realize that on a cathode-ray-tube display there is a gradient of brightness in the radial dimension, highest brightness occurs near the centre of the display and it decreases towards its periphery. Optimum brightness should therefore be set at the radial position of greatest importance to the task.

The experiments described by Lawson and Uhlenbeck (1950) have shown that only a slight change in threshold level is produced by a 10^6 variation of the trace intensity on both, the short persistence P1 and long persistence P7, screens. Modern radars use long persistence P19 and very long persistence P26 screens which were unavailable at the times of the described experiments. It was shown further that as long as the receiver gain is such as to produce, on an A-scope, an average noise deflection in excess of 0·5 mm, the threshold power is independent of the average deflection and hence receiver gain. It was observed that diminishing noise deflection causes the signal threshold power to rise; the signal threshold power becomes inversely proportional to the receiver gain when the noise deflection falls below 0·2 mm. Limiting, which is used in PPI displays to prevent defocusing of the CRT spot by large signals, may affect threshold signals to a great extent.

There is no fundamental difference in signal perception on A-scopes having different trace directions. However, lack of adequate training in the appreciation of a vertical sweep may cause a difference of several dB in the threshold signal level.

The dependence of the minimum detectable signal and hence the signal threshold on receiver bandwidth is given by eqn. (4.8). Since radar signals are in the shape of pulses, the energy of which is given by the product of the pulse amplitude and pulse length, the threshold signal pulse energy is proportional to the product of the i.f. bandwidth B and pulse length τ. Experiments involving some 10 000 individual observations have shown that minimum signal threshold power occurs when the product $B\tau$ is approximately equal to 1·2. When $B\tau \gg 1{\cdot}2$ the curve representing the dependence of the threshold signal power on the above product approaches asymptotically a line parallel to the curve showing the receiver noise power–receiver bandwidth relation. For $B\tau \ll 1{\cdot}2$ the signal threshold power is proportional to $1/B\tau$. Similar results were obtained for both the A-scope and PPI displays.

Due to the cross-modulation of noise components in the i.f. output signal, the video noise spectrum differs from that of the i.f. noise and therefore video bandwidth variations affect signal threshold power differently from that found for i.f. bandwidth variations. Since the highest significant video frequency at the output of the second detector is numerically equal to the i.f. bandwidth, an increase of the video bandwidth beyond this value will not produce any noticeable effect on the signal threshold power. Experimental evidence shows that increasing the value of the product of the video bandwidth and pulse length beyond 0·5 will not affect the above threshold. Similarly, the lowest threshold power is obtained for a trace length (the product of sweep speed and pulse length) of 1 mm which represents a 0·2° subtended angle at the eye from a viewing distance of 30 cm. For larger and smaller values of trace length an increase in threshold signal power was observed.

Defocusing in a direction perpendicular to the sweep was found to have practically no effect on threshold signals as long as the spot size does not exceed the spatial size of the signal caused by the aerial beamwidth. However, defocusing in a direction parallel to the sweep causes the signal threshold power to rise. The rise sets in when the spot length is approximately 1 mm long and continues to rise with increasing defocusing.

The signal threshold power was found to be inversely proportional to the square root of the pulse repetition frequency. This finding is in agreement with earlier statements (see Sections 3.3.1 and 3.3.2) about the integration improvement in case of cathode-ray-tube display–human observer integration being proportional to the square root of the number of integrated pulses. The number of integrated pulses is a linear function of the pulse repetition frequency as will be shown later. Increasing the number of signal pulses reduces the detectable threshold level.

Since the signal threshold level, as it has just been mentioned, is inversely proportional to the square root of the number of integrated pulses and this, in turn, is inversely proportional to the aerial's scanning rate, it is justified to expect the signal threshold power to be directly proportional to the square root of the sweep repetition frequency. This expectation is borne out by experimental evidence for A-scope displays but only partially and only for single scans in the case of PPI displays provided that the scanning frequency is such that the signal presentation time over which effective integration may take place does not exceed a maximum of several seconds and that at least one but preferably a few signal pulses are present during the aerial's transit. If the persistence of the CRT phosphor is such that the signal is present over several scans, scan-to-scan integration may occur; however, since increasing the scanning rate reduces proportionately the number of hits per scan, the number of pulses remains constant so that the signal threshold power in case of high (above 6 to 10 r.p.m.) scanning rates is almost independent of the scanning frequency.

Payne-Scott (1948) investigated, amongst others, the effect of the pulse repetition frequency and aerial scan rate on visual detection on PPI displays and found that for optimum visibility the highest pulse repetition frequency, which with the chosen pulse duration will not cause the permissible valve dissipation to be exceeded, and as low an aerial scan rate as possible is to be used.

Lawson and Uhlenbeck (1950) have shown that the minimum detectable signal power is directly proportional to the square root of the sweep repetition frequency and inversely proportional to the first power of the pulse repetition frequency. They quote results of experiments which show that for a constant pulse repetition frequency and varying signal presentation time the signal threshold power for an A-scope display is inversely proportional to the square root of the signal presentation time provided that this equals or exceeds 0·1 second. For shorter presentation times the threshold is inversely proportional to the first power of this time and for presentation times in excess of about 10 seconds the threshold becomes almost independent of it.

In the PPI case the signal presentation time depends on the aerial's scanning frequency, its horizontal beamwidth and the pulse repetition frequency. At a constant pulse repetition frequency the signal presentation time depends directly on the aerial's horizontal beamwidth and inversely on its scanning rate. The threshold signal power, on the basis of the A-scope experience, is expected to vary with the inverse of the square root of the signal presentation time and hence the inverse of the square root of the horizontal beamwidth. This is substantiated by experimental evidence over almost the whole range of the used beamwidths of 0·18° to 360°. At beamwidths below 1° the length of the displayed spot is less than 1 mm causing the signal threshold power to rise in accordance with what was said earlier. Similarly, for beamwidths in excess of 45°, where the operator cannot view the entire arc simultaneously, the signals are seen less efficiently.

Changes in the aerial's beamwidth produce changes in the number of hits per scan and possibly in the number of integrated pulses; the conclusions on the dependence of the signal threshold power on beamwidth, based on experimental data, contradict the erroneous assumption that wider arcs of broader aerials are easier to distinguish from a uniform noisy background. Experiments in which the signal presentation time and pulse repetition frequency were held constant and beamwidth variations were simulated by variations of the scanning frequency have shown that the signal threshold power is constant for all practical purposes over the whole range of beam angles. In radar practice where the pulse repetition frequency and aerial scanning rate are constant, dependence of the signal threshold power on the beamwidth is given implicitly by the dependence of the number of hits per scan on the aerial's beamwidth.

The signal threshold was found to rise with the number of positions at which signals may appear, and similarly on an A-scope display it was found to increase with the increase of the spacing of these positions. The lowest threshold was obtained for a 1 mm spacing. The position of the signal on a PPI display, as mentioned earlier, will affect its brightness; signals nearer the middle are brighter since they diminish in size. However, since the product of area and intensity remains constant, no appreciable range effect on the signal visibility was experienced (Baker, 1962) provided that the signals were restricted to ranges of less than 0·3 and more than 0·7 of the PPI's radius. Baker found that operators tend to concentrate on the observation of the annulus between the above dimensions and appear to ignore peripheral areas.

An operator anticipating the possible appearance of a signal will observe the display for a longer time than the duration of the signal's display. The interval of the display's observation was termed the attention interval. Sydoriak (see Lawson and Uhlenbeck, 1950) has shown experimentally that the signal threshold rises with an increase in the attention interval. The reason for this seems to be that the signal threshold power must rise in order to offset the increased chance of mistaking a large noise peak for a signal.

Occasionally the outputs of two or more receivers are mixed and signals from either receiver are observed. If, in the case of mixing only two outputs, the gains of the receivers are G_1 resp. G_2, the mixing ratio of the first receiver will be G_2/G_1, while that of the second one will be the reciprocal of this ratio. For unity mixing ratio the noise levels of the two receivers are equal. Sydoriak (Lawson and Uhlenbeck, 1950) has shown that the signal threshold power on an A-scope display increases with a rising mixing ratio. The reason for this lies in the manner in which an operator detects the presence of a signal by observing at the signal's position in an A-scope trace the occurrence of a small opening within which no noise fluctuations may be observed. At high mixing ratios the noise output of the other receiver may be observed in the gap due to a signal received by the first one.

While signal presentation on a PPI differs from that on an A-scope, similar dependence of the signal threshold power on the mixing ratio of the two channels was experienced. It manifests itself in the variation of the total light output of the signal paint which is a function of the linearity of the video amplifier, limiting and the dependence of the screen brightness on the exciting current.

It was mentioned earlier that with PPI displays it is necessary to use video limiting so as to prevent defocusing by large signals and reducing certain types of interference. Sydoriak (Lawson and Uhlenbeck, 1950) has studied the dependence of the signal threshold power on the limit level, i.e. the ratio of the limit level voltage to the average video voltage, and found that in the case of one channel limiting the signal threshold power rises sharply when the limit level falls below about 1 because at low limit levels the highly limited signal cannot override random noise fluctuations if the spot size is large compared with the signal pulse. Particular care must be exercised in limit level setting of the channels when video mixing is used, since the noise contribution of an unlimited channel would impair the occurrence of a well-defined upper limit.

Baker (1962) quotes experiments staged for the determination of the optimum PPI display size which have proved that for the search role a 12·5 to 17 cm (5 to 7 inch) diameter scope is superior to larger and also smaller ones.

The operator's performance may be affected by various environmental factors such as background noise, ambient temperature and illumination. All of these in excess of what the operator is used to will prove detrimental to his performance.

Tests conducted during the later years of the Second World War have shown, as reported by Baker (1962), that operators perform best in the early part of their watches. A group of 25 radar operators missed only 15 per cent of the signals in the first half hour of their watch as against 26, 27 and 28 per cent in the following three half hour periods respectively. Authorities which use radars employ, as a rule, several operators working simultaneously. Adding a second operator was proved experimentally to increase the probability of detection; however, no advantage of multiple operation can be expected unless each operator searches actively and continuously.

Frequent rest periods, e.g. alternation of a 30 minute watch with a 30 minute rest prevents a decrement in performance during an overall period of two hours. There is, however, suggestive evidence that if the resting observer sits back and watches his relief operator at work he derives no benefit from his rest interval.

It is interesting to note that radar operators at Adelaide's West Beach Airport, a relatively quiet airport, work seven hour shifts.

It was observed by Deese and Ormond, as reported by Baker (1962), that high signal frequency aids maintaining the radar operator's vigilance. Signal frequency and regularity are, in some situations, unknown quantities. However, injecting artificial signals, which should be indistinguishable from real signals, may be used to improve or, in some services, to check the vigilance of the operators.

The following method may be used for the latter task. An artificial target emerging from a partially screened trajectory is made to appear at an initial range on the display where power density is sufficient to produce a strong echo. The control operator, who knows when and where to expect the signal of the artificial target, may check then the vigilance of the operator from the range appearing for this target on the plotting board where it is entered on the instruction from the operator.

The screened trajectory of the above method may be replaced by the track of a real, but to the operator unknown, target which passes through the radar's area of responsibility. The control operator alerted to the arrival of the above target may determine the operator's vigilance by comparing the plotting board range with the range of the target's first appearance on the PPI screen, as before.

Judging by what was said about the influence of various factors on the radar operator's performance, it transpires that its determination requires many, often unavailable, data which change from operator to operator and often, during the same watch, for the same operator. According to Skolnik (1962) an empirical operator efficiency factor

$$\eta_0 = 0{\cdot}7 P_d^2, \tag{4.20}$$

with P_d the single scan probability of detection, assuming a good operator observing a PPI display under good conditions, may be used in overall system performance calculations.

4.3.1.3 Propagation losses: Besides the decrease in the received signal power with distance accounted for by the R^4 term of the radar equation (eqn. (4.9)), further reduction of signal power due to the oxygen and water vapour content of the atmosphere occurs in all weather conditions. Rain, snow, hail and fog attenuate electromagnetic waves to a great extent and cause clutter which is a serious impediment to radar operation.

Electromagnetic waves are affected by oxygen and water vapour in two ways. Part of their energy is absorbed, i.e. transformed into other forms of energy, and another part is scattered by their exciting the encountered particles into oscillations which radiate energy extracted from the radar radiation. The additional attenuation reduces the radiated energy thus causing a lower target illumination and consequently a reduction in the level of target return.

Absorption of electromagnetic waves by atmospheric gases was studied by Van Vleck (1947*a*, *b*) who derived a formula for absorption by oxygen based on the approximations of collision broadening theory which assumes that, although the electromagnetic energy is freely exchanged between the incident field and the molecules, some of the electromagnetic energy is converted to thermal energy during molecular collisions and so a part of the incident electromagnetic energy is absorbed. Others, like Meeks and Lilley (1963), Bean and Abbott (1951) and Reber, Mitchell and Carter (1970), confirmed experimentally Van Vleck's theory.

The following sections are based on their findings.

While it is appreciated that there is a general connection between atmospheric absorption and refractive index, the influence of the latter on electromagnetic wave propagation, together with the phenomenon of clutter due to meteorological causes, must be delegated to the appropriate sections of this treatise since it is attenuation that is of interest at present.

(*a*) *Absorption by oxygen*: Absorption of electromagnetic radiation by gases is due to the interaction of molecular dipole moments with the said radiation. The absorption of microwaves by oxygen was predicted by Van Vleck during his war-time studies of microwave propagation through the atmosphere. He predicted a millimetre wave spectrum for oxygen which consisted of many lines near a wavelength of 0·5 cm and a single line at 0·25 cm. This was later confirmed by measurements on the basis of which Van Vleck (1947*a*) was able to make a complete analysis of the millimetre wave spectrum of oxygen.

Oxygen molecules were found to be electrically non-polar; however, they posses a permanent magnetic moment of two Bohr magnetons and since Maxwell's equations are symmetrical as far as the electrical and magnetic fields are concerned, absorption of electromagnetic energy by oxygen must be due to the interaction of the magnetically polarized oxygen molecules with the radiation.

A Bohr magnetron, $\beta = eh/4\pi mc$, where e and m are the charge resp. the mass of an electron, $h = 6{\cdot}625 \times 10^{-27}$ ergsec is Planck's constant and c the velocity of propagation of electromagnetic waves in free space, is only about one hundredth of a Debye unit of electric polarity. Hence, the interaction of oxygen molecules with electromagnetic radiation is much weaker than if they possessed electric dipole moments.

The magnetic dipole moment of oxygen is due to the unpaired spins of two electrons. It interacts with the 'end-over-end' rotation of the oxygen molecule to form a 'rho-type' triplet, the fine structure of which yields microwave resonances. The selection principle, according to which the rotational quantum number for the molecule cannot change by more than one unit, prevents transitions between components removed by two units from being magnetically active. Otherwise the near coincidence of these components would cause strong absorption even at frequencies around 3 GHz.

The spacings of the components were determined by measuring the lines of atmospheric absorption bands of oxygen in stellar spectrograms.

A general formula for the absorption coefficient γ in dB/km as a function of frequency in GHz, pressure P in mm Hg and temperature T in Kelvin, based on Van Vleck's work was found by Meeks and Lilley (1963) to read

$$\gamma(f, P, T) = CPT^{-3} f^2 \, \Sigma \, S_N \exp\,(-E_N/kT), \tag{4.21}$$

where the summation over rotational states, specified by the quantum number N, must include all states with appreciable population at the temperatures encountered in the atmosphere.

In the above expression the term summed over odd values of N is

$$S_N = F_{N+}\,\mu_{N+}^2 + F_{N-}\,\mu_{N-}^2 + F_0\,\mu_{N0}^2$$

with

$$F_{N\pm} = \frac{\Delta f}{(f_{N\pm} - f)^2 + \Delta f^2} + \frac{\Delta f}{(f_{N\pm} + f)^2 + \Delta f^2},$$

where Δf, the line width parameter, specifies the half-width of the line at half the maximum absorption. It is a function of temperature and pressure and for air it is given by the empirical expression

$$\Delta f(P, T) = \alpha P[0{\cdot}21 + 0{\cdot}78\,\beta]\,(300/T)^{0{\cdot}85}.$$

The constant α determines the line broadening at unit pressure (1 mm Hg), while β specifies the relative effectiveness of N_2–O_2 collisons compared to O_2–O_2 collisions in producing pressure broadening.

Reber, Mitchell and Carter (1970) modified this last expression to read

$$\Delta f(P, T) = g(h)\,\frac{P}{P_0}\left(\frac{T}{T_0}\right)^{-1}$$

with $P_0 = 1013{\cdot}25$ millibars, $T_0 = 300$ K,

$$g(h) = \begin{cases} g_1 & 0 \leqslant h \leqslant h_1 \\ g_1 + \dfrac{g_2 - g_1}{h_2 - h_1}(h - h_1) & h_1 \leqslant h \leqslant h_2 \\ g_2 & h_2 < h \end{cases}$$

$g_1 = 0{\cdot}60$ GHz, $g_2 = 1{\cdot}17$ GHz, $h_1 = 8$ km and $h_2 = 25$ km.

The remaining parameters of the above expressions are as follows:

The non-resonant contribution F_0 in the expression for S_N is

$$F_0 = \frac{\Delta f}{f^2 + \Delta f^2}$$

further,

$$\mu_{N+}^2 = \frac{N(2N+3)}{N+1}$$

$$\mu_{N-}^2 = \frac{(N+1)(2N-1)}{N}$$

$$\mu_{N0}^2 = \frac{2(N^2+N+1)(2N+1)}{N(N+1)}.$$

The exponent in the Boltzmann factor is $E_N/kT = 2{\cdot}068\,44\,N(N+1)/T$ and for the absorption coefficient γ in dB and a normal concentration of O_2 in the air the constant $C = 2{\cdot}6742$.

The resonant frequencies f_{N+} and f_{N-} for odd values of N between 1 and 45 are given by Meeks and Lilley (1963) and also by Reber, Mitchell and Carter (1970).

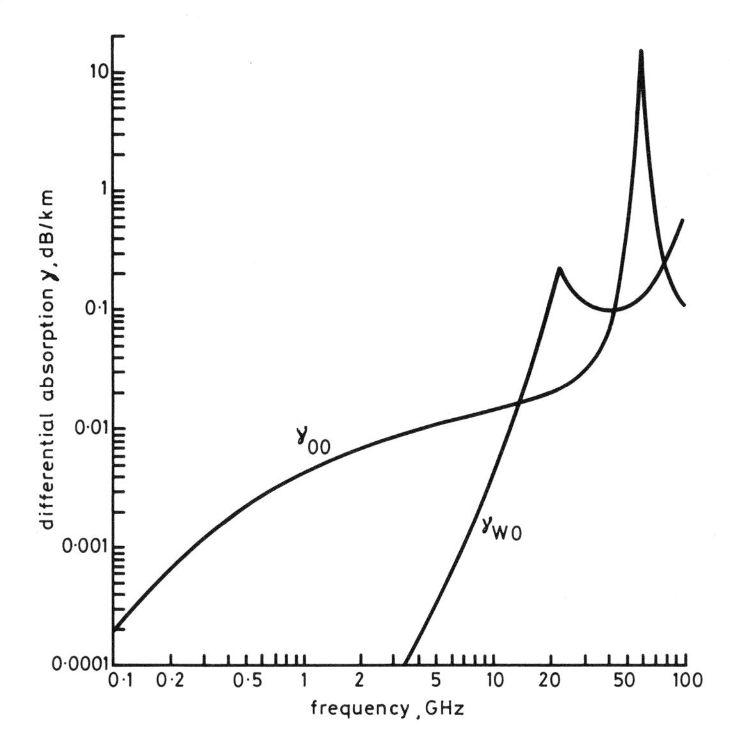

Fig. 4.6 *Surface values γ_{00} and γ_{W0} of absorption by oxygen and water vapour (From C.C.I.R., 1963, Vol. II)*

Pressure 760 mm Hg; Temperature 20° C;
Water vapour density 10 g/m^3 ≈ 7·6 g/kg

Figure 4.6 taken from the CCIR Report 234 (1963), shows the oxygen absorption coefficient at sea level for frequencies between 0·1 and 100 GHz. The absorption coefficient does not exhibit any peaks at frequencies below 10 GHz. At a frequency of approximately 0·1 GHz its value falls to about 0·0002 dB/km.

Approximate values of the absorption coefficient at the surface of the earth at a pressure of 760 mm Hg and a temperature of 20°C, for frequencies between 0·1 and 10 GHz, the range of frequencies used by most search radars, may be computed from the following expressions derived to fit the curves adopted by the CCIR for absorption by oxygen:

$$\gamma_{\text{dB/km}} = 10^{-[0 \cdot 648\,66 \log^2 f - 0 \cdot 673\,59 \log f + 2 \cdot 376\,75]},$$

$$0 \cdot 1 \leqslant f \leqslant 1\,\text{GHz} \qquad (4.22)$$

and

$$\gamma_{\mathrm{dB/km}} = 10^{-[0\cdot119\,67 \log^2 f - 0\cdot697\,82 \log f + 2\cdot376\,75]},$$
$$1\cdot0 \leqslant f \leqslant 10\,\mathrm{GHz}, \tag{4.23}$$

where f, the frequency, is to be inserted in GHz.

Most search radars operate in the above frequency ranges. Coefficients of absorption by atmospheric oxygen for higher frequencies may be read from the enclosed figure or obtained from other sources.

(*b*) *Absorption by uncondensed water vapour*: Contrary to absorption by oxygen, absorption by uncondensed water vapour is due to the mutual combination of closely spaced energy levels by electric dipole radiation. The transitions between nearby states occur according to the dipole selection rules of quantum mechanics which state that while all spectral lines represent transitions between distinct energy levels, all possible combinations of levels do not give observed lines. Certain selection rules must be obeyed. The most important of the roles for dipole radiation are Laporte's parity rule and the restrictions of the total rotational quantum number (Aller, 1953). The rotational energy levels of the water molecule are of four different symmetry types and the rotational quantum number is permitted to change only by one unit with restriction on the permitted combinations.

Among the levels of the water molecule which are excited at ordinary temperatures there is only one pair the number of which can combine and at the same time are so closely spaced as to resonate at microwave frequencies. This resonance occurs at about 22·22 GHz where absorption reaches a peak value of about 0·2 dB/km.

The absorption due to this line is given by Bean and Abbott (1957) as

$$\gamma = \frac{3\cdot5 \times 10^{-3}\rho}{\lambda^2}\left[\frac{\Delta f}{\left(\frac{1}{\lambda} - \frac{1}{1\cdot35}\right)^2 + \Delta f^2} + \frac{f}{\left(\frac{1}{\lambda} + \frac{1}{1\cdot35}\right)^2 + \Delta f^2}\right],$$

where ρ is the absolute humidity and Δf is the line width factor of the 22·22 GHz water vapour absorption line.

Additional absorption due to absorption bands above the mentioned one may be determined from

$$\gamma = \frac{1\cdot2 \times 10^{-2}\Delta f_1}{\lambda^2}\rho$$

with Δf_1 the effective line width of the higher absorption bands.

The approximations to the extrapolated CCIR (1963) curve at frequencies between 0·1 and 10 GHz read as follows:

$$\gamma_{\mathrm{dB/km}} = 10^{[0\cdot336\,67 \log^2 f + 2\cdot666\,67 \log f - 5\cdot494\,85]},$$

$$0\cdot1 \leqslant f \leqslant 1\,\mathrm{GHz} \qquad (4.24)$$

$$\gamma_{\mathrm{dB/km}} = 10^{[0\cdot619\,05 \log^2 f + 2\cdot490\,95 \log f - 5\cdot494\,85]},$$

$$1 \leqslant f \leqslant 10\,\mathrm{GHz}, \qquad (4.25)$$

where the frequency f is to be inserted in GHz.

It is interesting to note that although the troposphere is known to extend to a height of about 10 km above the surface of the earth, the effective distance for oxygen absorption, according to CCIR, is approximately the distance the radio wave travels in a constant density atmosphere extending in case of absorption by oxygen to a height of only 4 km, while for absorption by water vapour to only 2 km, with vacuum above.

Figure 4.6, taken from the mentioned CCIR document, shows attenuation caused by both gases, i.e. attenuation by oxygen, annotated γ_{oo}, and water vapour γ_{wo}, at an atmospheric pressure of 760 mm Hg, temperature of 20°C and water vapour density of 10 g/m^3. It may be observed that at frequencies up to about 13 GHz absorption by oxygen exceeds that caused by water vapour. From that frequency to about 43 GHz absorption due to water vapour becomes several times higher than absorption due to oxygen. This region contains an absorption peak exceeding 0·2 dB/km at the resonance frequency of 22·22 GHz. Beyond 43 GHz it is again oxygen absorption, rising towards a peak in excess of 10 dB/km at the 0·5 cm line, which is predominant in the total atmospheric absorption. This persists up to a frequency of about 80 GHz where water vapour absorption rising towards its second peak at about 175 GHz takes over.

The total atmospheric absorption in dB over a line of sight path is

$$L_a = \gamma_{oo} R_{eo} + \gamma_{wo} R_{ew},$$

where R_{eo} and R_{ew} are the effective path lengths for oxygen resp. water vapour.

Graphs of two-way atmospheric attenuation as a function of range, with frequency as parameter, for various ray elevation angles are given by Blake (1961).

4.3.2 *Typical values of search radar losses*

Since data required for the determination of losses are not always available, the following table lists typical values of the most common search radar losses applicable for search or surveillance radar performance assessment.

Loss	Typical values, dB
Aerial	3–5
Beam shape	2–5
Collapsing	1–4
Degradation	3
Integration	1–5
Limiting	1
Matching	0·5–1
Operator ($P_d = 0{\cdot}5$)	7·5
Plumbing	1–4
Scanning	0–6

Typical values for the sum of losses, excepting atmospheric absorption, lie between 12 and 20 dB.

4.3.3 Losses in the radar equation

Since losses in an engineering system are cumulative, it is the product, or sum if all the terms are expressed in dB, of all computed or estimated losses, excepting atmospheric attenuation, which is to be inserted into the radar equation (eqn. 4.9)).

The influence of atmospheric attenuation is accounted for by multiplying the above radar equation by an exponential term, similar to the generally used method of determining the influence of transmission line attenuation on the received power in communication networks.

The updated radar equation reads then

$$R^4 = \frac{PG^2\lambda^2\sigma}{(4\pi)^3 kT_0 BFL\, SN} \exp(-2\alpha R), \tag{4.26}$$

where L stands for the product of losses, α in the exponential term equals $0{\cdot}23\,\gamma_{\mathrm{dB/km}}$* and the factor 2 accounts for the two-way propagation of radar

* The number of units of attenuation due to propagation in a medium between points 1 and 2 is usually expressed, in Nepers, by the ratio of powers at these two points along the path as

$$\alpha = \ln(P_2/P_1).$$

The attenuation in dB is

$$\gamma = 10 \log(P_2/P_1) = 4{\cdot}343\,\alpha$$

so that

$$\alpha = 0{\cdot}23\,\gamma_{\mathrm{dB}}.$$

Hence, keeping to the communication engineering convention, atmospheric attenuation in the radar equation is expressed by the exponential term used above.

waves. The appearance of the range R in the exponential term complicates the solution of the radar equation. The approximate range is usually computed by neglecting initially the influence of the atmosphere and reducing this range by multiplying it by the exponential term in order to account for atmospheric attenuation.

4.4 Some implicit parameters in the radar equation

It was mentioned several times in what preceded that modern radars do not rely on single pulse detection but transmit a continuous sequence of pulses thus causing several pulses reaching the target and being returned by it to the radar during the aerial's dwell time on the target.

This definite number of pulses k, referred to as number of hits per scan, is a product of the radar's pulse repetition frequency (PRF) and of the aerial's dwell-time on the target given by

$$t = \frac{\theta}{\omega + v \sin \xi_0 / R}$$

with θ the aerial's horizontal 3 dB beamwidth, ω its angular scanning velocity, v the velocity of the target and ξ_0 the angle between the radius from the radar to the target at range R and the target's trajectory at the instant of interception as shown by Fig. 7.4.

The number of hits per scan is then

$$k = \frac{\theta\, PRF}{\omega + v \sin \xi_0 / R}$$

A judicious choice of the angle ξ_0 such that

$$\xi_0 = \left\{ \begin{array}{l} 2\pi \text{ for a radially receding} \\ \pi \text{ for a radially approaching} \end{array} \right\} \text{target}$$

when the aerial, viewed from above, rotates clockwise and vice versa, will cause the angular target velocity to be added to or subtracted from the aerial's angular scanning velocity depending on whether the target moves against or in the same sense as the aerial, thus decreasing or increasing the dwell-time on the target and changing the number of hits per scan accordingly. In case of a radial trajectory when $\xi_0 = \pi$ or 2π, the dwell-time reduces to θ/ω and the expression for the number of hits per scan reduces to the generally used expression

$$k = \frac{\theta\, PRF}{\omega}.$$

Assuming then for the initial approach that all k pulses are of equal amplitude, the signal-to-noise ratio per pulse, if k pulses are integrated, is one kth of the total signal-to-noise ratio required for detection with the specified probability of detection and false alarm. The parameter k, and with it the above three radar parameters, are thus introduced implicitly into the radar equation. The beam shape loss dealt with in Section 4.3.1.1 (*b*) corrects the inaccuracy introduced by the above assumption of all pulses being of equal amplitude.

It is interesting to examine the newly added parameters.

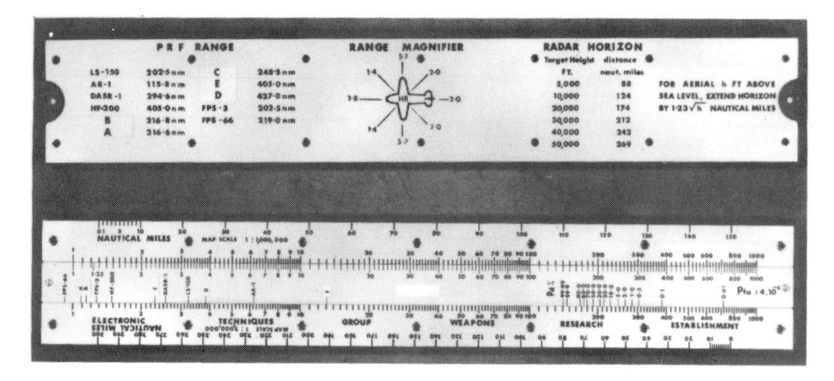

Fig. 4.7 *A radar slide-rule (After Rohan) (Reprinted with the permission of the Electronics Research Laboratory, Dept. of Defence Support, Commonwealth of Australia)*

4.4.1 Horizontal beamwidth

The coordinates of a target on a PPI screen are range, i.e. the distance of the target paint on the cathode ray tube from the centre of the beam sweep, and azimuth angle, i.e. the angular displacement of the said paint, usually, from the magnetic north.

The accuracy of the angular measure depends on the horizontal beamwidth of the radar's aerial which, in turn, is a function of the aerial's aperture relative to the used wavelength. A narrower beam, i.e. one due to a larger aperture to wavelength ratio, will yield a more accurate value of the measured angle.

The majority of modern search radars uses aerials having rectangular apertures with their larger side horizontal in order to achieve a narrow horizontal beamwidth.

The one-way amplitude pattern of a uniformly illuminated rectangular aperture is given by the $\sin x/x$ function where $x = \sin\theta$ with θ the off-axis angle and a, a function of the used wavelength and of the aerial's dimension in the plane of the angle θ, is so chosen as to yield the required half-power beamwidth (see Section 10.1.3.1 *b*).

It is generally accepted that the angular resolution of search radars should be comparable with the accuracy of position determination by equipment on board the targets of interest. The accuracy of position measurement in azimuth is primarily determined by the accuracy of synchronism between the aerial's rotation and the cathode ray tube's writing beam movement. Besides that, it is affected by the spread of the target paint in azimuth which may introduce inaccuracies in azimuth reading. The spread of the spot at small distances from the centre of the display depends on the spot size achievable with the used electron optics and on the horizontal beamwidth of the aerial. Based on the acceptable limits of azimuth determination accuracy, the horizontal beamwidth of most search radars is in the vicinity of one or two degrees.

The knowledge of the aerial's horizontal beamwidth and gain permits the determination of the vertical beamwidth by the application of eqn. (4.10).

4.4.2 Pulse repetition frequency (PRF)

The choice of the pulse repetition frequency is dictated not only by the aim of increasing detection probability but also by the contradictory requirements of a large unambiguous range and a high subclutter visibility.

The maximum unambiguous range of a radar is

$$R_u = \frac{c}{2\,\mathrm{PRF}},$$

where c is the velocity of propagation of electromagnetic waves.

However, due partly to technical causes and partly to prevent echoes of targets beyond the maximum displayed range (MDR) being shown in the interval of reception after the next transmitted pulse, the propagation time appropriate to the maximum displayed range must be shorter than 1/PRF, i.e. the maximum displayed range is generally shorter than the maximum unambiguous one. It is given by

$$\mathrm{MDR} = \frac{c}{2}\left(\frac{1}{\mathrm{PRF}} - t_d\right) = R_u - \frac{ct_d}{2},$$

where t_d is the 'dead' time, i.e. the time during which the receiver is inoperative, the time during and immediately following the pulse transmission. A 'dead' time of about 0·66 ms yields an approximately 100 km safety margin.

Should echoes of targets outside the unambiguous range be received after a pulse following the one causing the echo has been transmitted, e.g. during anomalous propagation conditions, a false, shorter than true range appropriate to the interval between the transmission of this pulse and the arrival of the echo of the preceding pulse would be indicated.

A low PRF is called for to yield a large unambiguous range.

Subclutter visibility (SCV) indicates how much weaker target returns embedded in clutter may be in order to be just detectable (Mücke and Röhrich, 1961).

Clutter attenuation limited by amplitude or phase fluctuations caused by scanning is, according to Grisetti, Santa and Kirkpatrick (1955)

$$\mathrm{SCV_{dB}} = 10 \log \frac{2\int_{-\infty}^{\infty} |f(\theta)|^2\, d\theta}{\int_{-\infty}^{\infty} |f(\theta + \Delta\theta) - f(\theta)|^2\, d\theta},$$

which, for a Gaussian approximation of the aerial's horizontal radiation pattern within the half-power beamwidth, reduces to

$$\mathrm{SCV_{dB}} = 10 \log \frac{1}{1 - \exp\left[-(1{\cdot}18/k)^2\right]}$$

showing that the number of hits per scan and hence the pulse repetition frequency should be high if good clutter rejection is to be achieved. The function $f(\theta)$ is the aerial's pattern factor in the horizontal plane and $\Delta\theta$ is the angle of the aerial's rotation during the interpulse interval.

The choice of the pulse repetition frequency then becomes a compromise between the two outlined requirements.

4.4.3 *Aerial scan rate*

The basic purpose of search or surveillance radars is the detection and observation of target movement which necessitates a high information flow rate resulting from the aerial's scanning. This requirement is supported also by the limited persistence of the currently used PPI screens which lies between 100 ms and 1 second when persistence is defined as the time interval during which the intensity of luminence falls to the 10 per cent level of its initial value.

Higher scanning rates diminish the number of hits per scan and thus reduce subclutter visibility. They reduce, even if to a small extent only, the radar range which, due to the $\sqrt{k}$ integration improvement factor for CRT-operator integration (Section 3.3.1), is a function of $(\sqrt{k}\,)^{1/4}$.

These considerations limit the aerial's scanning rate and thus the achievable rate of information flow.

The required clutter suppression depends on the radar's purpose. A short range radar, such as used at airfields, must have a more efficient clutter suppressing facility than long range radars as clutter echoes from nearby objects are more numerous and stronger than those from remote ones the illumination of which falls rapidly with an increasing distance due to the decrease of power density, increasing attenuation and due to the earth's curvature.

A subclutter visibility of, e.g. 25 dB, would require, on the basis of the above considerations, about 20 hits per scan which, with a 3 dB horizontal beamwidth of 2° and a pulse repetition frequency of 400 pulses per second, leads to a scan rate of about 7 revolutions per minute and a 375 km unambigious range.

4.5 A radar slide rule

The last shown form of the radar equation is the most often used expression for the quick assessment of search radar performance.

Neglecting temporarily atmospheric attenuation, the radar equation may be divided into three groups of terms by writing it as

$$R = \sqrt[4]{\frac{PG^2\lambda^2}{(4\pi)^3 kT_0 BFL}}\ \sigma^{1/4}\ \frac{1}{\sqrt[4]{SN}}\ . \tag{4.27}$$

The first group contains all the used radar parameters and is reasonably constant for a given radar. It may be viewed as a 'figure of merit' since it determines in broad terms the radar's detection range.

The second group, the target echoing area, is extrinsic and, in the first approximation, it may be taken, for the sake of generality, to be equal to 1 m^2, while the third one, the required signal-to-noise ratio, is determined by the specified probability of detection and false alarm. This term is the same for all radars using an identical method of detection, i.e. all radars relying on an operator–CRT combination for target detection with a given probability of detection and false alarm require the same signal-to-noise ratio for detecting a target of echoing area σ at a range R.

The atmospheric attenuation, once determined for the known path length, may be introduced as another loss by which the obtained range must be modified in order to account for the atmosphere's influence.

Rewriting eqn. (4.27), one then obtains

$$R = \text{const.}\frac{1}{\sqrt[4]{SN}}. \tag{4.28}$$

Taking logarithms on both sides, this expression becomes

$$\log R = \log \text{const.} + \tfrac{1}{4} \log 1/SN$$

which was used by Rohan for the construction of a probability of detection slide rule shown in Fig. 4.7.

The bottom scale of the slider (lower figure) carries a probability scale on the right and a number of markers denoted by various radar type designations on the left. The corresponding fixed scale has three logarithmic decades for ranges from 1 to 1000 nautical miles. The two upper scales, i.e. the one on the slider and its corresponding fixed one, are also normal logarithmic scales which are to be used as a conventional slide rule. The edges of the slide rule carry the two most often used map scales, namely the 1:1 000 000 and the 1:2 000 000 scales. They may be used for measuring distances on maps drawn in corresponding scales.

The reverse side of the slide rule, shown in the upper part of the figure, shows various data of interest, e.g. the maximum unambiguous range of the listed radars, the distance to the horizon in nautical miles for a number of aerial heights in feet and the 'range magnifier', i.e. the fourth root of the highly smoothed radar cross-section multiplier of an aircraft for a number of aspect angles.

To determine the range at which detection of an aircraft with a specified probability may be achieved, one places the appropriate radar's marker on the slider above the value of the 'range magnifier', read for the anticipated interception aspect angle from the sketch on the reverse side of the slide rule, on the bottom scale and reads on that same scale the range below the above probability of detection. The obtained range is then corrected by multiplying it by the target cross-section using the upper two logarithmic scales as a slide rule. Atmospheric attenuation for this range and considered radar determined from the graph supplied with the slide rule and shown by Fig. 4.8 may be used for further correcting the previously determined range.

The distance of the radar markers from the probability scale is proportional to the constant in eqn. (4.28). The most powerful of the radars shown is furthermost on the left; the others extend to the right in a diminishing order of their 'figure of merit'. Radar E is the least powerful of all radars marked on the slide rule.

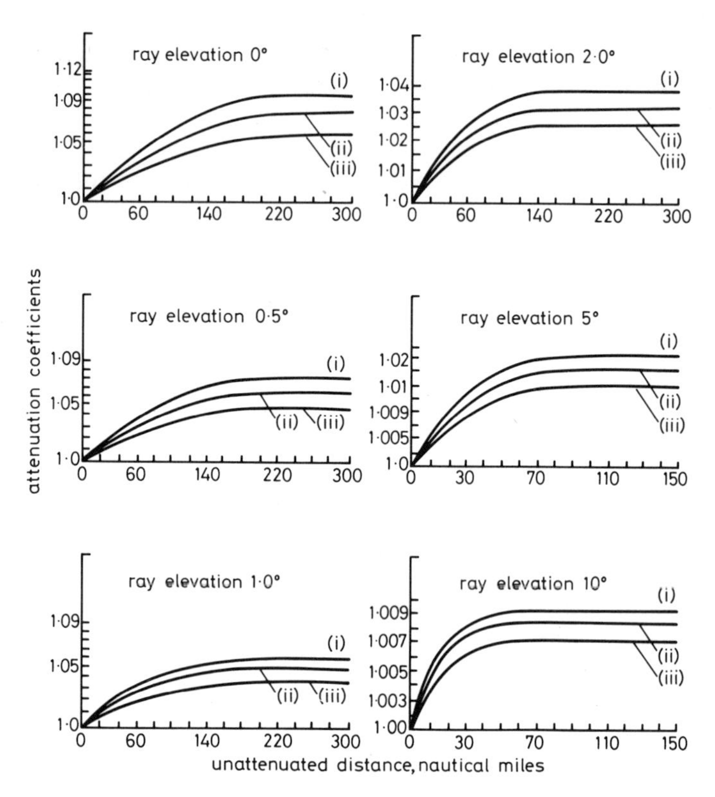

Fig. 4.8 *Atmospheric attenuation coefficients. (For use with the radar slide-rule only) (After Blake, 1963, adapted by Rohan) (Reprinted with the permission of the Electronics Research Laboratory, Dept. of Defence Support, Commonwealth of Australia)*

I	LS-150,	AR-1, DASR-1,	HF-200,	C
II	FPS-3,	FPS-66.		
III	E	D	B	A

The slide rule was designed for a probability of false alarm of 4×10^{-4} determined on the basis of a statistical study of operational logs of a number of surveillance radars and operators collected over several months.

Other radar slide-rules have been developed. Most of these compute the free-space detection ranges of specific radars for unity signal-to-noise ratio by evaluating the simplest form of the radar equation given earlier.

A slide-rule developed by Hovanessian is an example of a more sophisticated device suitable for the computation of the probability of detection of non-scintillating and 'exponentially' scintillating targets for given values of signal-to-noise ratio and probability of false alarm, besides the detection ranges for unity signal-to-noise ratio and the signal-to-noise ratio at other ranges.

The advent of programmable calculators capable of performing, besides the basic arithmetic operations, exponentiation, computing logarithms, taking roots of numbers, etc., made slide-rules obsolete. Simple programs for the evaluation of the radar equation containing a small number of programming steps are easy to develop. However, the necessity of entering programs every time a radar range computation is required leaves room for slide-rule applications.

4.6 The energy form of the radar equation

Expressing the radar equation in terms of the average power defined by

$$P_{av} = P\tau(\text{PRF}) = P\frac{1{\cdot}2}{B}(\text{PRF}),$$

where P is the peak transmitter power, τ the pulse length, B the bandwidth and (PRF) the pulse repetition frequency, yields the range

$$R^4 = \frac{P_{av}B}{1{\cdot}2(\text{PRF})}\,\frac{G^2\lambda^2\sigma}{(4\pi)^3 kT_0 BFL\,SN}\exp(-2\alpha R)$$

or

$$R^4 = \frac{P_{av}G^2\lambda^2\sigma}{(4\pi)^3(\text{PRF})kT_0(B\tau)FL\,SN}\exp(-2\alpha R)$$

since the recommended value of the product of the bandwidth and pulse length is equal to 1·2.

The energy content of the pulse waveform is

$$E = \frac{P_{av}}{(\text{PRF})},$$

so that the radar equation, expressed in terms of energy, becomes

$$R^4 = \frac{EG^2\lambda^2\sigma}{(4\pi)^3 kT_0(B\tau)FL\,SN}\exp(-2\alpha R). \tag{4.29}$$

However, taking the pulse length–bandwidth product to be equal to unity, an approximation which introduces only about a 0·2 dB error into the computed range, the radar equation becomes

$$R^4 = \frac{EG^2\lambda^2\sigma}{(4\pi)^3 kT_0 FLSN}\exp(-2\alpha R), \tag{4.30}$$

which, although originally based on the energy content of the pulse, is devoid of any reference to the pulse dimensions and is usable therefore for the computation of the detection range of any, including CW, radars provided that the energy term is understood to be

$$E = \int_0^T P(t)\, dt$$

with T representing the duration of the transmitted effective signal power $P(t)$, which would be equal to, e.g. the aerial's dwell-time on the target or range cell.

4.7 Shortcomings of the radar equation

It was mentioned earlier that since the simplest form of the radar equation neglects propagation and other effects, such as reflection, etc., while applicable for the comparison of radars, yields only free-space ranges which may differ to a great extent from those experienced in practice. This inadequacy applies also to the more complete form of the radar equation as given by eqns. (4.26) and (4.30).

This may be remedied to some extent by modifying the computed free-space range due, for simplicity's sake, to a ray of electromagnetic radiation travelling directly between the radar and the target and back, by the influence of another ray which on its path from the radar is reflected by the ground before reaching the target and again on its return path before arriving at the radar.

Other effects neglected by the radar equation are unwanted echoes, clutter, from natural terrain features, sea waves and man-made obstacles, meteorological pheomena, target behaviour, jamming, etc., and the influence of various technical methods for their combatting.

A short account of propagation is given in the following chapter while the other effects will be considered where appropriate.

Chapter 5

Radar wave propagation

5.1 Introduction

While there are many publications dealing with the propagation of electromagnetic waves, it is thought that a short introduction serving as a basis for the derivation of the mathematical expressions used later for the assessment of radar performance is appropriate at this juncture.

5.2 Maxwell's equations and geometrical optics

Maxwell's equations, as applicable to a homogeneous, isotropic dielectric (Stratton, 1941), yield the wave equation, the integral of which

$$E_y = f_1(x - v_0 t) + f_2(x + v_0 t) \tag{5.1}$$

consists of two waves, one travelling away from the source and one, a reflected wave, travelling towards it. Both waves have a velocity $v_0 = 1/\sqrt{\mu\epsilon}$, where μ and ϵ are the permeability resp. dielectric constant of the medium of propagation. The second wave is due to reflection at a discontinuity in the medium. The variable x is a function of the distance from the source.

Propagation at radar frequencies is usually dealt with by the principles of geometrical optics for which the scalar wave equation is required. This leads to an expression showing the dependence of the electric field on the modified index of refraction $n = (c/v_0)^2$ and on the position of the considered point in space. The constant c is the velocity of light.

Kline and Kay (1965) show that the wavefronts of the electric field in space are surfaces of constant phase and their normals, given by the gradients of these surfaces, determine the direction of the rays or orthogonal trajectories of the wavefronts which, for a position dependent n, are curved.

Freehafer (in Kerr, 1951) derived the differential equation of the ray trajectory.

The validity of the geometrical optics approach to problems of propagation is restricted to cases where the relative change of the modifed index of refraction and of the distance of neighbouring rays over a path length of a wavelength is much smaller than $2\pi n$. Caution is to be exercised in applying a geometrical optics approach to cases where the rays converge or diverge, i.e. where neighbouring rays cross to form foci or caustics. The above first restriction depends then on the frequency of the radiation and the structure of the medium of propagation, while the second one on the type of wave the ray pattern is to represent. Both restrictions are irrelevant in conventional radar problems and the use of the geometrical optics approach for their solution is generally accepted.

5.3 Radar wave propagation in the troposphere

The lowest layer of the atmosphere varying in height from about 6 km above the poles to 18 km above the equator is the troposphere. It is mostly in this layer of the atmosphere that radar wave propagation takes place. Since the troposphere is not homogeneous it will affect the said propagation.

Refraction of electromagnetic waves in the troposphere is caused by the varying density of the medium and the varying amount of moisture vapour in it.

An electromagnetic wave induces a dipole moment in each gas molecule thus causing an increase of the dielectric permittivity proportional to the number of induced dipoles per unit volume. Any spatial variation of the density implies a corresponding spatial variation in the velocity of propagation. At a first order approximation, the air may be considered to be a perfect gas in which the density and hence the refractive index variation is directly proportional to the pressure P and inversely proportional to the temperature T.

An electromagnetic wave interacting with the permanent dipole moment of water molecules aligns these so that they contribute to the polarization of the medium in proportion to the density of water molecules in it. Rising temperature causes depolarization due to an increase in random thermal motions. The contribution of the moisture to the refractive index is then directly proportional to the partial pressure of water vapour e in the medium and indirectly proportional to the temperature.

The index of refraction of a mixture of gases generally obeys the additive rule so that the total value equals the sum of the above contributions of the individual gases weighted in proportion to their partial pressures. The refractive index of the troposphere $n = \sqrt{\epsilon\mu}$, where ϵ is the dielectric permittivity and μ the permeability, is then the function of atmospheric pressure, temperature and specific humidity; however, the generally used unit in calculations is the refractivity N related to the refractive index by

$$N = (n-1) \times 10^6$$

and its dependence on the above properties of the atmosphere is

$$N = \frac{77\cdot6}{T}\left(P + \frac{4810\,e}{T}\right), \tag{5.2}$$

where the pressure P and the partial pressure of water vapour e are in millibars (mb) and the temperature T in Kelvin. The first term of this expression is often referred to as the 'dry' term, while the second one as the 'wet' term.

On average, the above three quantities decrease with an increasing altitude as follows:

> the pressure P by about 1·2 mb per 10 m, the temperature T by about 1°C per 200 m and humidity by about 1 mb per 300 m.

While various laws of variations with altitude are ascribed to refractivity, the most often used law of variation in radio engineering calculations is the linear one due to an artificial 'standard' atmosphere which provides an approximation applicable to the first few kilometres of the troposphere. The C.C.I.R., in 1959, proposed for adoption a 'fundamental reference' atmosphere (David and Voge, 1969), which, while equivalent to the 'standard' atmosphere in the first km of altitude, yields reasonable approximation to the real atmosphere over a more extensive region of propagation. Its refractive index dependence on the altitude h is

$$n(h) = 1 + 289 \times 10^{-6} \exp(-0{\cdot}136\, h), \tag{5.3}$$

where the altitude above sea level is to be taken in km.

Neither the 'standard' nor the 'fundamental reference' atmosphere is a true representation of prevailing atmospheric structures since the regular variations characterizing these atmospheres are rather exceptional especially at altitudes below about 500 m. Convection currents due to temperature and humidity variations of the ground which cause exchanges between the air and ground ultimately determine the air temperature and create turbulence of the air which brings about a vertical air movement accompanied by more or less adiabatic compressions and expansions, i.e. compressions and expansions without the exchange of heat. Besides that, the temperature and humidity distribution of the atmosphere is affected by advection, i.e. horizontal movement of air masses having different heat properties.

Based on what was said so far, it transpires that a ray propagating through the earth's atmosphere encounters refractive index variations along its trajectory which cause the ray path to bend. The refractive index, always slightly greater than unity near the surface of the earth, approaches unity with an increasing altitude. Vertical refraction, i.e. downward curvature of the ray, is closely proportional to $-dN/dh$, the negative vertical gradient of the refractivity.

5.3.1 The effective earth radius

The solution of problems associated with electromagnetic wave propagation is greatly facilitated by assuming straight line propagation which, in order to keep dimensions in proportion, necessitates the use of a modified earth's radius, i.e. an earth radius increased in length so as to straighten the ray paths. The magnitude of the increase depends on the gradient of the atmospheric refractive index which, as has been outlined above, accounts for the bending of the rays and which, for simplicity's sake, is taken in the following to be constant with height above the surface of the earth.

A correction factor for the true earth radius permitting the calculation of the effective earth radius, which makes the straight line propagation assumption acceptable, may be obtained, according to Doluchanow (1956), in the following way:

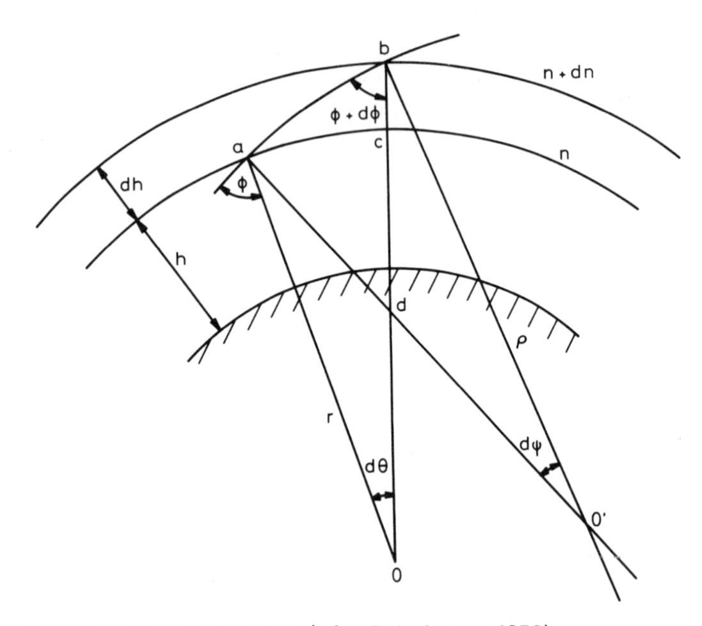

Fig. 5.1 *The radius of the ray curvature (After Doluchanow, 1956)*

Curves n and $n + dn$ at heights h resp. $h + dh$ above the surface of the earth represent in Fig. 5.1 shells of constant values of the refractive index. A ray $\overline{ab}$ enters the lower shell at an angle ϕ with respect to the line $\overline{0a}$ connecting the centre of the spherical earth 0 with point a on the shell n. Due to atmospheric refraction, the ray bends and reaches the upper shell at an angle $\phi + d\phi$. Normals erected in the points of intersection a and b of the ray with the two shells intersect at $0'$. The radius of curvature is ρ.

If the angles between the ray and lines $\overline{0a}$ and $\overline{0b}$ were equal, the angles at a and b in the triangles $0ad$ and $0'db$ would be equal and therefore the angles $d\psi$ and $d\theta$ would be also equal. However, since the angle at b is larger by $d\phi$, the angle $d\psi$ is also larger by the same amount, or

$$d\psi = d\theta + d\phi.$$

The length of the path $\overline{ab}$ is

$$\overline{ab} = \rho d\psi = \rho(d\theta + d\phi). \tag{5.4}$$

Expressing then $d\theta$ on the basis of the triangle, $0ab$, while considering dh to be so small that $\overline{ab}$ is a straight line, one obtains

$$d\theta = \frac{\overline{ab}\sin\phi}{r+h}, \tag{5.5}$$

which, when inserted into eqn. (5.4) yields

$$\overline{ab} = \frac{\rho\, d\phi}{1 - \rho\dfrac{\sin\phi}{r+h}}. \tag{5.6}$$

Assuming that the medium between the earth and the first shell has a refractive index n_1 and that between shells a refractive index n_2, when using Snell's law one may write

$$n_1 \sin\phi_1 = n_2 \sin\phi_2', \tag{5.7}$$

where ϕ_2' is the angle between sides $\overline{0a}$ and $\overline{ab}$ in the triangle $0ab$. Using the law of sines and writing $\phi_2 = \phi + d\phi$ one obtains

$$\frac{\sin\phi_2}{r+h} = \frac{\sin\phi_2'}{r+h+dn}$$

or

$$\sin\phi_2' = \frac{r+h+dh}{r+h}\sin\phi_2,$$

which, when inserted into eqn. (5.7) leads to

$$n_1(r+h)\sin\phi_1 = n_2(r+h+dh)\sin\phi_2. \tag{5.8}$$

Writing $n_2 = n_1 + dn_1$, performing the multiplication on the right-hand side of eqn. (5.8) and neglecting higher order infinitesimal terms, one arrives, after omitting the subscript of the refractive index n and angle ϕ, at the approximate expressions

$$d\phi \approx -\tan\phi\,\frac{n\,dh + dn(r+h)}{n\,[dh + (r+h)]}$$

or, since $dh \ll r+h$,

$$d\phi \approx -\tan\phi\left[\frac{dn}{n} + \frac{dh}{r+h}\right],$$

which, when inserted into eqn. (5.6) for the path length $\overline{ab}$, yields

$$\overline{ab} = -\frac{\rho\left[\dfrac{dn}{n} + \dfrac{dh}{r+h}\right]}{1 - \dfrac{\rho}{r+h}\sin\phi}\tan\phi. \tag{5.9}$$

But since dh was assumed such that $\overline{ab}$ is a straight line and therefore $dh = \overline{ab}\cos\phi$, the above expression becomes

$$\frac{dh}{\cos\phi} = -\frac{\rho\left[\dfrac{dn}{n} + \dfrac{dh}{r+h}\right]}{1 - \dfrac{\rho}{r+h}\sin\phi}\tan\phi,$$

which simplifies to

$$\rho = \frac{n}{-\dfrac{dn}{dh}\sin\phi}. \tag{5.10}$$

The refractive index of air is very near to unity. Furthermore, only rays which are nearly horizontal are of primary interest in radar and communication. For these $\sin\phi \doteq 1$ so that the above expression reduces, with good accuracy, to

$$\rho = \frac{1}{-\dfrac{dn}{dh}},$$

which shows that the curvature of the ray depends on the gradient of the refractive index. It is taken to be positive when the refractive index decreases with altitude.

The curvature of the earth of radius r is the reciprocal of this radius and the relative curvature of the considered ray to that of the earth is the difference between the two curvatures, or

$$\frac{1}{r} - \frac{1}{\rho} = \frac{1}{r} + \frac{dn}{dh}.$$

Equating this to the relative curvature of an imaginary earth having such a radius kr that the curvature of a ray propagating in its atmosphere is zero, or its radius of curvature is finite, yields

$$\frac{1}{r} + \frac{dn}{dh} = \frac{1}{kr}$$

so that

$$k = \frac{1}{1 + r\dfrac{dn}{dh}} \tag{5.11}$$

yields the coefficient k by which the radius of the true earth is to be modified in order to permit a straight line approximation of the almost horizontal ray trajectory. If the ray is entering the earlier mentioned atmospheric shells at such angles that the approximation $\sin\phi \doteq 1$ is no longer acceptable, this term, as shown in eqn. (5.10), must be included in eqn. (5.11).

The Central Radio Propagation Laboratory (CRPL) of the U.S. National Bureau of Standards uses an exponential distribution of the refractivity N expressed as

$$N = N_s \exp\left[-c_e(h - h_s)\right], \tag{5.12}$$

where N_s is the value of refractivity at the earth's surface at height h_s and

$$c_e = \ln\frac{N_s}{N_s + \Delta N}$$

with ΔN the difference between N_s and the value of N at a height of 1 km above the earth. The exponential distribution is a close representation of the average refractivity structure of the atmosphere to a height of about 3 km; it agrees with the 'standard' atmosphere up to a height of about 1 km.

Since the values of the ratio dn/dh in eqn. (5.11) are not readily available, one has to resort to using the formula (Bean and Dutton, 1966)

$$k = \frac{1}{1 - \dfrac{r}{n_s} c_e N_s \times 10^{-6}},$$

which, due to

$$n_s = 1 + N_s \times 10^{-6} \approx 1,$$

reduces to

$$k \doteq \frac{1}{1 - c_e N_s r \times 10^{-6}} \tag{5.13}$$

for the computation of the earth's radius modifying coefficient and which with

$$c_e = 0{\cdot}135\,788\,7/\text{km} \quad \text{and} \quad Ns = 289{\cdot}036\,27$$

obtained by interpolation between values given in the above reference yields, with the average earth's radius of 6371·228 km, a value of $k = 4/3$ for the 'standard' atmosphere.

5.4. The radar–target geometry

Having justified the straight-line approximation of the ray trajectory one arrives at a simple configuration for the radar–target geometry, shown in Fig. 5.2, on the basis of which one may derive most expressions for the computation of the geometrical quantities required for the relevant propagation studies.

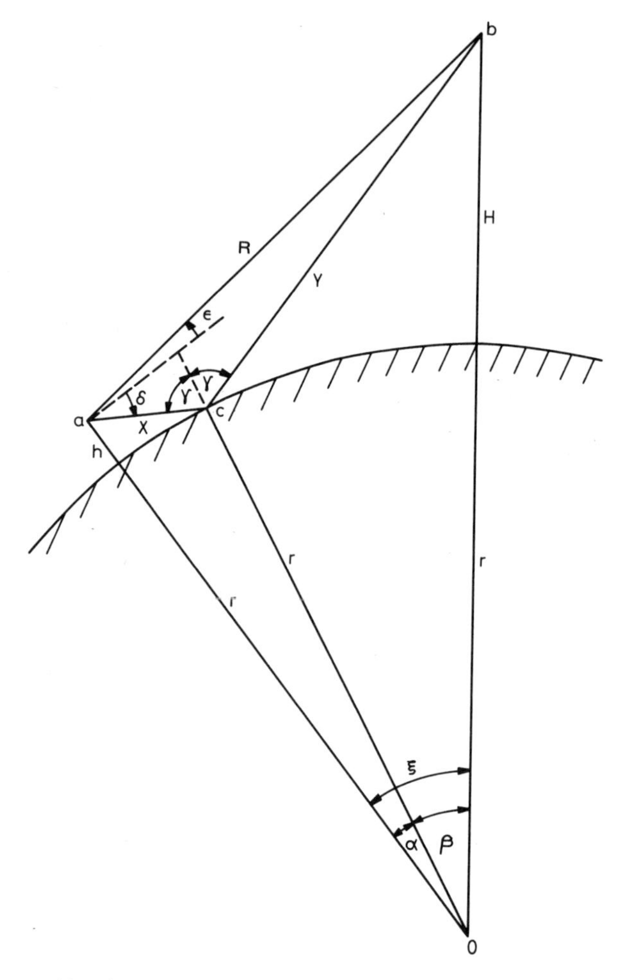

Fig. 5.2 *The considered geometry*

While the earth's radius is designated by r, it is to be accepted that it represents in what follows the modified earth radius given by the value kr which, e.g. in the case of the 'standard' atmosphere is $4r/3$. The lengths h and H are the radar aerial resp. the target height and R is the slant range of the target from the radar aerial which range is either specified or determined from the radar equation depending on the nature of the problem at hand.

Denoting then the sum of the modified earth radius and the aerial height by A and that of the modified earth radius and target height by B one may express, provided that both heights are known, the angle ξ between A and B as

$$\xi = \sqrt{\frac{R^2 - (h-H)^2}{AB}} \tag{5.14}$$

by using, in the triangle $0ab$, the law of cosines and the approximation

$$\cos \xi = 1 - \frac{\xi^2}{2}, \tag{5.15}$$

which, being accurate to five decimal places for angles up to about 6°, is applicable with this accuracy to ground ranges of almost 900 km. In cases where the target height is not known it may be computed by the law of cosines using the given range R, the length A and the angle ϵ at which the slant range is inclined with respect to the local horizontal at the radar's aerial.

Three methods of computing the angles α resp. β required for determining the lengths X and Y of the reflected ray following the path acb have been examined. The first one, an attempt of finding an exact solution, equates the sines of the angles at c expressed on the basis of the law of sines from the triangles $0ac$ and $0cb$

$$\frac{A \sin(\xi - \beta)}{\sqrt{A^2 + r^2 - 2Ar\cos(\xi - \beta)}} = \frac{B \sin \beta}{\sqrt{B^2 + r^2 - 2Br\cos\beta}}. \tag{5.16}$$

Using then the approximation of eqn. (5.15) and setting

$$\sin(\xi - \beta) \approx \xi - \beta \tag{5.17}$$

an approximation accurate to four decimals places up to an angle of about 4°, or a ground range of almost 600 km, one arrives, after some algebraic manipulation, at the biquadratic equation

$$\begin{aligned} &\beta^4 ABr(h-H) - \beta^3 2ABr\xi(h-H) \\ &\quad + \beta^2 r\{(h-H)[AB(\xi^2+2) + 2r^2 + r(h+H)]\} \\ &\qquad -\beta 2A^2H^2\xi + A^2\xi[B^2 - r\xi(B+H)] = 0. \end{aligned} \tag{5.18}$$

This, however, does not lend itself to easy evaluation.

The second method, a computation of the angle α, assumes that the direct ray ab and the reflected ray cb are parallel so that the angle between the reflected ray and the local horizontal at c is $\alpha + \epsilon$.

Writing the law of sines for the triangle $0ac$ and using the mentioned approximations of the sine and cosine functions one obtains the equation

$$r\left(1 - \frac{\alpha^2}{2}\right)\cos\epsilon - r\alpha\sin\epsilon = A\cos\epsilon(1 - 2\alpha^2) - 2A\alpha\sin\epsilon,$$

which may be solved for α to yield

$$\alpha \doteq \frac{h}{A \tan \epsilon} - \frac{3h^2}{2A^2 \tan^3 \epsilon} + \frac{9h^3}{2A^3 \tan^5 \epsilon}. \tag{5.19}$$

The accuracy of α computed from this expression is acceptable for higher values of the elevation angle ϵ; it is not good for close targets at low altitudes.

The third method uses the expressions

$$\begin{aligned} p &= \frac{2}{\sqrt{3}} \sqrt{r(h+H) + \left(\frac{r\xi}{2}\right)^2} \\ \phi &= \cos^{-1} \frac{2r^2 \xi (h-H)}{p^3} \\ \beta &= \frac{\xi}{2} + \frac{p}{r} \cos \frac{\phi + \pi}{3} \end{aligned} \tag{5.20}$$

given by Fishback in Kerr (1951) for the computation of the angle adjacent to the greater of the extended radii A and B. The changes to the notation used in and the modifications of Fishback's expressions have been made to suit the present treatise.

The angle α becomes then

$$\alpha = \xi - \beta. \tag{5.21}$$

The lengths of parts X and Y of the reflected ray computed from the law of cosines in the two previously used triangles and the earlier introduced approximation to the cosine function become

$$X = \sqrt{h^2 + rA\alpha^2}$$

and

$$Y = \sqrt{H^2 + rB\beta^2} \tag{5.22}$$

respectively.

The angle δ between X and the local horizontal at the radar's aerial computed by using the law of sines in the triangle $0ac$ and the approximation of the angle α as in eqn. (5.17) is given by

$$\delta = \cos^{-1} \frac{r\alpha}{X} \tag{5.23}$$

and the angle of incidence γ, from the sum of the angles in the same triangle by

$$\gamma = \frac{\pi}{2} - |\delta - \alpha|. \tag{5.24}$$

A terrain inclination at point c will modify the angle γ and hence the reflection coefficient and phase shift on reflection and also the position of point b where the direct and reflected rays intersect. An inclination towards the radar will reduce γ and hence also the distance of b from the radar and vice versa. The behaviour of the reflection coefficient and phase shift is described in Chapter 6.

The ground range from the radar's aerial to the 'point' of reflection is

$$\alpha r. \tag{5.25}$$

In reality reflection occurs from elliptically shaped areas, the Fresnel zones, the dimensions of which may be computed from eqns. (6.12), (6.13), (6.14) and (6.15).

5.5 Reflection and interference

It was mentioned earlier that the electromagnetic field due to an aerial's radiation in case of multipath propagation is the resultant of at least two components. One represented by the vector $\bar{E}_1$, travels directly from the aerial to the point of interest in space, where it arrives with the phase ψ_1, while $\bar{E}_2$ the other component reaches that same point, after reflection from the ground, with phase ψ_2.

The resultant field is the vector sum

$$\bar{E} = \sqrt{\bar{E}_1^2 + \bar{E}_2^2 + 2\bar{E}_1\bar{E}_2 \cos(\psi_1 - \psi_2)}.$$

It is to be noted that the magnitudes of the above two vectors are

$$E_1 = EG(\epsilon) \exp(\alpha R)$$

and

$$E_2 = EG(\delta)\rho \exp[(R + \Delta R)\alpha]$$

respectively, where E is the amplitude of the aerial input energy, $G(\epsilon)$ and $G(\delta)$ are the aerial gains in the direction ϵ resp. δ, ρ is the magnitude of the composite reflection coefficient, as described in Chapter 6, and the exponential terms are atmospheric attenuations suffered by the two fields propagating over the path R and $R + \Delta R$ respectively. The quantity ΔR is the difference between the direct and reflected path length.

Considering a single ray representing the above direct component radiated by an aerial at height h above the ground and another ray, the reflected component, assumed to be emanating from the aerial's mirror image at depth h below the ground, one arrives, when reflection occurs from a plane earth, at the path difference

$$\Delta R = 2h \sin \epsilon \tag{5.26}$$

and the corresponding phase difference between the direct and reflected ray then becomes

$$\psi = \frac{2\pi}{\lambda} \Delta R. \tag{5.27}$$

The phase shift μ occurring on reflection which is computable from the complex coefficient of reflection must be added to ψ in order to obtain the total phase shift ξ between the two rays so that

$$\xi = \mu + \frac{2\pi}{\lambda} \Delta R. \qquad (5.28)$$

Chapter 6 shows that the phase shift on reflection for a horizontally polarized radiation is very close to π radians for the whole range of grazing angles of interest in radar and communication so that for the two rays mentioned earlier to meet in phase the phase shift due to the path length difference, at horizontal polarization, must equal an odd multiple of π radians. Maxima of the resultant field occur then for

$$2h \sin \epsilon = (2n-1)\pi \qquad (5.29)$$

while minima, at horizontal polarization, occur when the total phase difference is an odd multiple of π radians or the phase shift due to path length difference added to that occurring on reflection must be an even multiple of π radians, i.e. when

$$2h \sin \epsilon = 2n\pi. \qquad (5.30)$$

Using then the identity $\lambda/2 = \pi$, the elevation angles ϵ at which maxima of the field occur are given by

$$\epsilon = \sin^{-1}(2n-1)\frac{\lambda}{4h}, \quad n = 1, 2, \ldots, k \qquad (5.31)$$

while those of the minima by

$$\epsilon = \sin^{-1}\frac{n\lambda}{2h}, \quad n = 1, 2, \ldots, k. \qquad (5.32)$$

A lobing pattern is then created in the vertical plane containing the aerial when the angle ϵ varies from 0 to $\pi/2$ radians.

The number of minima may be obtained by evaluating n from eqn. (5.32) for $\epsilon = \pi/2$ yielding

$$n = 2h/\lambda,$$

while the number of maxima, obtained in the same way from eqn. (5.31), is the nearest smaller integer to

$$n = \frac{4h + \lambda}{2\lambda}.$$

The number of lobes and minima increases while the angles of their directions decrease with an increasing aerial height resp. frequency.

Numbering of the lobes and minima usually starts at the bottom with the first lobe being at an angle ϵ obtained for $n = 1$ from eqn. (5.31).

Since the phase shift on reflection at vertical and circular polarization changes with grazing angle variations to a greater extent than that for horizontal one, as shown in Chapter 6, the above expressions for the angles of the maxima and minima are not applicable for vertical and circular polarization for which the phase difference of the two rays required for lobe or minimum formation have to be computed for the appropriate phase shifts on reflection.

The approximate difference in height of two extremes, i.e. maxima or minima, may be computed from

$$\Delta H \doteq R\,\Delta\epsilon$$

with R their mean range and $\Delta\epsilon$ the difference of their elevation angles so that the difference in heights of an adjacent maximum and minimum, at horizontal polarization, using eqns. (5.31) and (5.32) for an identical n is

$$\Delta H \doteq R\lambda/4h. \tag{5.33}$$

The approximate difference in range ΔR of two extremes at a constant height is

$$\Delta R = R_1 \frac{\sin(\epsilon_2 - \epsilon_1)}{\cos \epsilon_2} \tag{5.34}$$

where R_1 is the range to the more remote extreme at an angle ϵ_1 and ϵ_2 is the angle of the closer extreme.

Since ΔR is a function of the range, the frequency of field variations from maxima to minima and vice versa increases as a target flying at a constant height approaches the radar.

Reflection from rough surfaces is dealt with in Chapter 6 which surveys the relevant literature on the subject. Besides that it defines the reflection coefficients for horizontal, vertical and same sense and opposite sense circular polarization, with same sense meaning that both the transmitted and received waves rotate in the same sense, while in circular polarization of the opposite sense the two waves rotate in opposite sense. The composite reflection coefficient applicable to rough convex surfaces is also studied there.

5.6 Meteorological effects on radar wave propagation

Besides the always present attenuation by the oxygen and water vapour content of the atmosphere, described in Section 4.3.1.3 on propagation losses, radar performance is also affected by such meteorological phenomena as rain, snow, clouds, hail, etc.

Rain, snow, clouds and hail affect radar performance in two ways. They attenuate waves propagating through them and they cause unwanted echoes known under the cumulative name of clutter. It is attenuation that is of interest at present; clutter will be considered later.

5.6.1 Meteorological conditions at various heights

Since weather manifests itself in different ways at the various heights above the earth, its influence on radar performance against targets flying at various heights will differ. Further, since the character of the weather pattern changes with geographical location, the following few lines will examine the influence of conditions prevailing in the area of the radar subjected to more thorough scrutiny.

5.6.1.1 Clouds and precipitation: Water vapour, generally an invisible gas, may reach only a certain density in the atmosphere depending on the air temperature. If this level of density is exceeded, be it due to an increased water content of a thermally stable air, e.g. by evaporation of water surfaces or of the ground, or due to cooling of the air through nightly radiation in the vicinity of the ground or due to an upward motion of the air, then saturation is reached and the excess of water vapour is released by condensation.

Condensation is the cause of all precipitation. It occurs either when the temperature of the air is reduced while the volume remains constant and the air is cooled to dew point, or when the volume of the air is increased without an increase in its temperature, or when the simultaneous change of temperature and volume reduces the moisture holding capacity of the air below its existing moisture content.

Cloud formation depends on atmospheric instability and vertical motion. The international classification of clouds (Blüthgen, 1964) with respect to the heights of their bases distinguishes

high clouds (CH) at heights between 5000 and 13 000 m,
middle clouds (CM) between 2000 and 7000 m and
low clouds (CL) between the ground and 2000 m.

Precipitations are classified according to the process involved in the cooling of moist air masses.

The convective type of precipitation, due to high summer ground-temperatures, associated with cumulus and cumulo-nimbus clouds, develops when lower upper-atmospheric temperatures cause the release of convective instability. When the air at upper levels is less moist than that at lower levels the whole layer is forced upwards; the dryer air column follows a dry adiabatic rate until, eventually, the condensation level is reached after which the lower layers of the rising air mass cool at the saturated adiabatic rate. Thus, the actual lapse rate of the total thickness of the raised layer increases and, if this new rate is greater than that of the saturated adiabatic, the air layer becomes unstable.

Precipitation of this type, occurring often with hail, is of the thunderstorm type though not necessarily accompanied by thunder and lightning. It occurs as a heavy downpour over limited areas of 20 to 50 km^2 and lasts for about $\frac{1}{2}$ to one hour.

Showers of rain or soft hail pellets may form a streaky distribution of precipitation when cold, moist, unstable air is forced by wind to pass over a warmer surface. This type of precipitation is widespread, though of brief duration, in any one locality.

Cyclonic precipitation is tied to frontal phenomena between different air masses caused by the ascent of air through horizontal convergence of air streams in areas of low pressure. Due to temperature convection components in middle latitudes this type of precipitation is prevalent in the cooler seasons but may occur during any season. It manifests itself as a widespread constant rain lasting six to twelve hours.

Cold air accompanying cold fronts forces warm air to rise in a turbulent process which produces extensive cloud formation with a low-lying condensation base. This releases a heavy, rapidly changing, but regionally restricted rain along the cold front which occasionally follows immediately after the warm front accompanying precipitation. However, it soon breaks up into a series of short showers which, in contrast to the thermal convection phenomena, pass rapidly with a diminishing intensity since the oncoming cold air contains less humidity and tends towards a rapid stabilization. The rapid movement causes the vast amounts of water from the shower clouds to be distributed over larger areas than those which occasionally persist for hours over the same region.

Precipitation generally originates at heights below about 3000 m depending on the season and geographical latitude of the location of interest.

5.6.1.2 Attenuation by precipitation: Electromagnetic radiation at microwave frequencies propagating through precipitation is attenuated partly by scattering of energy by drops of water and partly by the conversion of the absorbed energy into heat. A spherical particle of radius a removes, according to Goldstein (see Kerr, 1951), the total energy

$$W = SQ(a, \lambda)$$

from the radiation at wavelength λ. The symbol S in this expression stands for the incident Poynting vector and $Q(a, \lambda)$ is the total cross-section of the rain-drops, i.e. the sum of the absorption and scattering cross-section, or

$$Q(a, \lambda) \approx \lambda^2 \left(\frac{2\pi a}{\lambda}\right)^3 \left[1 + \left(\frac{2\pi a}{\lambda}\right)^3\right]$$

if $n(a)$ is the number of drops per m^3 with radius a in the range da and l is the thickness of the precipitation layer, the attenuation is given by

$$S = S_0 \exp\left[-434 \int_0^l \int_0^\infty n(a)\, Q(a, \lambda)\, da\, dl\right],$$

where the constant before the integration signs converts the dimension of the attenuation constant to the generally used dimension of dB/km.

In general, there are large differences in precipitation rates even at places only a few hundred metres apart; furthermore, current measurements of drop sizes are very crude and there is no conclusive evidence to show that rain of a known rate of fall has a uique drop size distribution although a probable drop size distribution may be assigned to various precipitation rates. The observed drop sizes range from 0·01 to 0·6 cm. However, the larger drops tend to break up when falling and the actual drop size depends on the strength of the air currents. Temperature affects the dielectric properties of water and hence the total attenuation due to precipitation, fog and clouds.

Ryde (1946) computes attenuation at a temperature of 18°C for frequencies between 3 and 30 GHz as

$$\alpha_{\mathrm{dB/km}} = kp,$$

where p is the precipitation rate in mm/h and k is a function of the wavelength given by Ryde in graph form which may be expressed analytically as

$$k = 0{\cdot}002\,747\,\lambda^2 - 0{\cdot}042\,408\,\lambda + 0{\cdot}149\,679 \qquad (5.35)$$

with the wavelength λ in cm.

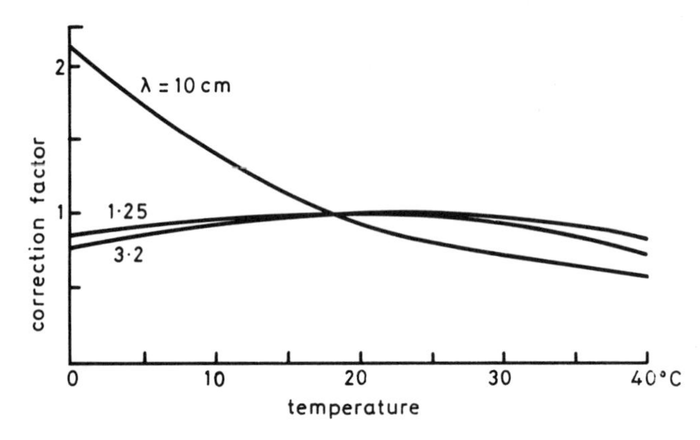

Fig. 5.3 *Temperature correction for k*

The difference in the value of k for precipitation rates between 1 and 100 mm/h is negligible but k may double when the temperature drops from above 18°C to zero. Figure 5.3, based on figures published by Bean, Dutton and Warner (see Skolnik, 1970), shows the variation of the average multiplicative correction factor of k for wavelengths of 1·25, 3·2 and 10 cm with temperature between 0°C and 40°C; k equals unity at 18°C.

While rain attenuation is negligible at frequencies below 3 GHz, it rises with frequency to exceed the combined oxygen and water vapour attenuation.

Attenuation caused by hail is only about one hundredth of that due to rain.

Ryde (1946) found that clouds of ice-crystals and snow do not cause any appreciable attenuation even if the rate of fall is over 125 mm/h. However, attenuation by spheres consisting of concentric layers of different dielectric constant, e.g. a melting ice sphere, may cause attenuation approaching that caused by rain.

5.6.1.3 Attenuation by fog and clouds: Attenuation by fog depends on its density which is usually judged by the distance in metres over which a prominent dark object may be identified in daylight by the naked eye against the sky at the horizon or over which a known moderately unfocussed light source may be just seen by night (Bean, Dutton and Warner in Skolnik, 1970).

Although visibility does not depend entirely on the liquid water content of the atmosphere it is generally used to express it approximately and also to estimate electromagnetic wave attenuation.

Cloud droplets are water or ice particles having radii smaller than 0·01 cm. Attenuation by clouds depends on their liquid-water content M usually expressed in g/m³ and given by

$$M = \frac{4\pi\rho}{3} \sum_{i=1}^{N} r_i^3,$$

where ρ is the density of water in g/m³, N the number of drops per unit volume and r_i their radii.

The attenuation in dB/km, according to Gunn and East, as quoted by Bean, Dutton and Warner, is

$$K = 0{\cdot}4343 \, \frac{6\pi}{\lambda} \, \mathrm{Im} \left(- \frac{m^2 - 1}{m^2 + 2} \right),$$

where m is the complex index of refraction of water and Im( ) denotes the imaginary part of the expression in the bracket.

Figure 5.4, based on figures given by Saxton and Hopkins, as quoted by Bean *et al.*, shows the attenuation by fog and clouds at 0°C for the three previously used wavelengths of 1·25, 3·2 and 10 cm.

5.6.2 Depolarization by precipitation

Besides attenuation precipitation depolarizes electromagnetic waves due to the distortion of drops of water by winds (Ananasso, 1980). Photographic observations have shown that a drop of water having a radius larger than about 1 mm acquires the shape of an asymmetrical oblate spheroid when falling through the air. Oguchi (1960) assumed an ellipsoidal drop shape and determined the difference in attenuation of waves polarized along its two main axes, while Chu (1974) found that horizontally polarized waves are attenuated 3 to 5 dB more than vertically polarized ones. Further, it was observed that circular polarization is transformed into an elliptical one and that drop inclination rotates the polarization to a certain extent.

5.7 Anomalous propagation

A survey of radar wave propagation would not be complete without some reference to 'anomalous' propagation.

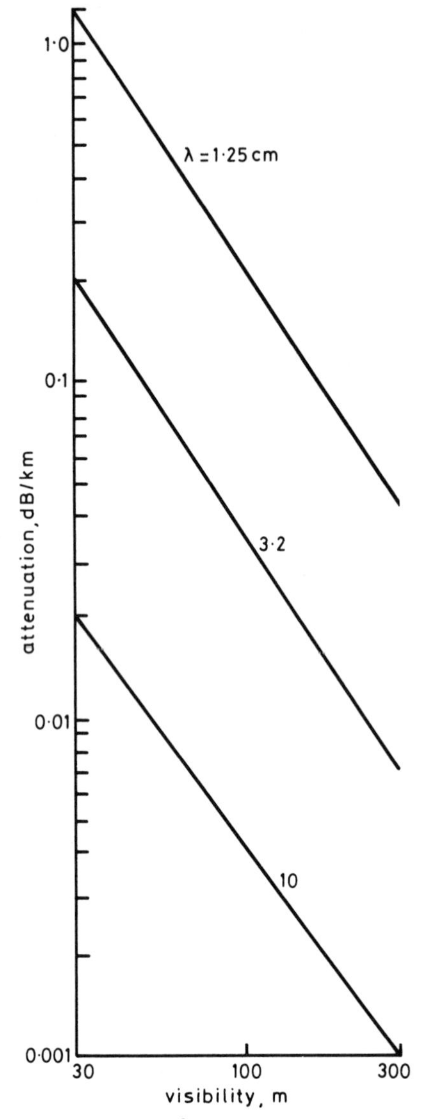

Fig. 5.4 *Attenuation by cloud or fog at 0° C*

The term anomalous propagation refers to propagation under such meteorological conditions that, in case of super-refraction, the curvature of rays is either equal or greater than that of the earth's surface so that the rays follow it or are forced back towards it, they are ducted, while, in case of subrefraction, they bend away from the surface of the earth and may be lost in space.

Super-refraction is a frequent phenomenon while sub-refraction is rather rare.

5.7.1 Meteorological conditions supporting super-refraction

There are various causes of ducting of electromagnetic waves which are surveyed briefly in the following.

At night when the surface temperature falls faster than that of the air, temperature inversions may occur near the ground at heights below about 300 m and produce conditions conducive to ducting.

Similar conditions may be produced by an advection process by which a mass of warm dry air blows from the land over a cooler sea causing evaporation which increases the humidity near the surface thus decreasing the temperature.

In a subsidence inversion the air mass at the centre of an anticyclone descending from higher levels is dry. Water vapour from moist surface may ascend to the base of the inversion by convection or turbulence and intensify the humidity contrast. The resulting humidity inversion may cause ducting.

Ducting is less pronounced and extends over smaller areas over land since hills, woods, buildings, etc., disturb conditions suitable for ducting. Over sea, however, where conditions favourable to ducting may extend over vast areas, propagation by ducting may occur over ranges of several hundred km, as reported by Booker (1946).

Ducting is a diffraction problem which has been described by several authors among others by Kerr (1951), Budden (1961*b*), and Matthews (1965).

The boundaries of a tropospheric surface duct are the ground and an elevated layer of stratified air, while stratified air layers form both boundaries of an elevated duct. Propagation within a duct is similar to that in a waveguide. However, since one or both boundaries are 'soft' the effective plane of reflection depends on the angle of incidence of the wave into the boundary formed by the stratified layer. A duct can support only certain modes of propagation the wavelengths of which are shorter than the critical wavelength.

$$\lambda_{max} = 2{\cdot}296 \left(\frac{\Delta n}{\Delta h}\, d^3\right)^{1/2}$$

as given by Kerr (1951) for a simplified model of wave propagation in an atmospheric duct of width d. The ratio $\Delta n/\Delta h$ is the gradient of the refractive index. All quantities are to be measured in the same units.

The condition for ducting may be derived from the earlier obtained expression for the curvature of the rays

$$\rho = -\frac{1}{dn/dh}$$

from which, due to

$$n = 1 + N \times 10^{-6},$$

one may express the gradient of the refractive index as

$$\frac{\Delta N}{\Delta h} = -\frac{1}{\rho} \times 10^{6}.$$

If ducting is to occur, the radius of curvature of the ray must equal or be smaller than that of the earth's surface, i.e.

$$\frac{\Delta N}{\Delta h} \leqslant -\frac{10^{6}}{r}, \tag{5.36}$$

where r is the earth's radius. Using the average value of $r = 6371{\cdot}228$ km yields

$$\frac{\Delta N}{\Delta h} \leqslant -157\,N\text{-units per km}$$

as the condition for ducted propagation.

5.7.2 Linear refraction index profile
Budden (1961 *b*) considers the propagation of horizontally and vertically polarized electromagnetic waves in a duct with a linear decrease of the refractive index and other profiles. The following is based largely on his treatise.

5.7.2.1 Horizontal polarization: The modified refractive index may be defined by

$$\mu = 1 + M \times 10^{-6},$$

where

$$M = [(n-1) + h/a] \times 10^{6}$$

with n denoting the hitherto used refractive index, h the height above a perfectly conducting ground and $a = kr$ the modified earth radius. Its rate of change with height is

$$\frac{d\mu}{dh} = \frac{1}{a} - \kappa,$$

where κ is the lapse rate of the refractive index having an approximate value of 3×10^{-5}/km under nornal turbulent conditions. A value of $\kappa > 1/a$ causes μ to decrease with height and consequently the rays to bend towards the ground thus fulfilling the requirements for ducting.

Assuming that the lapse rate κ of the modified refractive index is positive and constant, the modified refractive index becomes

$$\mu = \mu_0(1 - \kappa h), \tag{5.37}$$

where μ_0 is the value of μ at the ground, at $h = 0$. Since the values of κh used in practice are small, they do not exceed about 10^{-4}, the above expression may be written, with negligible error, as

$$\mu^2 = \mu_0^2(1 - 2\kappa h). \tag{5.38}$$

While profiles described by the last two expressions may be valid to a height of several hundred metres when temperature inversion exists, they cannot occur in practice since at great enough heights the curvature of the earth always yields a negative lapse rate of the refractive index.

Assuming that a wave leaving the earth with a normal at an angle θ and entering an elevated layer at a height h and distance d with a normal at an angle ψ with the horizontal undergoes a phase shift

$$\exp\left[-j\,\frac{2\pi}{\lambda}\,\mu(d\cos\psi + h\sin\psi)\right]$$

which, due to Snell's law

$$\mu\cos\psi = \mu_0\cos\theta,$$

becomes

$$\exp\left[-j\,\frac{2\pi}{\lambda}\,(\mu_0 d\cos\theta + \mu h\sin\psi)\right] \tag{5.39}$$

and since its second term is a function of the height, (5.39) may be written as

$$\exp\left[-j\,\frac{2\pi}{\lambda}\,\left(\mu_0 d\cos\theta + \int_0^h \mu\sin\psi\,dh\right)\right]. \tag{5.40}$$

Using Snell's law and eqn. (5.38) one arrives at

$$\begin{aligned}\mu^2\sin^2\psi &= \mu^2 - \mu^2\cos^2\psi\\ &= \mu^2 - \mu_0^2\cos^2\theta\\ &= \mu_0^2(\sin^2\theta - 2\kappa h)\end{aligned}$$

or, since the value of μ at the ground is approximately unity,

$$\mu^2\sin^2\psi = \sin^2\theta - 2\kappa h.$$

The virtual height of reflection is that height h_r at which the vertical component of μ, i.e. $\mu\sin\psi$, is zero or

$$h_r = \frac{\sin^2\theta}{2\kappa}.$$

The first term of (5.39) does not depend on μ; the phase shift is given by the second term of (5.40) and the total phase shift is twice this value or

$$\exp\left[-2j\,\frac{2\pi}{\lambda}\int_0^{h_r}\mu\sin\psi\,dh\right]$$

for a ray travelling from the ground to the height h_r and back.

Substituting

$$\mu\sin\psi = (\sin^2\theta - 2\kappa h)^{1/2}$$

one arrives at

$$\exp\left[-2j\,\frac{2\pi}{\lambda}\int_0^{h_r}(\sin^2\theta - 2\kappa h)^{1/2}\,dh\right].$$

Budden's solution of the integral thus yields the reflection coefficient

$$\rho = j\exp\left[-\tfrac{2}{3}j\,\frac{2\pi}{\lambda}\,\frac{\sin^3\theta}{\kappa}\right] \tag{5.41}$$

referred to level h. For a horizontally polarized wave it is the ratio of the electric field in the ascending and descending waves.

Since the ground was assumed to be a perfect conductor, the horizontal component of the electric field must be zero so that the reflection coefficient of the ground must be

$$\rho_0 = -1$$

as it will be shown also in Chapter 6, and then

$$\sin^3\theta = (n - \tfrac{1}{4})\,\frac{3\lambda\kappa}{2} \tag{5.42}$$

with n an integer denoting the order of the propagation mode, which cannot be zero or negative since $\sin\theta = \mu\sin\psi$ when $h = 0$, i.e. the direction of the wave normal to the ground which must have a positive real part. When n is positive $\sin\theta$ is real and positive since the other roots of eqn. (5.42) have negative real parts and the modulus of the reflection coefficient is always unity.

All modes are locked modes with most of the energy confined below the height

$$h_r = \frac{1}{2\kappa}\left[(n - \tfrac{1}{4})\,\frac{3\lambda\kappa}{2}\right]^{2/3}, \tag{5.43}$$

which is called the 'track width' for the mode of order n. It increases with n and λ but it is the height over which the anomalous lapse rate prevails that determines the number of possible modes. The angle θ for these modes is small so that only rays launched near the normal may be trapped.

The wave equation which the horizontal field component E_y must satisfy for vertical propagation is the Stokes equation (Budden, 1961 *a*)

$$\frac{d^2E_y}{dh^2} + \left(\frac{2\pi}{\lambda}\right)^2 (\sin^2\theta - 2\kappa h)E_y = 0 \tag{5.44}$$

the solution of which reads

$$E_y = Ai\left\{\left[2\kappa\left(\frac{2\pi}{\lambda}\right)^2\right]^{1/3}(h-h_r)\right\} \tag{5.45}$$

with $Ai(\xi)$ and Airy integral function shown in Fig. 5.5 (Budden, 1961 *b*).

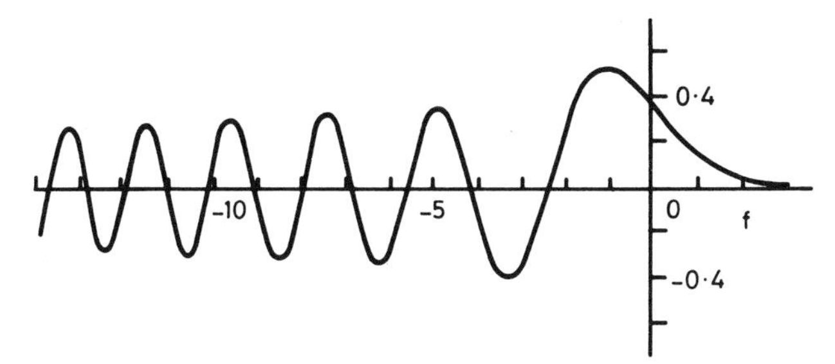

Fig. 5.5 *The Airy integral function Ai(ξ) (After Budden, 1961b)*

Figure 5.6 (Budden, 1961 *b*) shows the dependence of the field components on the height h in an atmosphere with a linear decrease of the squared refractive index. The upper three curves, with the modes indicated by the number at the top, are for a horizontal polarization or TE modes, while the lower ones are for vertical polarization or TM modes; h_r is the virtual height of reflection.

For higher order modes or longer wavelengths the track width may extend into regions where the assumption of a linear decrease of the modified refractive index, according to eqn. (5.37), is no longer valid. The modes are no longer locked then; the higher order modes may lose energy from the top and the guide is termed, therefore, to be leaky.

5.7.2.2 Vertical polarization: In the case of vertical polarization it is more convenient to consider the behaviour of the magnetic field of a wave which has only a horizontal component H_y so that all modes are TM modes.

Since the variation of H_y with height is very small for all values of h and μ^2 of interest, the wave equation reduces to

$$\frac{d^2H_y}{dh^2} + \left(\frac{2\pi}{\lambda}\right)^2 H_y(\sin^2\theta - 2\kappa h) = 0$$

which is of the same form as eqn. (5.44) arrived at in the case of horizontal polarization.

The reflection coefficient referring here to the magnetic field H_y is given by eqn. (5.41) which, assuming the ground to be a perfect conductor for which

$$\frac{dH_y}{dh} = 0 \quad \text{when } h = 0,$$

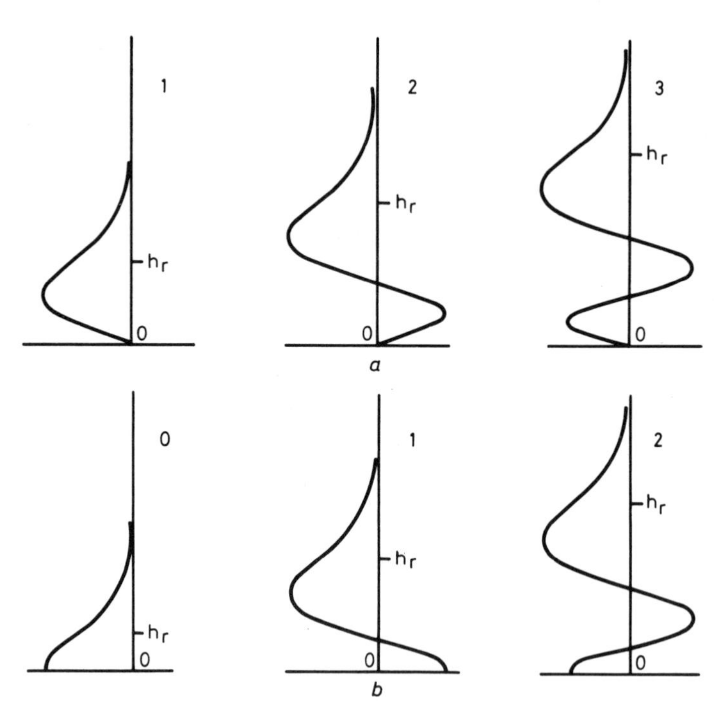

Fig. 5.6 *The dependence of the field components on the height h (After Budden, 1961b)*
(*a*) Horizontal polarization
(*b*) Vertical polarization

leads to

$$\rho = j \exp \left[-\tfrac{2}{3} j \frac{2\pi}{\lambda} \frac{\sin^3 \theta}{\kappa} \right] = 1$$

for the reflection coefficient of the ground, from where

$$\sin^3 \theta = \frac{3\kappa\lambda}{2} (n + \tfrac{1}{4}) \tag{5.46}$$

with n zero or an integer.

The virtual height of reflection or the 'track width' in case of vertical polarization is

$$h_r = \frac{1}{2\kappa}\left[\frac{3\kappa\lambda}{2}(n+\tfrac{1}{4})\right]^{2/3}, \tag{5.47}$$

an expression similar to that for horizontal polarization.

Both expressions, i.e. eqn. (5.42) and (5.46), for the mode conditions are approximate only since they are based on asymptotic approximations to the Airy integral function and here too, as in the case of horizontal polarization, a more accurate result is obtained by solving the given wave equation to read

$$H_y = Ai\left\{\left[2\kappa\left(\frac{2\pi}{\lambda}\right)^2\right]^{1/3}(h-h_r)\right\}.$$

Due to the assumption of the ground being a perfect conductor and using the expression $h_r = \sin^2\theta/2\kappa$ one arrives at

$$Ai\{-\sin^2\theta\,(\pi/\kappa)^{2/3}\} = 0$$

as the requirement which satisfies the boundary condition $dH_y/dh = 0$ when $h = 0$.

The error due to the approximation is of the order of 10% for the lowest propagation mode but diminishes for higher modes.

It may be observed that in case of horizontal polarization there are n half-waves in the field profiles shown by Fig. 5.6, while there are $n+\frac{1}{2}$ half-waves in those shown for vertical polarization.

Assuming that the lapse rate of the modifed index of refraction in the duct is $\kappa = 10^{-4}$/km, the angles of the wave normal elevations at ground level and track widths for the first four modes for horizontal and vertical polarization and a frequency of 1·3 GHz are given in Table 5.1.

Table 5.1 Wave normal elevation angles at ground level and track-widths for ducting

Mode	Horizontal polarization		Vertical polarization	
	θ degrees	h_r metres	θ degrees	h_r metres
0	–	–	0·118	21·1
1	0·169	43·8	0·201	61·6
2	0·225	77·1	0·245	91·2
3	0·261	104·2	0·277	116·5

5.7.3 Other refractive index profiles

The hitherto considered profile with a linearly decreasing refractive index with height does not occur in practice. It is useful, however, for studying the conditions conducive to ducting as was done in the preceding sections.

In reality, the square of the refractive index μ^2 is some slowly varying function of height and the reflection coefficient is given by the phase integral

$$\rho = j \exp\left[-2j \frac{2\pi}{\lambda} \int_0^{h_r} q \, dh\right],$$

where q, the so-called Booker variable, is

$$q^2 = \mu^2 - \mu_0^2 \cos^2\theta$$

with μ the real part of the refractive index and θ the angle between the wave normal and the horizontal, as given by Budden (1961 *b*).

For waves having a horizontal electric vector, i.e. TE modes, the mode condition may be written as

$$\int_0^{h_r} q \, dh = \frac{\lambda}{2}\left(n - \tfrac{1}{4}\right) \tag{5.48}$$

when assuming the ground to be a perfect conductor. The coefficient n is a positive but non-zero integer. Solutions of eqn. (5.48) yielding real values of θ for higher frequencies and heights were $d\mu^2/dz$ is negative lead to locked modes, while those giving complex values of θ belong to leaky modes.

Within the duct, i.e. for $h < h_r$, the field of the locked modes is the sum of the fields of two crossing waves and is given by

$$E_h = \frac{2E_0}{\sqrt{q}} \exp(j\pi/4) \exp\left(-j\frac{2\pi}{\lambda} x \cos\theta\right) \cos\left(\frac{2\pi}{\lambda} \int_h^{h_r} q \, dh - \pi/4\right).$$

x is the direction of propagation and E_0 depends on the type of radiator launching the wave. This expression breaks down near the reflection height h_r where a linear variation of μ^2 with h may often be assumed. The field then, for $h \approx h_r$, is

$$E_h = 2E_0 \pi^{2/3} (\lambda\kappa)^{-1/6} Ai\left[-\left(\frac{3}{2}\frac{2\pi}{\lambda}\int_h^{h_r} q \, dh\right)^{2/3}\right]$$

Rotheram (1973) considers besides super-refraction at ranges below that of the radio horizon the phenomenon of super-diffraction, occurring beyond this range far from the transmitter, where the field is dominated by the leaky waveguide modes with their attendant properties of excitation and leakage by secondary radiation, to study the propagation of various modes in an evaporative duct by determining the height-gain function for each terminal in a communication set-up and an excitation factor from second-order differential equations for the individual propagation modes. The height-gain function shows how the signal varies with the height h, while the excitation factor determines the amplitude excited in a given mode by a source. He ascribes discrepancies between theory and experimental results to scattering by rough surfaces at the bottom surface of the duct and atmospheric turbulence within the duct. In an effort to overcome the rough surface effect he modifies the surface impedance so as to consider the energy lost by scattering to be absorbed by the earth and treats atmospheric scattering in a similar manner to modify the refractive index to have a lossy component.

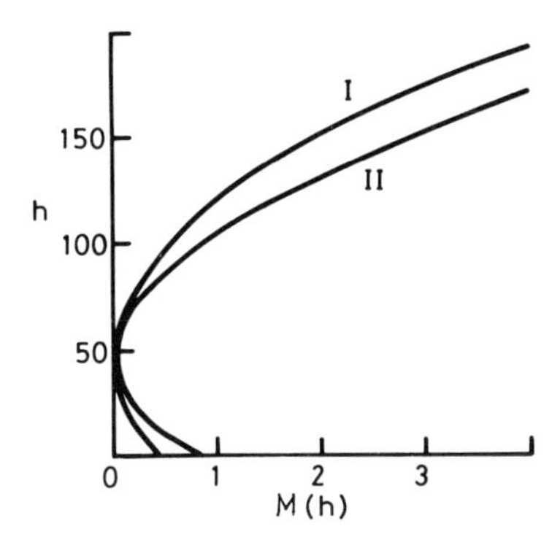

Fig. 5.7 *The considered modified refraction index profiles (After Fok et al., 1958)*

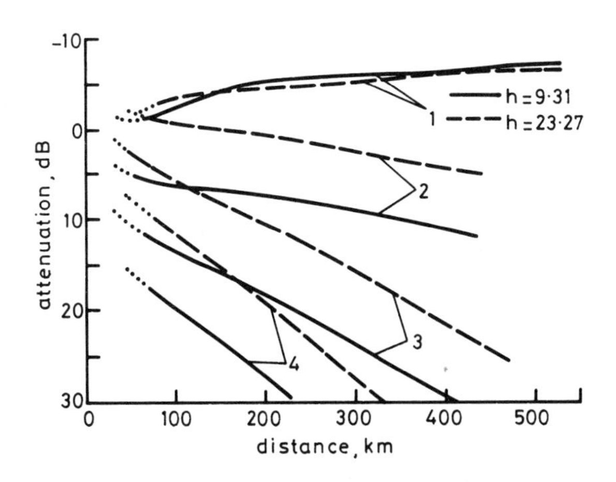

Fig. 5.8 *Attenuation vs. distance for profile I (After Fok et al., 1958)*

Fok, Vainshtein and Belkina (1958) compute the attenuation coefficient for waves travelling in surface trospheric ducts exhibiting two different profiles of the modified index of refraction M shown in Fig. 5.7. Both profiles have the same inversion height but profile II is twice as strong as profile I.

Figure 5.8 shows the dependence of the attenuation on range when both, the transmitter and the receiver, are within the duct of profile I, while Fig. 5.9 shows it for profile II. The full curves apply to the case when the height of the aerials is 9·31 m; the height of the aerials for the attenuation shown by the dashed curves is 23·27 m. Curves 1 are for a frequency of 10 GHz, curves 2 for 3 GHz, curves 3 for 1 GHz and curves 4 for 0·33 GHz.

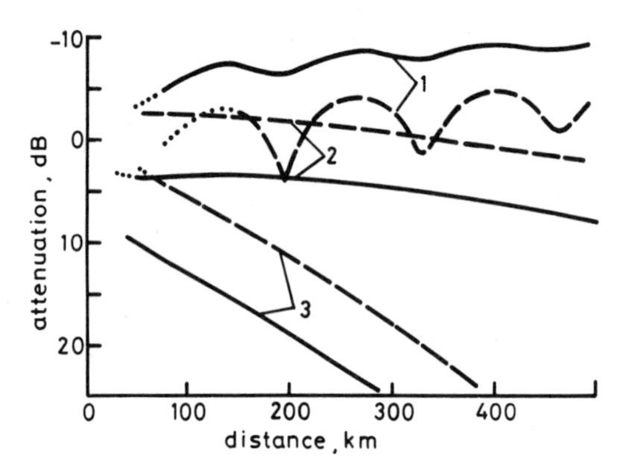

Fig. 5.9 *Attenuation vs. distance for profile II (After Fok et al., 1958)*

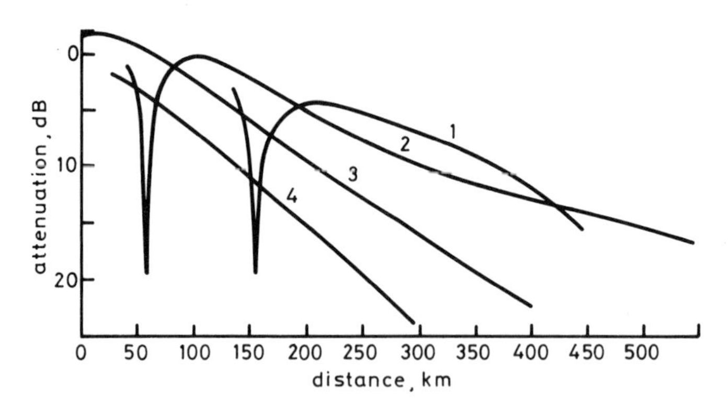

Fig. 5.10 *Attenuation vs. distance. One terminal above the inversion layer (After Fok et al., 1958)*

It is interesting to note that in the case of curves 1 the attenuation diminishes with the distance. This is attributed by the authors to the waves being cylindrical in character and the reflecting layer being such that the trapped wave does not penetrate beyond its limits.

Figure 5.10 shows the dependence of the attenuation on the distance for profile I when one of the terminals is high above the inversion layer, while the other one is at one-fifth of the inversion height. The numbering of the curves is identical with those of the previous figures.

5.7.4 Elevated ducts

Only ducts having the ground for their lower boundary have been considered so far. Elevated ducts with both boundaries formed by discontinuities in the atmosphere are also experienced in practice.

The atmospheric conditions for this type of duct to occur are such that μ, the modulus of the modified refractive index, increases, near the earth, with height in the normal manner but at an elevated height it starts decreasing. A duct supporting a waveguide mode of propagation occurs in the region of the above abnormal variation of μ. Within this range there is a level at which μ reaches a maximum and viewed from this level μ decreases for both a decreasing and an increasing height with the consequence that an upward travelling ray is bent downwards and a downward travelling wave is bent upwards.

Assuming that the mentioned maximum designated here by μ_1, occurs at a height h_1 and that the elevation of the wave normal of the wave travelling upwards at this level is θ, then, on the basis of what preceded, the Booker variable may be expressed as

$$q^2 = \mu^2 - \mu_1^2 \cos^2\theta .$$

There are two levels, h_0 below and h_{00} above h_1, at which $q = 0$, or where

$$\mu = \mu_1 \cos\theta .$$

The reflection coefficient ρ_1 referred to level h_1 for a wave travelling upwards is

$$\rho_1 = j \exp\left(-2j \frac{2\pi}{\lambda} \int_{h_1}^{h_{00}} q \, dh\right),$$

while ρ_2, i.e. that for a downward travelling wave, is

$$\rho_2 = j \exp\left(+2j \frac{2\pi}{\lambda} \int_{h_1}^{h} q \, dh\right).$$

The mode condition, based on the necessary premise that a wave travelling along a waveguide should, after two reflections, remain identical with the original wave entering the first reflection, i.e. that the total phase shift should be $2n\pi$, or that $\rho_1\rho_2 = 1$, which when applied to the considered duct propagation is

$$\exp\left(-2j \frac{2\pi}{\lambda} \int_{h_0}^{h_{00}} q \, dh\right) = -1$$

from where

$$\int_{h_0}^{h_{00}} q \, dh = \frac{\lambda}{2}(n - \tfrac{1}{2})$$

with n a positive integer. A negative value of n would yield the same series of modes for a wave travelling in the opposite direction, i.e. if n positive applies to an upgoing wave, a negative n would appertain to a downgoing one. The above mode condition is applicable to both, the TE and TM modes, with the same propagation constant $2\pi \cos \theta/\lambda$.

The two levels h_0 and h_{00} are referred to as the bottom and top of the track, while their difference is the track width.

The above approximate expressions are not applicable to cases where the $q = 0$ levels are close together or to the ground or the top of the track. If the upper reflection level is close to the top of the track a leaky mode may develop.

A knowledge of the variation of the refractive index μ across the duct is necessary for the determination of the heights h_0 and h_{00}. Approximations of the μ distribution are acceptable for small height differences. However, these must be more accurate for large differences especially if higher order modes are to be determined.

5.7.5 Some observed duct propagations and their effect on radar coverage

Several experimenters have reported duct propagation at metre and decimetre waves over great distances.

Booker (1946) describes observations made during the Second World War by operators seeing returns from Mt. Etna and other terrain features in Sicily and echoes from Greece at ranges of about 600 km on the PPI display of radars operating at a frequency of 600 MHz in Malta and also those made by operators of radars working at a frequency of 200 MHz in Bombay seeing the far side of the Arabian coast at distances of about 2500 km.

Smith studied conditions conducive to ducting by observing PPI displays at two operational radars in Australia and correlating their occurrence with meteorological data prevailing at the same time. Smith and Soden (1974) describe the observations aimed at finding the probability of occurrence of super-refraction at microwave frequencies at Darwin, Northern Territory, with a view of predicting it on the basis of meteorological data. The appearance of a number of radar returns from terrain features beyond the unambiguous radar range of the particular radar was taken as a measure of the presence of super-refraction. It was found that there is a season-dependent diurnal variation of the occurrence of these echoes. The greatest likelihood of ducting at Darwin by surface ducts below about 2·5 km is during the tropical dry season and the dry-to-wet season transition periods, while, as far as the diurnal dependence is concerned, the occurrence of ducting for paths over the sea and over the continent is most likely during the early morning and late evening hours, depending on the season.

A correlation of 0·82 between the observed anomalous propagation and meteorological data obtained from radiosonde measurements for the wet season, i.e. January and February, was obtained. The correlation then drops gradually during the wet-to-dry and dry season to reach about half the above value during the dry-to-wet transitional period, after which it rises again.

Smith and Soden conclude that since, at microwaves, the refractive index of the lower atmosphere near Darwin is very variable it is not practicable to predict the occurrence of ducting on the basis of meteorological data alone. However, some rough estimates may be made on the basis of the collected statistical data. A reasonably successful prediction is possible in more moderate climates.

Generally, oceanic evaporation ducts are found nearly all the time over all oceanic areas. Ducting phenomena influence radar performance and are of particular importance for shipborne and coastal radars.

5.7.6 Sub-refraction

Sub-refraction is a rare phenomenon and it has received but little attention so far. However, a few words on this topic are required to round-off the treatise on anomalous propagation.

Sub-refraction may develop when the ground temperature in a nocturnal inversion falls below the dew-point temperature and the water in the lowest layers of the air condenses. The heat of condensation is released directly to the air, under conditions of radiative fog formation, the humidity lapse tends to counteract any temperature inversion. If the humidity inversion is sufficiently strong substandard refraction having a positive gradient may ensue.

Little mention of sub-refraction has been found in the available relevant literature. Bean and Dutton (1966) conclude that although sub-refraction is normally neglected, it is potentially a very important refractive factor for distances of less than about 40 km. Even though the percentage occurrence of subrefractive layers may be as high as 6 per cent, as observed over a five year period in Washington D.C., it is frequently offset by the simultaneous occurrence of an adjacent super-refractive layer.

Chapter 6

Rough surface reflection – a survey

6.1 Introduction

The propagation of electromagnetic radiation between a radar and a target is a multipath phenomenon.

When investigating the field strength due to a source located near the surface of the earth, it is customary to determine it from the vector sum of the field due to a direct ray between the source and the point of interest in space and a ray which arrives at that same point by a path including a reflection from the ground.

Reflection from a smooth plane earth depends on the electrical properties of the ground at the clearly defined intersection of the ground with a ray joining the image of the source of radiation with the said point. However, since the surface of the earth is not smooth but has irregularities, the smooth ground reflection coefficient is usually modified to account for scattering by the roughness of the ground. Besides that, since the surface of reflection is not plane but convex, a further modification of the smooth earth reflection coefficient, by the so-called divergence factor, is necessary.

It is intended to give a brief survey of the current state of knowledge of the above multipath problem.

6.2 Specular reflection

Specular reflection occurs only on a plane, smooth, surface where the laws of geometrical optics apply. The resultant field strength at the position of a target, in the radar case, or at the receiving aerial, in case of communication, is the vector sum of the free-space field strength of the transmitter at the target resp. aerial and the free-space field strength of the image of the transmitter at the same point.

Denoting then the first of these fields at the transmitter aerial by E_1 and the second one at the image of the transmitters aerial by E_2, the total field at the target is

$$E = E_1 G(\epsilon) + E_2 \rho G(\delta) e^{-j\zeta}, \tag{6.1}$$

where $G(\epsilon)$ is the transmitter aerial gain in the direction of the target, $G(\delta)$ is the aerial gain in the direction of the point of reflection, ρ is the magnitude of the Fresnel reflection coefficient and ζ is the total phase shift given by

$$\zeta = \frac{2\pi}{\lambda} \Delta R + \mu$$

with ΔR the difference between the reflected and direct path lengths, λ the wavelength of the radiation and μ the phase-shift suffered by the reflected ray on reflection. Generally, the Fresnel reflection coefficient is

$$\rho = \rho_{v,h} e^{-j\mu_{v,h}} \tag{6.2}$$

with

$$\rho_v e^{-j\mu_v} = \frac{\epsilon_c \cos\theta - \sqrt{\epsilon_c - \sin^2\theta}}{\epsilon_c \cos\theta + \sqrt{\epsilon_c - \sin^2\theta}} \tag{6.3}$$

for vertical polarization, and

$$\rho_h e^{-j\mu_h} = \frac{\cos\theta - \sqrt{\epsilon_c - \sin^2\theta}}{\cos\theta + \sqrt{\epsilon_c - \sin^2\theta}} \tag{6.4}$$

in the case of horizontal polarization. The polarization is referred to as being vertical when the magnetic vector is parallel to the boundary surface and the electric vector is parallel to the plane of incidence, and horizontal in the case when the electric vector is parallel to the boundary surface or perpendicular to the plane of incidence (Jordan, 1950). In the above expressions

$$\epsilon_c = \epsilon_1 - j60\,\sigma\lambda \tag{6.5}$$

is the complex dielectric constant with $\epsilon_1 = \epsilon/\epsilon_0$ the relative dielectric constant of the reflecting medium, σ its conductivity, λ the wavelength of the radiation, and θ the angle of incidence of the radiation, i.e. the complementary angle of the grazing angle. Specular reflection occurs at a point within the Fresnel zone, a highly elongated ellipse with its major axis in the intersection of the plane of incidence with the surface of the ground.

The coefficient of reflection for vertical polarization is the ratio of the reflected to the incident magnetic field or the ratio of the vertical components of the electric field. The ratio of the horizontal components yields the negative value of the coefficient of reflection. The coefficient of reflection for horizontal polarization is the ratio of the reflected to the incident electric field or the ratio of the corresponding vertical components of the magnetic field. Here again, the ratio of the incident and reflected horizontal components of the magnetic field yields the negative value of the coefficient of reflection since their direction differs by π radians.

Since the complex dielectric constant of most reflecting media encountered in radar practice is much larger than unity, eqns. (6.3) and (6.4) simplify to read

$$\rho_v e^{-j u_v} \approx \frac{\sqrt{\epsilon_c}\cos\theta - 1}{\sqrt{\epsilon_c}\cos\theta + 1}$$

and

$$\rho_h e^{-j h_u} \approx \frac{\cos\theta - \sqrt{\epsilon_c}}{\cos\theta + \sqrt{\epsilon_c}}$$

thus yielding approximate expressions which permit the observation of the variation of the magnitude and phase components with the variation of the grazing angle $\pi/2 - \theta$.

When θ tends to $\pi/2$, i.e. the grazing angle approaches zero, the magnitude of both, vertical and horizontal, reflection coefficients tends to -1. The coefficient of reflection for horizontal polarization then decreases gradually with the increase of the grazing angle from 0 to $\pi/2$. The dependence of the magnitude of the coefficient of reflection for vertical polarization is more complicated. When the angle of incidence reaches a value making $\cos\theta = 1/\sqrt{\epsilon_0}$, the magnitude of the coefficient ρ_v falls sharply to its minimum to rise again with a decreasing angle of incidence or increasing grazing angle. The angle of incidence corresponding to the minimum value of ρ_v is the pseudo-Brewster angle.

The magntidue of ρ_v would be zero at the Brewster angle of incidence if reflection from a perfect conductor occurred.

Having passed this minimum, ρ_v increases with rising values of the grazing angle to reach a value equal to that of the horizontal coefficient of reflection at $\theta = 0$ or a grazing angle of $\pi/2$ radians.

The phase shift on reflection is also more complicated in case of vertical polarization. Considering an increasing grazing angle, the phase shift μ is π for both polarizations for a zero grazing angle and while with an increasing grazing angle the phase component for horizontal polarization rises slightly above this value, the phase shift for vertical polarization falls abruptly near the pseudo-Brewster angle to remain at a low value for the rest of the higher values of the grazing angle.

Figure 6.1 shows the magnitude and phase components of the vertical and horizontal coefficient of reflection at a frequency of 3.0 GHz for ground having a relative dielectric constant of 10 and a conductivity of 2 mmho/m, while Fig. 6.2 shows the same for reflection from sea water which has a relative dielectric constant of approximately 69 and conductivity of approximately 6·5 mho/m at that frequency. Both illustrations were computed and plotted by the later described radar models.

The constant amplitude electric field vector of a circularly polarized radiation rotates about the axis of propagation. A clockwise rotating electric field vector, when viewed in the direction of propagation, is termed righthand circularly polarized, while one rotating counterclockwise is left-hand circularly polarized. Circularly polarized waves reflected by spherical objects are opposite sense circularly polarized. Hence, circular polarization is often used for rain clutter reduction since an aerial transmitting circularly polarized waves of one sense can receive, without

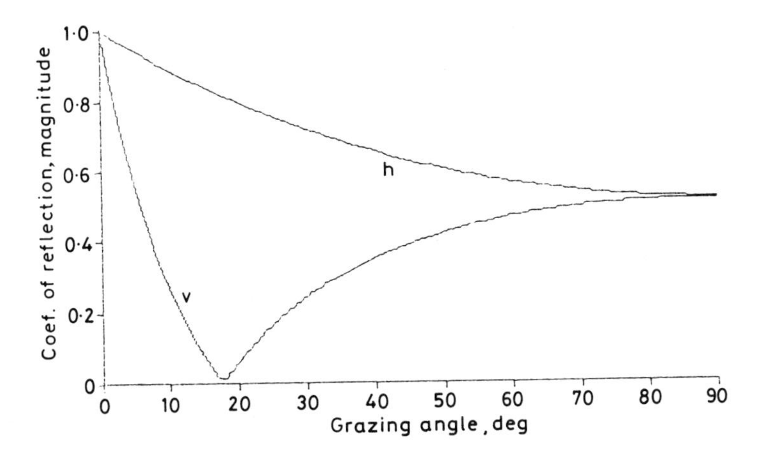

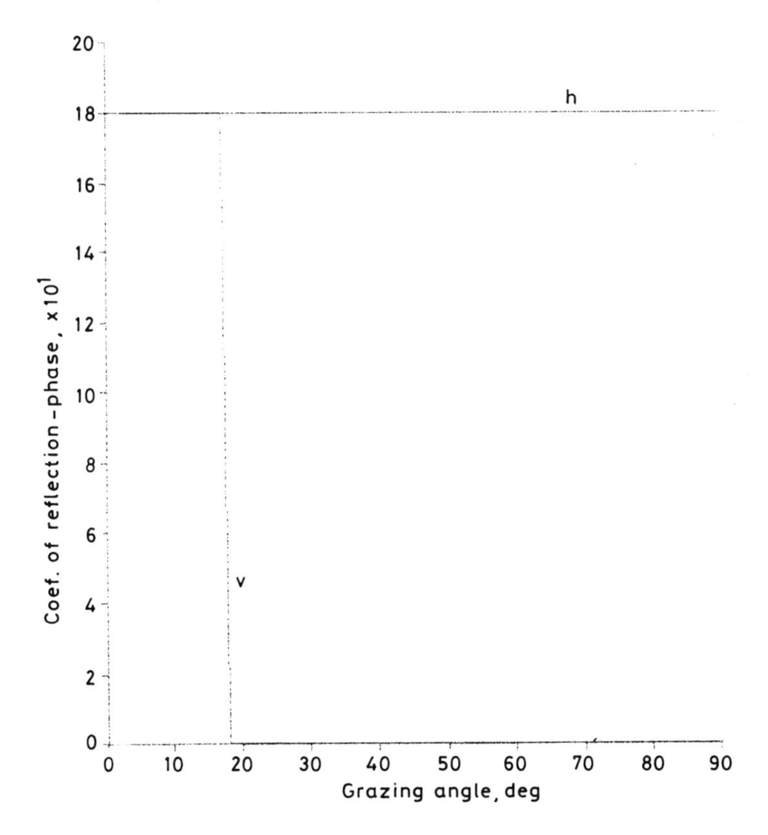

Fig. 6.1 *Reflection coefficient, magnitude and phase vs. the grazing angle. Plane, smooth earth ($\epsilon_r = 10$, $\gamma = 2$ mmho/m, $f = 3$ GHz)*
h = horizontal polarization
v = vertical polarization

excessive attenuation, only waves of the same sense. While targets of interest might contain features which reverse the sense of polarization, they reflect some energy in a sense acceptable to the aerial.

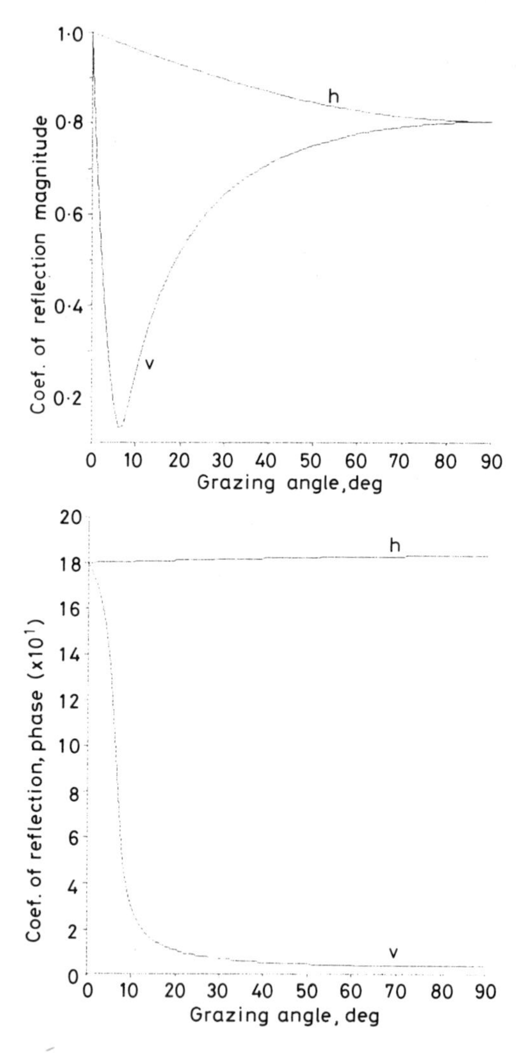

Fig. 6.2 *Reflection coefficient, magnitude and phase vs. grazing angle. Smooth sea ($\epsilon_r = 69$, $\gamma = 6{\cdot}5$ mho/m, $f = 3$ GHz)*
h = horizontal polarization
v = vertical polarization

A circularly polarized wave consists of a vertical and a horizontal component with a $\pi/2$ radians phase difference between them. Each of these components is reflected according to its appropriate coefficient of reflection. Due to the difference in the magnitude of the two coefficients, the originally circularly polarized wave becomes elliptically polarized on reflection.

The reflection coefficient in the case of receiving same sense circularly polarized waves is, according to Katz (1963), the vector sum of the reflection coefficients of the vertical and horizontal components, i.e.

$$\rho_{cs} = \tfrac{1}{2}\sqrt{\rho_v^2 + \rho_h^2 + 2\rho_v\rho_h \cos(\mu_h - \mu_v)}, \tag{6.6}$$

while that for an opposite sense circularly polarized wave it is

$$\rho_{co} = \tfrac{1}{2}\sqrt{\rho_v^2 + \rho_h^2 - 2\rho_v\rho_h \cos(\mu_h - \mu_v)}. \tag{6.7}$$

The phase shift on reflection for same resp. opposite sense circular polarization is

$$\mu_c = \tan^{-1}\frac{\rho_v \sin\mu_v \pm \rho_h \sin\mu_h}{\rho_v \cos\mu_v \pm \rho_h \cos\mu_h} \tag{6.8}$$

Figure 6.3 shows the dependence of the magnitude and phase of the two circular reflection coefficients on the grazing angle when using identical parameters with those used in Fig. 6.1.

When reflection of same sense circularly polarized waves occurs from a smooth surface the magnitude of their reflection coefficient near zero grazing angle incidence is close to those of linearly polarized waves; it falls off gradually to reach a very low value at normal incidence. The magnitude of the reflection coefficient of an opposite sense circularly polarized wave is zero at zero grazing angle and rises with an increasing grazing angle to the same value as that of linearly polarized waves under identical conditions.

Finally, it should be mentioned that the ground is not homogeneous. The conductivity and dielectric constant change both in the horizontal and vertical direction due to geological stratification. Besides that, the above quantities exhibit seasonal variations. Josephson and Blomquist (1958) found wide variations in the conductivity of the ground when frozen earth was subjected to heavy thawing. Field strength variations of up to 14 dB were recorded at v.h.f. and vertical polarization over the same path at different seasons which, when related to the dielectric constant corresponds to its variation from about 3 to 30. The above two authors have determined that the value of the effective dielectric constant depends mainly on the water content of the ground and is relatively independent of the type of ground. They used the empirical relation

$$\epsilon = 0{\cdot}78W + 2{\cdot}5 \tag{6.9}$$

for the calculation of the dielectric constant from measured v.h.f. field strengths.

A more general formula

$$\epsilon = kW + 2{\cdot}0, \tag{6.10}$$

where k is 0·5 for sand-like ground, 1·0 for average soil and 1·5 for clay and electrically similar media, has been proposed by Biggs (1970), who quotes Albrecht's

(1965) dependence of the conductivity on moisture:

$$\sigma = 7{\cdot}7 \times 10^{-5}\,(0{\cdot}73W^2 + 1)(1 + 0{\cdot}03t), \tag{6.11}$$

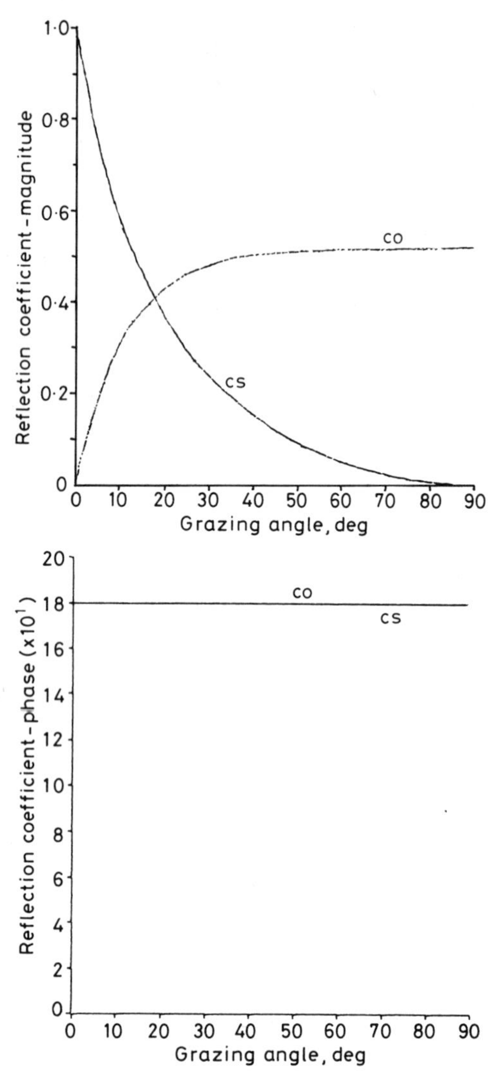

Fig. 6.3 *Reflection coefficient, magnitude and phase vs. grazing angle. Plane smooth earth ($\epsilon_r = 10$, $\gamma = 2$ mmho/m, $f = 3$ GHz)*
cs = same circular polarization
co = opposite circular polarization

where t is the ground temperature in °C. The term W in the above expressions is the percentage of weight of water in a given volume to the weight of water which could occupy the same volume.

The above two quantities constitute the complex dielectric constant given by eqn. (6.5). However, in practice the reflection coefficient is influenced more by the roughness of the ground than by its electrical constants. The use of the reflection coefficient as a function of the angle of incidence, as given by Fresnel's relations (6.3) and (6.4), is justified only when the ground is smooth, i.e. when the irregularities are much smaller than half a wavelength.

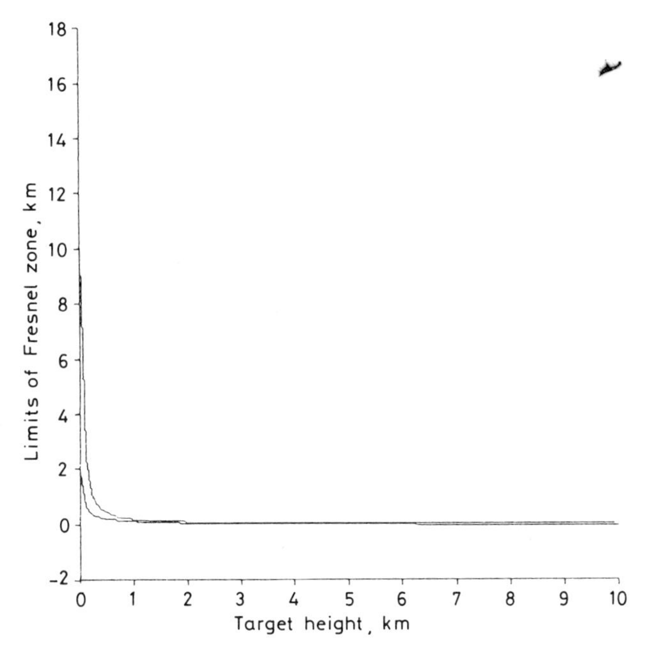

Fig. 6.4 *Limits of the first Fresnel zone. Aerial height = 15 m, ground range 10 km, λ = 10 cm*

6.3 Fresnel Zones

While it is generally accepted that, in plane earth approximation, reflection from a single point is used for field strength calculations in radar and communication practice, it is understood that a large part of the ground is illuminated by the transmitter and the incident radiation is reradiated in all directions. The phase relations of most components of the reradiated, indirect energy are such that they cancel, leaving only the energy reradiated from a comparatively small elliptical patch of the ground to combine with that energy which travels to a point of interest directly from the transmitter without the intermediate reflection. Due to the longer path traversed by the reradiated energy and its phase shift on reflection, given by eqns. (6.3) and (6.4), the direct and reradiated energy combine with various phases thus causing the development of the well known lobing pattern of the vertical coverage diagram.

The above elliptical reflecting area is called a Fresnel zone since it corresponds to the Fresnel zones encountered in the diffraction theory of light.

The surface at which the phase of the indirect radiation satisfies the phase requirements for reinforcing the direct radiation may be shown (Kerr, 1951) to be a family of ellipsoids of revolution, the collinear major axes of which lie on the line connecting the aerial with the point of interest in space, i.e. the target in case of radar, and having the transmitting aerial and the target in their foci. The previously mentioned elliptical patch on the ground is the intersection of the ellipsoids with the reflecting surface. It is obvious that there will be a family of ellipses each corresponding to the intersection of an appropriate ellipsoid with the reflecting surface. The energy reradiated from zones bounded by adjacent ellipses are in phase opposition. However, since the amplitudes of the reradiated energy change gradually, complete cancellation does not occur. The individual ellipses are those responsible for the appropriate lobes mentioned earlier, the innermost ellipse corresponding to the lowest lobe, the main contributor of reradiated energy.

Formulae for the calculation of the dimensions of the Fresnel zones have been published by various authors (Norton and Omberg, 1947; Beckmann and Spizzichino, 1963; Kerr, 1951).

According to Kerr, the centre of the nth ellipse x_{0n} may be determined from the expression

$$x_{0n} = \frac{d}{2}\left\{1 - \frac{\dfrac{H^2 - h^2}{d^2}}{\left[\dfrac{\delta}{c} + \sec\theta\right]^2 - 1}\right\}, \tag{6.12}$$

where d is the ground range between transmitter and the target resp. receiver, H is the target resp. receiving aerial height, h is the transmitter aerial height, $\theta = \tan^{-i}$ $(H - h)/d$ and

$$\delta = \frac{2Hh}{d} + n\frac{\lambda}{2}$$

with λ the wavelength and n the Fresnel zone number ($n = 1$ for the first, i.e. innermost, zone).

The minor axis intercepts are given by

$$y_{1,2} = \pm\frac{d}{2}\sqrt{\left(\frac{\delta}{d}\right)^2 + \frac{2\delta}{d}\sec\theta}\sqrt{1 - \frac{\left(\dfrac{H+h}{d}\right)^2}{\left[\dfrac{\delta}{d} + \sec\theta\right]^2 - 1}} \tag{6.13}$$

while those of the major axis may be computed from

$$x_{1,2} = x_0 \pm y_{1,2} \sqrt{1 + \frac{1}{\left[\frac{\delta}{d} + \sec\theta\right]^2 - 1}}. \tag{6.14}$$

Figure 6.4 shows the extent of the first Fresnel zone at a frequency of 3 GHz, an aerial height of 15 m at a ground range of 10 km from a target varying in height from 0 to 10 km, while Fig. 6.5 shows the distance of the idealized point of reflection for the same conditions.

The area of the Fresnel zones is

$$A = \frac{\pi d^2}{4} \frac{\left(\frac{\delta}{d} + \sec\theta\right)\left[\left(\frac{\delta}{d}\right)^2 + 2\frac{\delta}{d}\sec\theta\right]}{\sqrt{\left[\frac{\delta}{d} + \sec\theta\right]^2 - 1}} \left[1 - \frac{\left(\frac{H-h}{d}\right)^2}{\left[\frac{\delta}{d} + \sec\theta\right]^2 - 1}\right]. \tag{6.15}$$

The location of the Fresnel zones and their extent is of interest in radar siting where it might be important to know the area and location of smooth surface required to support lobing (see Chapter 9). The criterion of smoothness will be given below.

6.4 Rough surfaces

When the reflecting surface is not smooth, specular reflection diminishes with the effect of reducing the second term of eqn. (6.1). The reduction is a function of the surface roughness and is accounted for by multiplying the Fresnel reflection coefficient by a factor

$$\rho_s = \exp\left[-2\left(\frac{2\pi\sigma_h\cos\theta}{\lambda}\right)^2\right] \tag{6.16}$$

derived by Pekeris and, independently, by MacFarlane but published by Ament (1953). The term σ_h is the standard deviation of the surface height about the mean height in the first Fresnel zone which, when referring to the sea surface, is related to the significant sea wave height, given by Burling and quoted by Nathanson (1969) as

$$H_{1/3} = 4\sigma_h. \tag{6.17}$$

It is defined as the average of the highest third of the waves in the wave record.

Figure 6.6 shows the average scattering coefficient given by eqn. (6.16) as a function of the grazing angle with heights of surface irregularities of 0 cm, 5 cm and 10 cm as parameters. Figure 6.7 shows the same for ten times higher surface irregularities. Both figures were computed for a wavelength of 10 cm. Fig. 6.8 shows the dependence of the above coefficient on the ratio

height of terrain irregularity/4λ

with four values of the grazing angle as parameters.

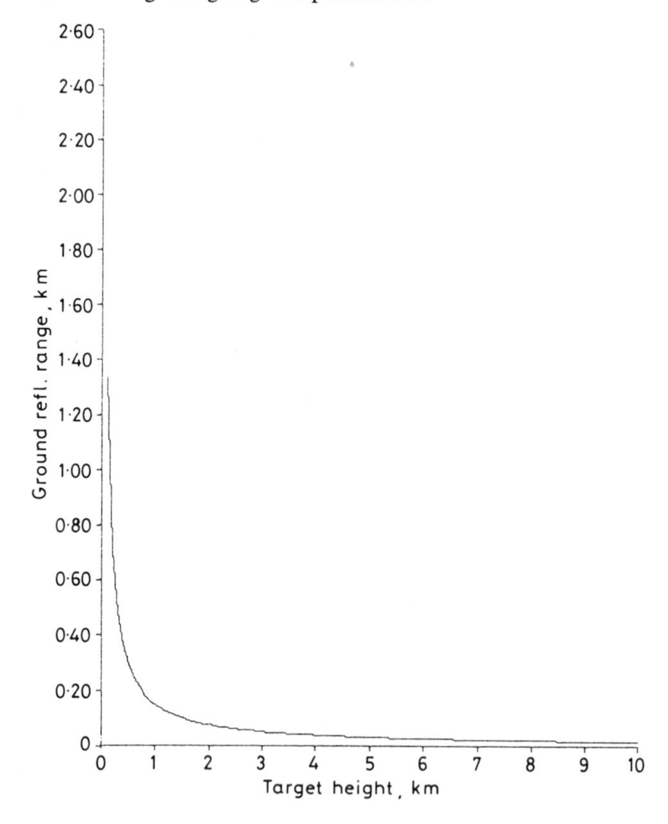

Fig. 6.5 *Distance of the 'point' of ground reflection from the radar aerial as a function of target height. Aerial height = 15 m, ground range 10 km, λ = 10 cm*

The derivation of the factor ρ_s is based on expressing the phase difference between rays reflected from the lowest and highest level of the protuberances of a rough surface. If the height of these is h, the path difference of the rays is $\Delta R = 2h \cos\theta$ and their phase difference

$$k\,\Delta R = \frac{2\pi}{\lambda} 2h \cos\theta = \frac{4\pi h}{\lambda} \cos\theta. \tag{6.18}$$

For small phase differences the effect of terrain roughness is negligible and the reflecting surface appears to be smooth. When the phase difference is π radians, the two rays are in phase opposition and cancel, i.e. the specular component of the radiation disappears and the radiation is dispersed into other directions.

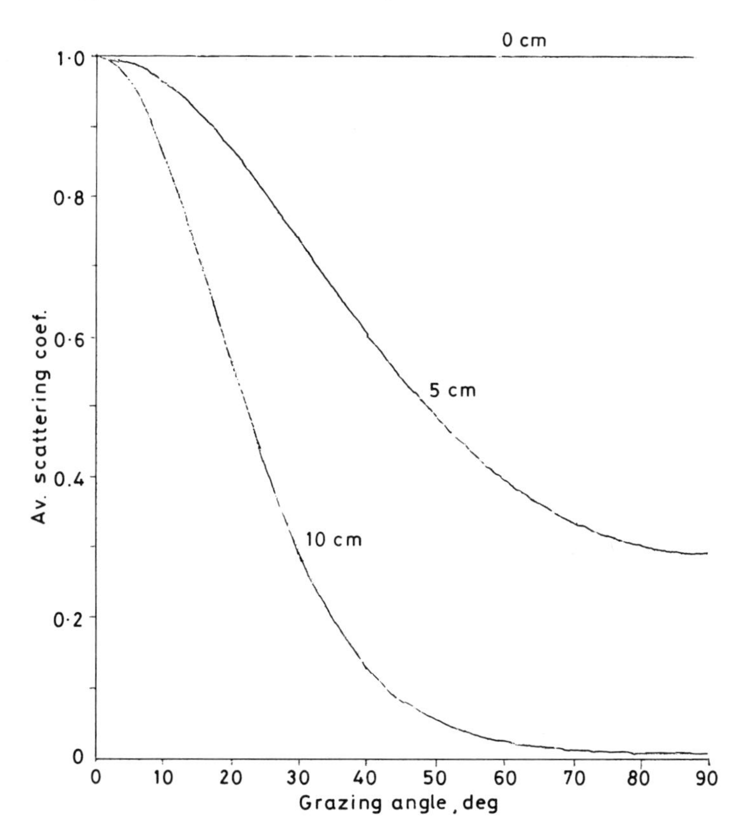

Fig. 6.6 *The average scattering coefficient as a function of the grazing angle with heights of surface irregularities as parameter, f = 3 GHz*

The phenomenon is due to a critical roughness at the used grazing angle. Between the two extremes, i.e. specular reflection and no specular reflection, there is a value of phase difference which divides surfaces into smooth and rough ones. Rayleigh suggested that this phase difference be $\pi/2$ radians when

$$h \cos \theta = \frac{\lambda}{8}. \tag{6.19}$$

Calling this height of the terrain irregularities the critical height h_c at a particular angle of incidence θ, the terrain is considered to be smooth for

$$h \leqslant h_c$$

and rough otherwise. Rough surfaces suppress lobing. The critical height decreases with a decreasing angle of incidence or with an increasing grazing angle.

Ament's roughness factor (6.16) is in good agreement with rough sea measurements reported by Beard, Katz and Spetner (1956), Beard (1961) and Taylor and Glover (1975).

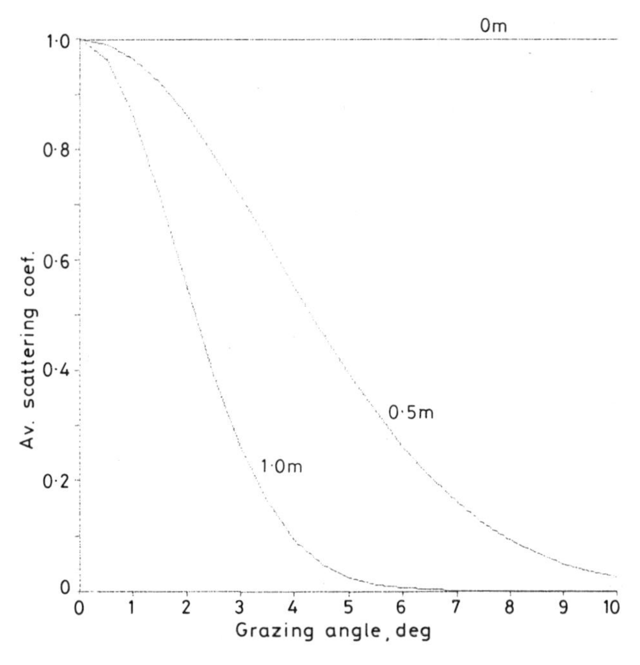

Fig. 6.7 *The average scattering coefficient as a function of the grazing angle with heights of surface irregularities as parameter. $f = 3\,GHz$*

While the arbitrarily chosen Rayleigh criterion of surface roughness has been generally accepted, other values for the critical phase difference, namely $\pi/4$ and even $\pi/8$, have been suggested as being more realistic (Norton and Omberg, 1947; Kerr, 1951). However, in view of the crudity of the assessment of surface roughness, there seems to be no justification in being more precise; in the end the choice of the criterion depends on convention.

Beckmann (1963) shows that the field scattered by a rough surface may be determined by the sum of a specular and a diffuse component. The specular component constitutes the steady part, while the diffuse one the fluctuating part of the reflected signal.

Since terrain irregularities, except man-made ones and possibly sand-dunes, are random, they must be considered on a statistical basis. Various authors proposed various statistical distributions.

A review of the various approaches to the problem of rough surface reflection is given by Beckmann (1963).

The theory of scattering by periodic surfaces yields insight into the general behaviour of rough surfaces. The case of normal incidence onto a sinusoidal surface was investigated by Rayleigh who postulated the field to be the sum of plane waves and found the unknown terms of the sum by satisfying the exact boundary conditions at the surface.

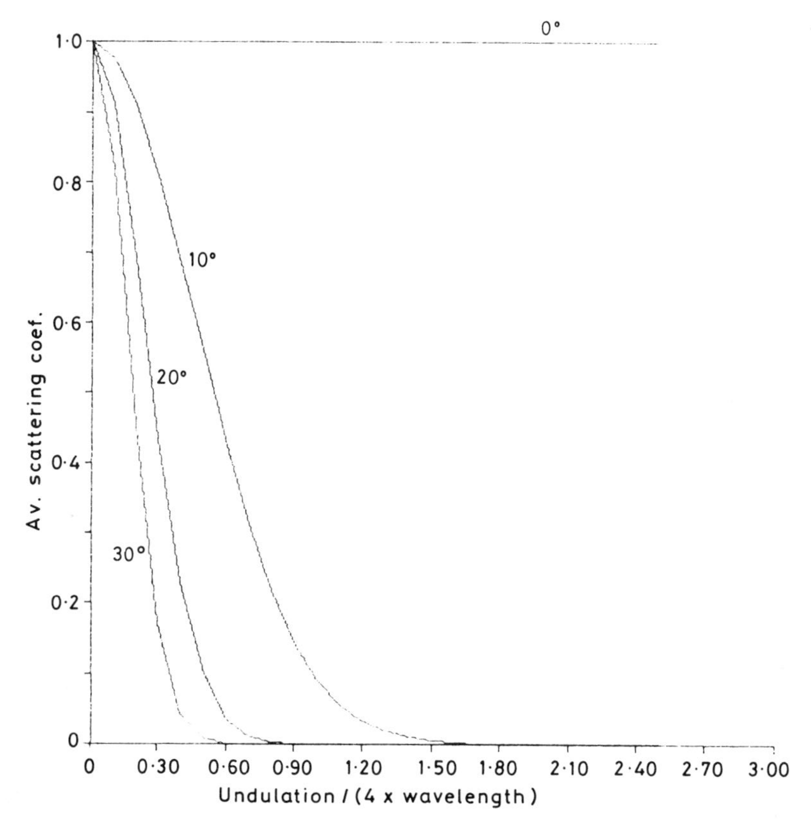

Fig. 6.8 *The average scattering coefficient as a function of the ratio H/4λ with the grazing angle as a parameter. λ = 10 cm*

Kirchhoff's method consists of postulating or approximating the boundary conditions on the surface and finding the corresponding field by the application of the Helmholtz integral. There are various ways of approximating the boundary conditions, e.g. estimating the surface current distribution from the incident magnetic field or expressing the total field by means of local reflection coefficients. Basically, these approximations are only different versions of the methods of geometrical optics.

Twersky (1957) proposed a rough surface model consisting of a random distribution of arbitrary boss-like protuberances on a ground plane and computed the reflection coefficient and differential scattering cross-section per unit area in terms of the scattering amplitude g of a single protrusion of an almost arbitrary shape, their average number per unit area and a given incident radiation. Scattering amplitude, as defined by Twersky, is the far field response in the direction of reflection when the scatterer is excited by a plane wave. In principle, it is equivalent to the scattering coefficient.

The far field due to a finite three-dimensional scatterer, according to Twersky, is

$$E_2^{v,h} = \frac{\exp\left(j\dfrac{2\pi}{\lambda}R_0\right)}{j\dfrac{2\pi}{\lambda}R_0} g^{v,h}(\theta_1, \theta_2)|E_1|, \tag{6.20}$$

where R_0 is the distance from the scatterer, θ_1 and θ_2 is the angle of incidence resp. reflection, $|E_1|$ given by

$$E_1^{v,h} = e^{jk_1 \cdot \bar{r}} \tag{6.21}$$

is the exciting plane wave with $\bar{k} = (2\pi/\lambda)/(\bar{k}_1/k_1)$ a propagation vector and $\bar{r}$ a radius vector. The relation between $g^{v,h}(\theta_1, \theta_2)$ and the customary scattering coefficient ρ, in case of a three-dimensional scatterer, is

$$\rho^{v,h}(\theta_1, \theta_2) = -\frac{\pi}{k^2 \cos\theta_1} g^{v,h}(\theta_1, \theta_2), \tag{6.22}$$

where $k = 2\pi/\lambda$.

Although Twersky's model is limited to bosses protruding from a plane and having large separations from each other, it is the only model permitting the exact theoretical investigation of polarization problems.

Ament (1953, 1956) applied noise theory statistics to the specification of a rough, perfectly reflecting, surface which when illuminated by a plane wave reflects statistically predictable fields. His rough surface consists of perfectly conducting infinite half planes tilted at a specified angle β from the normal to the ground plane.

His reflection coefficients are (Ament, 1960)

$$\rho_h = \frac{k\cos\theta_1 - \mu}{k\cos\theta_1 + \mu} \tag{6.23}$$

$$\rho_v = \frac{k\cos\theta_1 - \mu\sin(\theta_1 + \beta)}{k\cos\theta_1 + \mu\sin(\theta_1 - \beta)}$$

where μ must satisfy

$$k^2 - \mu^2 - k^2\sin^2\theta_1 + 2Nj\sec\beta\sqrt{[k^2 - (k\sin\theta_1\sin\beta + \mu\cos\beta)^2]} = 0$$

with $k = 2\pi/\lambda$ and N the average number of half-planes per unit length.

In the limiting case, when $\beta = \theta = 0$, this expression reduces to

$$\sqrt{k^2 - \mu^2}\,[\sqrt{k^2 - \mu^2} + 2Nj] = 0$$

so that the correct value of μ for this case, as computed from the last factor, becomes

$$\mu = +\sqrt{k^2 + 4N^2}.$$

Rice (1951) generalized Rayleigh's approach by applying random Fourier coefficients to the rough surface profile and expressing the variance of the surface in terms of its statistical properties. This method was found to be better suited to the investigation of ground wave propagation problems than to scattering problems and is outside the scope of this treatise.

Beckmann (1963) considers the mutual phase interference of elementary waves scattered by a rough surface described by a Markov chain with a finite number of states and discrete moments of transition to obtain a realistic field. His rough surface model consists of plane elements of random slope, each of which reflects the incident beam with the Fresnel reflection coefficient in the locally specular direction only. He adds then the elementary waves reflected in the same direction with respect to their phases to form one resultant wave in that direction.

DuCastel and Spizzichino (see Beckmann, 1963) generalized and refined Beckmann's approach from an infinite, one-dimensional, surface with plane facets to a finite, two-dimensional, surface with continuous radii of curvature. Elementary waves scattered into certain directions originate, according to this model, at a number of 'brilliant points'. Determining then the number of these points per unit area permits the determination of the r.m.s. value of the scattered field in any direction.

Since the normal distribution was accepted to be the most important and typical for the distribution of protuberances of a rough surface, most formulae used in periodic rough surface scattering are based on it (Beckmann, 1963).

Beckmann (1973) investigates scattering by non-Gaussian surfaces and derives a two-dimensional density function and a correlation coefficient by using an expansion in orthogonal polynomials. He uses the results for obtaining a general method for finding the field scattered by a rough surface by both physical and geometrical optics when the rough surface is generated by a stationary not necessarily normal random process.

In the physical optics approach he assumes that the illuminated part of the surface is much larger than the square of a wavelength, that it is generated by a stationary, mean-square, continuous random process and that its geometrical and electrical characteristics are such that physical optics is applicable.

In the geometrical optics method, it is accepted that the field scattered into a given direction is determined by the probability of the inclination of the surface elements being such as to reflect the incident wave into the given direction.

Beckmann (1973) concludes that for very low values of the grazing angle of the incident wave or for a very slight surface roughness, the scattering characteristics of the surface do not depend significantly on its probability distribution. In all other cases, i.e. for very rough surfaces or higher values of the grazing angle, the scattering characteristics are strongly influenced by the probability distribution of the surface.

Results derived for normally distributed surfaces are not applicable to other surfaces except in the cases mentioned above. Measurements have established that the surface of a fairly smooth sea may be taken to be normally distributed.

6.5 Scattered field and its phase distribution

Beckmann (1963) investigates the field scattered by rough surfaces by considering the resultant field to be the vector sum of a number of elementary waves in mutual phase interference, which, due to scattering by random surfaces, is found to be also random.

The scattered field E_2 at a distance d from the point P is given by the Holmholtz integral

$$E_2(P) = \frac{1}{4\pi} \iint_s \left(E \frac{\partial \psi}{\partial n} - \psi \frac{\partial E}{\partial n} \right) dS, \tag{6.24}$$

where

$$\psi = \frac{\exp(jk_2 d)}{d}$$

is a solution of the three-dimensional wave equation and E is the voltage on the surface S.

The mean scattered power is generally

$$\langle P_2 \rangle = \tfrac{1}{2} Y_0 \langle E_2 E_2^* \rangle = \tfrac{1}{2} Y_0 |E_{20}|^2 \langle \rho \rho^* \rangle \tag{6.25}$$

with $Y_0 = 1/120\pi$ the admittance of free space and

$$\langle \rho \rho^* \rangle = \frac{F_3^2}{A^2} \int_{-X}^{X} \int_{-X}^{X} \int_{-Y}^{Y} \int_{-Y}^{Y} \exp[jv_x(x_1 - x_2)$$

$$+ jv_y(y_1 - y_2)] \, dx_1 dx_2 dy_1 dy_2 \exp[jv_z(\xi_1 - \zeta_2)],$$

where

$$F_3(\theta_1; \theta_2, \theta_3) = \frac{1 - \cos\theta_1 \cos\theta_2 - \sin\theta_1 \sin\theta_2 \cos\theta_3}{\cos\theta_1(\cos\theta_1 + \cos\theta_2)}$$

is a three-dimensional factor dependent on the angle of incidence θ_1, reflection θ_2 and the angle θ_3 between the plane of specular reflection and scattered reflection, $\zeta_1 = \zeta(x_1, y_1)$ and $\zeta_2 = \zeta(x_2, y_2)$ random variable functions of the coordinates of the considered points on the surface, v_x, v_y and v_z are the orthogonal components of the vector $\bar{v} = \bar{k}_1 - \bar{k}_2$ with $\bar{k}_1$ and $\bar{k}_2$ given earlier and $A = 4XY$ the projection of the surface S onto the x–y plane.

However, it is the power scattered into the specular direction which is of greatest interest. Beckmann (1963) shows that for this case

$$\langle \rho\rho^* \rangle = \exp\left| -\left(\frac{4\pi\sigma \cos\theta_1}{\lambda}\right)^2 \right|$$

arriving thus at Ament's rough surface scattering factor given earlier by eqn. (6.16).

Prosin and Pavel'yev (1972) determine the scattered signal power allowing for shadowing in the Fresnel zone by using Kirchhoff's approximation and avoiding the simplifying assumptions that the reflections from areas of the surface are independent. They conclude that the effective scattering region of an uneven surface is much greater than that of a plane surface, and in case of glancing incidence it extends to infinity. Shadowing by some surface elements limits the magnitude of the reflection coefficient in the latter case.

Outside a narrow cone about the direction of specular reflection, the amplitude of the field scattered by a rough surface is, according to Beckmann (1963), always Rayleigh distributed. The amplitude of the scattered field has this distribution everywhere when reflection from a very rough surface occurs at other than grazing incidence.

Considering the phase, Beckmann (1963) defines an incoherent wave as one having a random, uniformly distributed phase over an interval of 2π and shows that the mean power densities of incoherent waves may be added algebraically. A wave having a constant phase is coherent; its total power density is the vector sum of individual fields. The total field of n coherent waves of equal amplitudes is n^2 whereas it is only n in case of incoherent waves.

The phase difference of two waves scattered from points of a surface defined by $\zeta(x_1)$ and $\zeta(x_2)$ is

$$\Delta\psi_{1,2} = \frac{2\pi}{\lambda}\left[(x_2 - x_1)(\sin\theta_1 - \sin\theta_2) + (\zeta_1 - \zeta_2)(\cos\theta_1 + \cos\theta_2)\right]. \tag{6.26}$$

Since for reflection in the non-specular direction, i.e. when $\theta_1 \neq \theta_2$, the phase difference $\Delta\psi$ of individual scattered waves will vary over many intervals 2π, independent of the distribution of ζ, the phases of the individual scattered waves are uniformly distributed over 2π and the scattered field is always incoherent and Rayleigh distributed.

When scattering in the specular direction is considered, i.e. when $\theta_1 = \theta_2$, the first term of the expression for $\Delta\psi$ disappears and only the random component

$$\psi = \frac{4\pi}{\lambda}\zeta(x)\cos\theta$$

remains for the phase of the individual components. For very rough surfaces, at other than grazing incidence, the waves scattered even in the specular direction will be incoherent and have a Rayleigh distributed field amplitude. For a smooth, plane, surface, when $\zeta(x) = 0$, the phases of all components are equally zero, i.e. phase coherent.

Using the theorems of probability theory, the mean square value of the scattering coefficient may be expressed as the sum of the variance of ρ and the product of the mean values of ρ and its complex conjugate

$$\langle \rho\rho^* \rangle = \sigma_\rho^2 + \langle \rho \rangle \langle \rho \rangle^*. \tag{6.27}$$

The variance of a smooth surface is zero; the reflected waves are coherent. Only incoherent waves are reflected by a very rough surface when the absolute value of ρ is Rayleigh distributed and $\langle \rho\rho^* \rangle = \sigma_\rho^2$. The transition between these extremes, namely between coherence and incoherence, for which the ratio $\langle \rho \rangle \langle \rho \rangle^* / \sigma_\rho^2$ changes from ∞ to 0, is gradual and continuous.

6.6 Curvature effect

It was mentioned earlier that due to reflection occurring from a convex, even if smooth, surface, an additional factor, the so-called divergence factor has to be introduced in order to account for the reduction of the energy reflected in a given direction.

Visualizing a conical bunch of rays intercepting an area of the reflecting surface, it is obvious that the cross-section of the reflected bunch will be more divergent when reflection occurs from a convex surface than when a plane surface were involved. The field strength is proportional to the square root of the ray density. The divergence factor is the square root of the ratio of the cross-section of the bunch of rays reflected from the plane surface to that reflected from a convex surface. It was derived by Van der Pol and Bremmer (1939) and is

$$D = \frac{r(X+Y)\sqrt{\sin\gamma\cos\gamma}}{\sqrt{[(r+H)X\cos\tau_2 + (r+h)Y\cos\tau_1](r+h)(r+H)\sin\xi}}, \tag{6.28}$$

where X and Y are the slant ranges of the transmitting resp. target or receiving aerial from the point of reflection, r is the earth radius modified by the appropriate atmospheric index of refraction, γ is the angle of incidence and reflection in the specular direction, h and H are the transmitting aerial resp. the target or receiving aerial height, τ_1 and τ_2 are the angles between X and Y and the lines connecting the centre of the earth with the transmitting resp. target or receiving aerial, and ξ the angle between the above two lines.

Accepting that the aerial resp. target heights are small when compared with the earth's radius, that the angle τ_2 is the complement of the grazing angle which is always very small in the region where the divergence factor has an appreciable effect and that the sum of the ranges X and Y is approximately equal to the ground range $r\xi$, the above expression reduces to (Beckmann, 1963)

$$D = \frac{1}{\sqrt{\left[1 + \dfrac{2XY}{r_1(X+Y)\cos\gamma}\right]\left[1 + \dfrac{2XY}{r_2(X+Y)}\right]}} \tag{6.29}$$

with r_1 and r_2 the radii of curvature at the intersection of the reflecting surface with two vertical planes perpendicular to the direction of propagation.

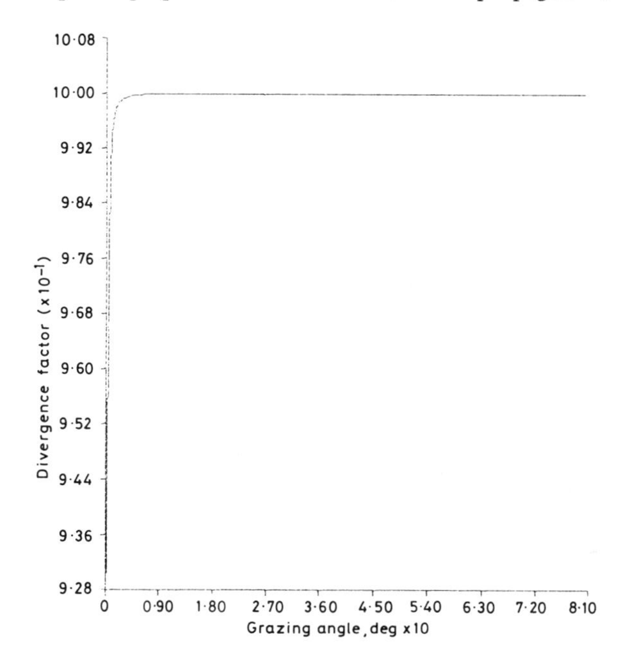

Fig. 6.9 *The divergence factor as a function of the grazing angle. Aerial height 15 m, target height 1000 m and range 1—100 km*

In practice, the earth is considered to be spherical with $r_1 = r_2 = r$, when

$$D = \frac{1}{\sqrt{1 + \dfrac{2XY}{r(X+Y)}\left[1 + \dfrac{1 + \dfrac{2XY}{r(X+Y)}}{\cos\gamma}\right]}}. \tag{6.30}$$

Figure 6.9 shows the divergence factor as a function of the grazing angle computed for an aerial height of 15 m, target height of 1 km and range of 1 to 100 km.

6.7 Glistening surface

Diffuse scattering is a non-directional phenomenon. The reradiating region, the glistening surface, is that region of the earth's surface from which power may be reradiated by surface irregularities having mean square slopes

$$\tan \beta_0 \leqslant \frac{2\sigma_h}{T}, \tag{6.31}$$

where σ_h is the standard deviation of the heights of the normally distributed surface irregularities about the mean height and T is the horizontal autocorrelation distance of the surface. The glistening surface may extend over the whole distance between the radar aerial and the intersection of the earth's surface with the line joining the target with the centre of the earth.

Spizzichino (see Beckmann and Spizzichino, 1963) shows that in case the aerials are strongly beamed so that the glistening surface is limited solely by the aerial radiation pattern, the mean absolute square of the scattering coefficient is

$$\langle|\rho|^2\rangle = \frac{1}{4}\frac{g_1(2\beta_0)}{g_1(0)} + \frac{1}{4}\frac{g_2(2\beta_0)}{g_2(0)}, \tag{6.32}$$

where $g_1(\alpha)$ and $g_2(\alpha)$ are the aerial gains in the direction making an angle α with the line connecting the transmitting and receiving aerials, or, in the radar case, the transmitting aerial and target.

The mean reflection coefficient for diffuse scattering, using Spizzichino's notation, is

$$\langle|R_d|^2\rangle = \langle|\rho_d|^2\rangle R_0^2 D, \tag{6.33}$$

where R_0 is the smooth, plane earth reflection coefficient and D the divergence factor which, however, in diffuse scattering, has a negligible effect on the reflection coefficient.

In the considered case, when the aerial patterns determine the extent of the glistening surface, diffuse scattering is determined mainly by the parameter

$$K_\alpha = \frac{1}{d}\sqrt{\left(\frac{h}{\alpha_2}\right)^2 + \left(\frac{H}{\alpha_1}\right)^2}, \tag{6.34}$$

where h and H are the transmitter resp. the receiver aerial or target heights and α_1 and α_2 are the 3 dB half beamwidths of the transmitting and receiving aerials.

If $K_\alpha > 0{\cdot}5$ for a symmetrical radar-ground-target geometry and $0{\cdot}5 < K_\alpha < 1$ for an asymmetrical one, then the reflection coefficient is negligibly small. The mean absolute square of the scattering coefficient increases with a decreasing value of K_α indicating that the scattered energy comes from an increasingly extensive area of the earth's surface. When $K_\alpha < 0{\cdot}1$, the reflected energy comes only from the areas near the transmitting and receiving aerials.

When one terminal is near the ground while the other one is very far, the mean absolute square of the scattering coefficient is

$$\langle|\rho_d|^2\rangle = \frac{1}{4\pi\beta_0^2}\int\frac{dS}{d^2}, \tag{6.35}$$

where d is the distance of the lower terminal to a point on the surface of the earth and S is the glistening surface.

The glistening surface boundary, as given by Spizzichino, is

$$y = \pm\frac{d_1 d_2}{d_1 + d_2}\left(\frac{h}{d_1} + \frac{H}{d_2}\right)\sqrt{\tan^2\beta_0 - \frac{1}{4}\left[\frac{h}{d_1} - \frac{H}{d_2}\right]^2} \tag{6.36}$$

with d_1 and d_2 denoting the ground ranges of the reflection point from the base of the radar aerial resp. from the intersection of the line from the target to the centre of the earth and the earth's surface; the other quantities have been defined earlier.

6.8 Depolarization

In what was said so far it was tacitly accepted that the polarization of the waves before and after reflection is identical. This survey would not be complete without a few words on the subject of depolarization.

Depolarization is the change of polarization from that of the incident wave to that of the reflected one. The meaning of the linear polarization terms of vertical and horizontal polarization were given above. Elliptical polarization occurs when two linearly polarized waves of unequal amplitude, phase and polarization planes but identical frequency and direction of propagation interfere. Linear and circular polarization are, from the mathematical point of view, only special cases of elliptical polarization.

According to Beckman (1968) an electric field strength $\bar{E}$, where the dash denotes a vector quantity, may be represented by the sum of its space components as

$$\bar{E} = E_h\bar{e}_h + E_v\bar{e}_v = E_h\left(\bar{e}_h + \frac{E_v}{E_h}\bar{e}_v\right), \tag{6.37}$$

where the ratio of the two complex quantities E_v and E_h, denoted by p, is the complex polarization factor which uniquely defines any polarization of an electromagnetic wave. Its magnitude is the ratio of the amplitudes of the two components, while its argument is their phase difference.

If the polarization of the incident wave is defined by p_1 and that of the scattered one by p_2 then

$$q = \frac{p_2}{p_1} \tag{6.38}$$

is the depolarization factor. It is equal to the ratio of the generalized scattering coefficients (Beckmann, 1963)

$$\chi_v = \frac{E_{v2}}{E_{v1}}, \qquad \chi_h = \frac{E_{h2}}{E_{h1}},$$

where the subscripts 1 and 2 refer to the incident resp. scattered values of the waves. The above generalized scattering coefficients reduce to those given by eqns. (6.3) and (6.4) when scattering from a properly oriented smooth plane occurs.

Other useful quantities for the study of depolarization are the degree of polarization

$$P = \frac{M^2 - N^2}{M^2 + N^2} \leqslant 1, \tag{6.39}$$

where M and N is the major resp. minor axis of the polarization ellipse, and the depolarization ratio

$$D = \frac{|E_c|^2}{|E_p|^2} \tag{6.40}$$

with the subscripts c and p denoting cross resp. parallel polarized components, i.e. components perpendicular resp. parallel to the direction of the incident linear polarization.

If the direction of interest in the above expression makes an angle γ with the axes of the polarization ellipse, the depolarization ratio becomes

$$D = \frac{N^2 + M^2 \tan^2\gamma}{N^2 \tan^2\gamma + M^2} \tag{6.41}$$

which, for the case of $\gamma = 0$, yields the cross-polarization ratio

$$D = \frac{N^2}{M^2}. \tag{6.42}$$

For scattering by a perfectly conducting plane in the specular direction the depolarization factor

$$q = \frac{\rho_v}{\rho_h} = -1 \tag{6.43}$$

indicating that a smooth plane will, in general, depolarize all incident polarizations except horizontal for which the complex polarization factor p is zero, and vertical with $p = \infty$.

At near grazing incidence onto a finitely conducting plane both vertical and horizontal reflection coefficients are identically -1 so that the depolarization factor will be unity indicating that no depolarization occurs. In general, a large, smooth, conducting plane surface, the depolarization factor of which is equal to -1, depolarizes incident radiation; it reverses the sense of rotational polarization and the orientation of the polarization ellipse, leaving, however, its shape unchanged.

Beckmann (1963) has shown that the ratio (6.43) applies equally to rough, perfectly conducting, surfaces, disregarding the distribution of protuberances, provided that the radii of curvature of these are much larger than the wavelength of the incident radiation. If the surface is not a perfect conductor, the depolarization factor q will be a random quantity since the reflection coefficients will vary over the surface in accordance with the local angle of incidence and, in case of a non-homogeneous scattering surface, also with the local electrical properties of the surface. In the plane of incidence the depolarization factor is

$$q = \frac{\rho_v(\delta)}{\rho_h(\delta)}$$

with

$$\delta = \frac{\theta_1 + \theta_2}{2},$$

where θ_1 and θ_2 are the specular angles of incidence resp. reflection.

If the radii of curvature of the surface irregularities are smaller than a wavelength, the local diffraction effects complicate matters and the depolarization factor must be determined from its definition given by eqn. (6.38).

The field scattered by very rough surfaces has a Rayleigh distribution

$$w(|E_2|) = \frac{2|E_2|}{s} e^{-|E_2|^2/s} \tag{6.44}$$

in all non-specular directions. In the above expression s is the mean square value of the field given by

$$s = \langle |E_2|^2 \rangle = \langle E_2 E_2^* \rangle = 120\pi \langle w \rangle, \tag{6.45}$$

where w, $|E_2|$ and s apply to both linear polarizations. Protuberances of different shapes depolarize radiation in different ways. If the configuration of the protuberances is such that they may be taken to be more or less symmetrically distributed about a plane parallel to the plane of incidence, and for reasons of symmetry not possess cross-polarized components, then the fields $|E_{v2}|$ and $|E_{h2}|$ may be assumed to be independent and

$$|q| = \frac{|E_{v2}|}{|E_{h2}|} \tag{6.46}$$

which, as shown by Beckmann (1963), has a probability distribution

$$P\{Q < Q_0\} = \frac{1}{2}\left[1 - \cos\left(2 \arctan \frac{Q_0}{Q}\right)\right] \tag{6.47}$$

where

$$G = \sqrt{\frac{\langle |E_{v2}|^2 \rangle}{\langle |E_{h2}|^2 \rangle}}.$$

The absence of depolarization of purely horizontally ($p_1 = 0$) and purely vertically ($p_1 = \infty$) polarized radiation in the plane of incidence applies also in the case of rough surfaces regardless of the conductivity and the distribution of the protuberances.

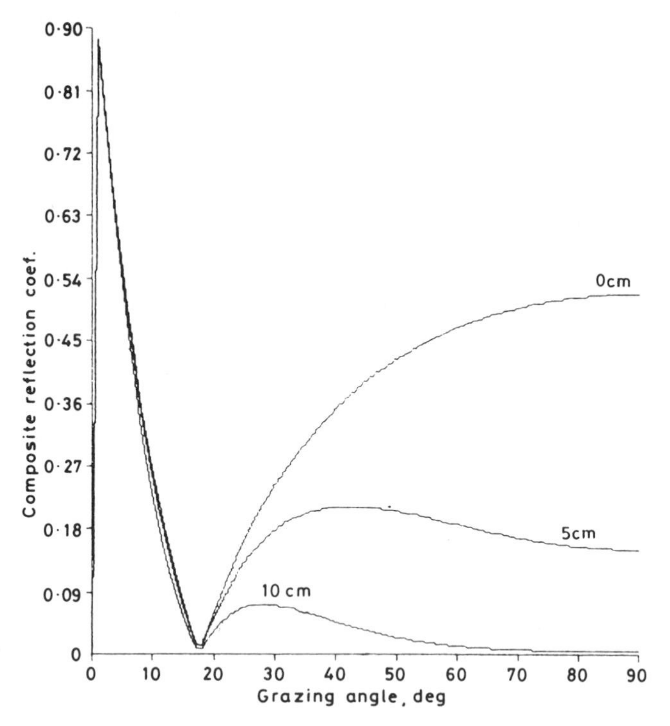

Fig. 6.10 *The vertical polarization composite reflection coefficient vs. grazing angle with the height of surface irregularities as parameter. $\lambda = 10\,cm$*

If the surface is oriented so that its normal at the point of incidence does not lie in the plane of incidence, an incident elementary wave will be scattered laterally out of the plane of incidence causing also the depolarization of purely horizontally and purely vertically polarized waves. In general, the depolarization from the polarization factor of the incident wave p_1 to that of the scattered one p_2, on scattering from a smooth conducting plane in other than specular directions is, according to Beckman (1963), given by

$$p_2 = \frac{p_1(1 - \tan\beta \tan\beta_2) + \tan\beta + \tan\beta_2}{p_1(\tan\beta + \tan\beta_2) + \tan\beta \tan\beta_2}, \tag{6.48}$$

where the polarization angle β, i.e. the angle between the incident field $\bar{E}_1$ and the intersection of the wave front with the reflecting plane, may be determined from

$$\sin\beta = \frac{\sin\theta_3 \sin\theta_2}{\sqrt{1-(\cos\theta_1 \cos\theta_2 - \sin\theta_1 \sin\theta_2 \cos\theta_3)^2}} \qquad (6.49)$$

with θ_1, θ_2 and θ_3 as defined earlier, and β_2, the angle between the reflected field $\bar{E}_2$ and the intersection of the wave front with the reflecting plane, from

$$\cos\beta_2 = \cos\beta\cos\theta_3 - \sin\beta\cos\theta_1 \sin\theta_3, \qquad (6.50)$$

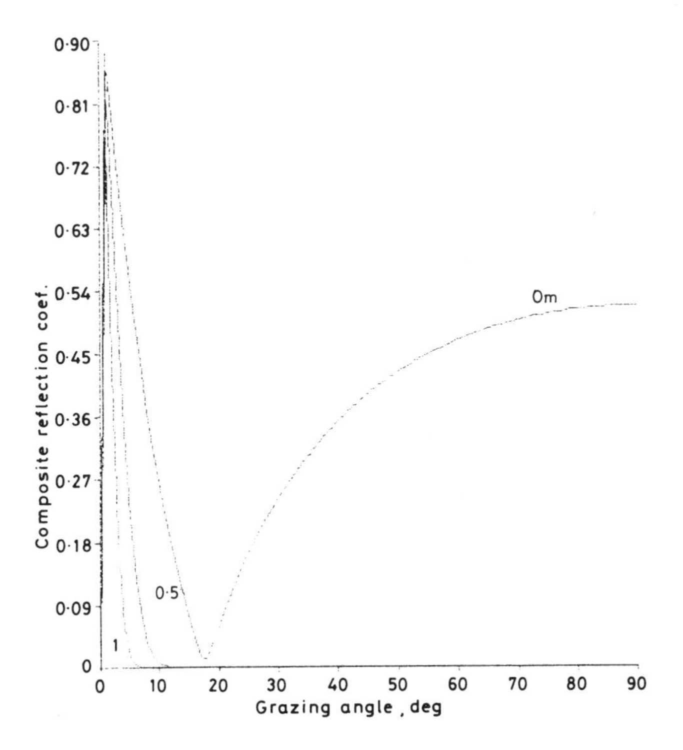

Fig. 6.11 *The vertical polarization composite reflection coefficient vs. grazing angle with the height of surface irregularities as parameter.* $\lambda = 10$ *cm*

while for scattering in the specular direction

$$p_2 = \frac{p_1(\rho_h \tan\beta\tan\beta_2 + \rho_v) - \rho_h \tan\beta + \rho_v \tan\beta}{\rho_h + \rho_v \tan\beta\tan\beta_2 - p_1(\rho_h \tan\beta_2 - \rho_v \tan\beta)}. \qquad (6.51)$$

A rough, locally flat, surface will depolarize in the same manner.

6.9 Composite reflection coefficient

On the basis of what was said so far it transpires that the reflection coefficients given by eqns. (6.3) and (6.4) are applicable to a smooth, plane, earth only. When reflection from a surface other than that specified above occurs, the reflection coefficients must be modified to take the roughness and convexity of the reflecting surface into consideration. This, as has been indicated earlier, is accomplished by the application of eqns. (6.16) and (6.30). The composite reflection coefficient then becomes

$$\rho = \rho_{v,h}\rho_s D, \tag{6.52}$$

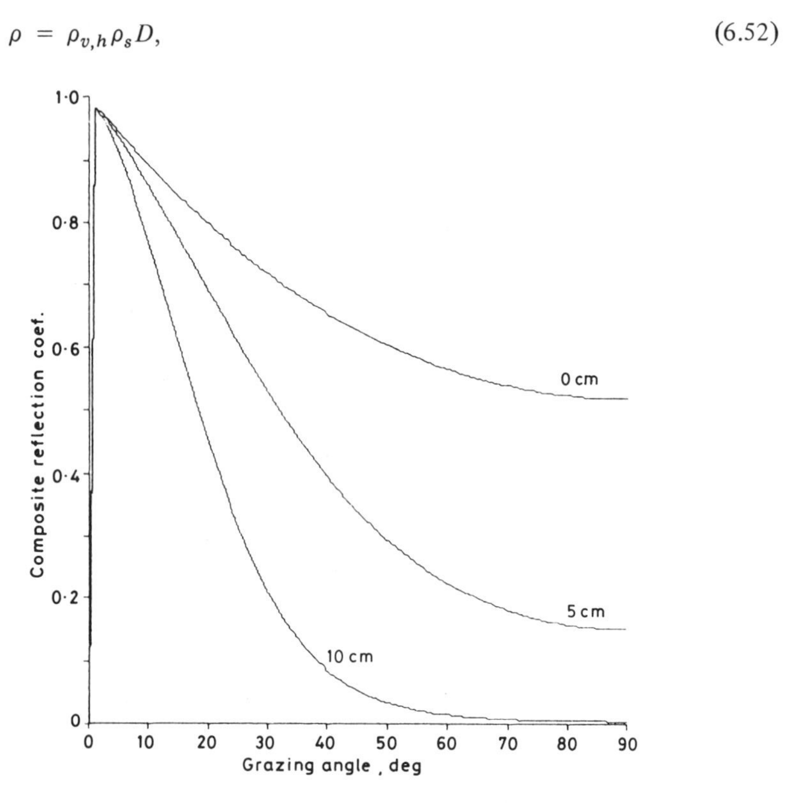

Fig. 6.12 *The horizontal polarization composite reflection coefficient vs. grazing angle with the height of surface irregularities as parameter.* $\lambda = 10\,cm$

where $\rho_{v,h}$ is the plane, smooth, surface reflection coefficient given by either eqn. (6.3) or eqn. (6.4) depending on whether vertical or horizontal polarization is being considered.

Figure 6.10 shows the composite reflection coefficient as a function of the grazing angle for the reflection of a vertically polarized wave of 10 cm wavelength from ground having a relative dielectric constant of 10, a conductivity of 2 mmho/m and heights of surface irregularities of 0, 5 and 10 cm.

Figure 6.11 shows the same for heights of surface irregularities of 0, 0·5 and 1·0 m, while Figs. 6.12 and 6.13 illustrate the variation of the composite reflection coefficient with the grazing angle for horizontal polarization and other conditions identical to those of Figs. 6.10 resp. 6.11.

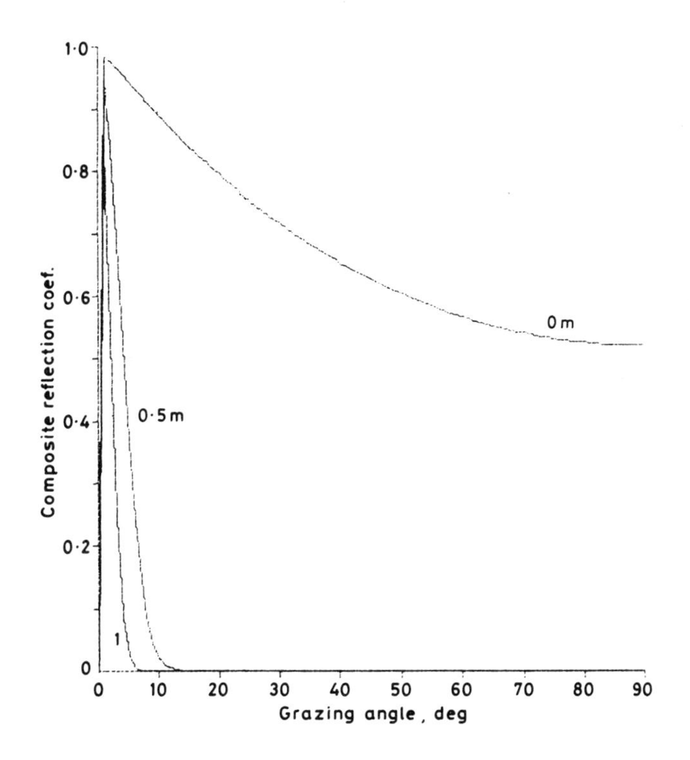

Fig. 6.13 *The horizontal polarization composite reflection coefficient vs. grazing angle with the height of surface irregularities as parameter. $\lambda = 10$ cm*

Comparing these with Figs. 6.1, 6.5, 6.6 and 6.8, one observes, as would be expected, that the influence of the divergence factor given by Fig. 6.9 is restricted to very low grazing angles. Due to this, the composite reflection coefficient is zero at zero angle of grazing incidence from whence it rises sharply to a value given by the product of the plane earth reflection coefficient and the scattering coefficient. The influence of the latter is noticeable particularly at higher values of the grazing angle where the scattering coefficient prevents the composite reflection coefficient reaching values due to reflection from a plane earth. The shape of the curves showing the composite reflection coefficient at small surface irregularities is similar to the plane earth ones; however, they differ markedly when the heights of the irregularities exceed the wavelength.

6.10 Conclusions

The above short study illustrates the importance of modifying the plane earth reflection coefficient by factors applicable to rough and convex surfaces if the results of field computations are to be realistic. The composite reflection coefficient will always be lower than that obtained for a smooth, plane, earth at the same grazing angle.

Chapter 7

The mathematical modelling of the target echoing area

7.1 Introduction

The target echoing area σ, is the least exactly defined parameter of the radar equation. It depends not only on the target's size and shape, but also on its attitude with respect to the radar aerial. It is known to vary greatly even with small changes in attitude. This variation is usually neglected in performance computations and an average value of the target echoing area is used as a rule, although some account of echoing area fluctuations is considered by Swerling's formulae used for the computation of the probability of detection.

7.2 Target echoing area, its definition and dependence on various factors

The scattering of electromagnetic energy from a target is a rather complicated phenomenon which depends on a number of factors. This section defines the radar target echoing area, also often called the radar cross-section, and indicates the influence of various factors on its value.

Far-field conditions, i.e. a range R large compared with the wavelength and with the dimensions of the scattering target, and a homogeneous and isotropic propagation medium, in which there is a linear relation between the field quantities at every point, are assumed, unless stated otherwise.

7.2.1 The definition of the echoing area

Radar cross-section investigations are based on the principles of electromagnetic theory which permit the definition of the echoing area by

$$\sigma = 4\pi \lim_{R \to \infty} R^2 \left|\frac{\bar{E}_s}{\bar{E}_i}\right|^2 = 4\pi \lim_{R \to \infty} R^2 \left|\frac{\bar{H}_s}{\bar{H}_i}\right|^2, \qquad (7.1)$$

where the range R is permitted to become arbitrarily large in order to make the cross-section σ, a scalar quantity, independent of the range. The quantities $\bar{E}_s$ and

$\bar{H}_s$ represent the scattered electric and magnetic fields respectively, while $\bar{E}_i$ and $\bar{H}_i$ stand for the incident fields.

7.2.2 Monostatic and bistatic echoing area

As has been indicated above, the radar echoing area of any object, other than a sphere, depends on the angle of incidence of the electromagnetic radiation and also on the angle of look by the receiver.

Most radar installations are monostatic so that the above angles are identical; with bistatic radars, however, the two angles have to be considered separately.

Since most methods of calculating echoing areas and most published measurements are for a monostatic case only these will be considered here and the reader interested in the computation of the bistatic echoing areas is referred to the publication of Crispin, Goodrich and Siegel (1959), Ruck *et al.* (1970), Kell (1965) and others.

7.2.3 Polarization and the scattering matrix

Radar transmission is either linearly or elliptically polarized. Linear polarization is either horizontal or vertical depending on whether the electric field vector, the magnitude of which varies periodically in time, is always in the horizontal or vertical plane. If the electric vector varies periodically in magnitude and direction, elliptical polarization ensues. Elliptical polarization is the general form, linear and circular polarization are special cases. An elliptical polarization is termed right-handed if the electric vector rotates clockwise when looking from the aerial in the direction of propagation. It is left-handed if the electric vector rotates in the opposite sense. Most targets reflect, at any particular aspect, differently polarized incident waves in different ways thus the targets have polarization selective properties. Further, for many shapes the polarization of the scattered field differs from the polarization of the incident field. This is referred to as 'depolarization' and is caused by, e.g. the phase change of only one of the field vectors on reflection from a conducting surface. In this way, a right elliptically polarized wave becomes, on reflection, left elliptically polarized and vice versa (see Section 11.2.1).

Since multiple scattering which affects polarization and phase, both neglected by geometrical optics, has been proved to be a major contributor to depolarization (Ruck *et al.* 1970), polarization selectivity of the echoing area disappears when the target is everywhere convex and smooth enough for a geometrical optics approach of echoing area determination to be used.

The definition of echoing area, eqn. (7.1), assumes that a radar is capable of transmitting and receiving both, horizontal and vertical, polarizations either simultaneously or independently in any order and that it can measure the amplitude and plane of the received signal. It implies that the echoing area is to be determined from the ratio of the magnitudes of the scattered and incident fields.

In many cases it is more convenient to represent the echoing area by a scattering matrix

$$[S] = \begin{bmatrix} a_{11} & a_{12} \\ a_{21} & a_{22} \end{bmatrix}, \tag{7.2}$$

where the coefficients a_{ij} are generally complex quantities with the subscripts 1 and 2 representing any two general orthogonal polarization components, e.g. horizontal and vertical, or, in the case of elliptical polarization, right resp. left elliptical polarizations.

All available information on the scattering properties of the target are contained in the scattering matrix. It permits one to compute the echoing area for a particular frequency and target and transmitting–receiving aerial orientation. Conversely, if the echoing area is known for several polarizations the scattering matrix may be uniquely defined.

If the target has more than two planes of symmetry intersecting along the radar's line-of-sight, the scattering matrix S reduces, for orthogonal polarization, to a scalar multiplied by a unit matrix.

The matrix may be truncated for target sizes comparable to a wavelength since in this case the scattering is well approximated by the dipole or quadrupole approximation. However, if the target is not axially symmetrical, the determination of the echoing area is complicated even in this case.

7.2.4 Frequency dependence of the echoing area

The dependence of the echoing area on the phase relation of the various contributing reflections from individual features of a composite target has been studied by Kell (1965). Phase change may be considered as a wavelength change.

The reflective properties of a target, to a first approximation, are dependent only on its size and not on its shape. In general, there are three forms of scattering (Weinstock, 1965):

(a) 'Rayleigh scattering', when the wavelength λ is large compared with the target's dimensions;
(b) 'Resonance region', when λ is of the same order as the dimensions; and
(c) 'Surface and edge scattering' occurring in the high frequency region when λ is small compared with the dimensions.

Variation of the wavelength will make the target echoing area go through all these forms of scattering.

In a fashion similar to that observed for polarization selectivity, the frequency dependence of the echoing area disappears when the conditions for geometrical optics are satisfied.

The target echoing area depends also on range. In order to interpret or apply results after analysing an electromagnetic scattering problem theoretically, it is convenient to denote the field in the neighbourhood of the source of radiation and scattering body as near field, which includes the Fresnel diffraction region, while the field in-between is usually referred to as far-field. Radar theory is interested mainly in the radar echoing area in the far-field when the range is greater than the so-called 'Rayleigh range'. This, in accordance with aerial theory, is defined as the range beyond a distance l^2/λ, where l is the maximum dimension of the aerial. However, since the reflected as well as the incident paths are to be considered, in the radar case, the above distance becomes $2l^2/\lambda$.

Figure 7.1 shows the minimum Rayleigh range for a number of typical radar target dimensions as a function of the frequency.

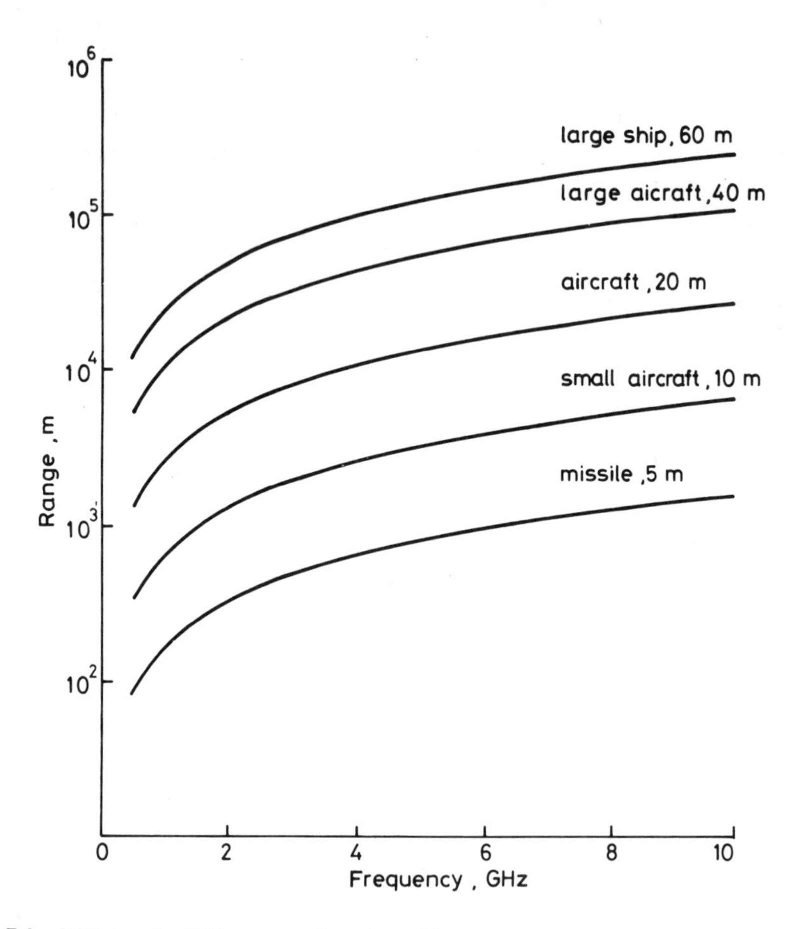

Fig. 7.1 *Minimum far-field ranges as functions of frequency*

7.2.5 Complex targets – statistical analysis

For some investigations it is advantageous to treat complex targets as either an assembly of independent random scatterers or a number of dominant scatterers with an assembly of smaller scatterers. Since the received signal power is directly proportional to the echoing area, these targets may be investigated with the aid of statistical methods used in signal detection theory (Goldstein, 1951) based on Rice's work on noise theory (1944, 1945).

7.2.5.1 Assembly of independent random scatterers: Based on the assumption that the mutual coupling between the individual scattering elements in a large target is negligible, the total field due to an assembly of scatterers is a linear combination of the fields of the individual scatterers.

Investigating the echoing area statistically shows that it has a Rayleigh probability distribution with a probable value equal to zero but with a finite probability of its reaching any value, no matter how large. A value of less than $\sigma_0/2$ will occur for 29% of the time and a value of more than $5\sigma_0$ for 1% of the time.

Similarly to signal detection theory, the first probability distribution shows the magnitude of the echoing area fluctuation, while the second probability distribution gives information about the range of fluctuation.

The first and second probability distributions describe the echo and hence the echoing area fluctuation completely.

7.2.5.2 One steady scatterer with an assembly of indepdent random scatterers: If E is the signal due to a steady scatterer and σ_j that due to the jth scatterer, the total echo in this case becomes

$$S = E + \sum_j \sigma_j, \qquad (7.3)$$

where the sum $\Sigma\sigma_j$ is the total contribution from the random scatterers. E is assumed to be constant in time.

On the basis of the mentioned dependence of the received signal on the echoing area, the first probability distribution of the echoing area may be expressed by an equation similar to that used for the power distribution of signal plus noise (eqn. (3.48)) originally derived by Rice (1944, 1945).

This, with the Bessel function I_0 (#) expanded into a power series, leads to a chi-square distribution.

For a more detailed treatment the reader is referred to the named references and to Rohan (1981).

It should be noted here that Swerling chose the chi-square distribution for his third and fourth case of fluctuating target models.

The second probability distribution is similar to that of an assembly of independent random scatterers.

7.3 The computation of the target echoing area

A number of analytical techniques for the exact determination of the echoing area is available. However, while in some cases one has to resort to asymptotic approximations of some exact expressions, in others some simplifying assumptions have to be made so that the final numerical solution will be only approximate.

7.3.1 Exact methods

Maxwell's equations form the basis of all exact methods and they have to be solved for the boundary conditions appropriate to the target.

The target from which scattering is to be investigated is considered to be in a charge-free homogeneous medium – free space – and the source of radiation is assumed to be far enough for the waves incident upon the target to be plane.

Detailed descriptions of the various methods are given by Ruck *et al.* (1970).

7.3.2 Approximate methods

Due to the difficulties in obtaining the input quantities required for exact solutions of the problem, researchers usually avail themselves of approximate methods of echoing area determination. These methods are often based on some physical assumption and are, in general, easier to apply. They provide numerical answers. A number of approximate methods is currently being used (Ruck *et al.*, 1970). Some of them use the concepts of geometrical optics, the geometrical theory of diffraction, or physical optics, while others base their approach on linear system theory, electromagnetic theory of fields induced on or near the surface of convex conducting bodies or travelling wave theory in seeking expressions for the field scattered by bodies of interest.

7.4 The echoing area of simple shapes

It has been pointed out earlier that complex radar targets may be considered to be assembled from components possessing relatively simple shapes. The echoing area of these may be used, if appropriately superposed, to calculate the echoing area of the complex target.

The term simple shapes in the context of this treatise means a configuration of a relatively simple geometry in contrast to the more complex geometry presented by the configurations of radar targets, e.g. aircraft, ships and missiles.

The echoing area of a 'smooth' body is approximately

$$\sigma \approx \pi R_1 R_2, \tag{7.4}$$

where R_1 and R_2 are the principal radii of curvature, generally large in terms of the wavelength, at the specular point of reflection.

Formulae for the product $R_1 R_2$, depending on the specification of the reflecting surface, are given by Crispin and Siegel (1968).

If the echoing area of an object is determined by reflection from two sources, each of which yields an echoing area σ_i, the received signal power is proportional to $(\sigma_1^2 + \sigma_2^2 + 2\sigma_1\sigma_2 \cos\psi)^{1/2}$, where ψ is the phase difference between the signals due to the two sources. In many cases one is interested in σ_1 and σ_2 only and one neglects the phase; the limiting values of the signal will be the out-of-phase minimum $(\sigma_1^{1/2} - \sigma_2^{1/2})^2$ and the in-phase maximum $(\sigma_1^{1/2} + \sigma_2^{1/2})^2$. If it is assumed that ψ is a statistical quantity with a uniform distribution between 0 and 2π, the average echoing area becomes $\sigma_{av} = \sigma_1 + \sigma_2$ and the expected range of values for the combined signal becomes

$$\sigma_{av} - 2\sigma_1\sigma_2 \quad \text{and} \quad \sigma_{av} + 2\sigma_1\sigma_2 .$$

A few simple shapes will be considered below.

The echoing area of an *ellipsoid* of dimensions a, b and c is

$$\sigma = \frac{\pi a^2 b^2 c^2}{(a^2 \sin^2\theta \cos^2\psi + b^2 \sin^2\theta \sin^2\psi + c^2 \cos^2\theta)^2}, \tag{7.5}$$

where θ in a spherical system of coordinates is measured from the Z-axis, while ψ, in the XY plane, from the X-axis.

The expressions for the echoing area of a prolate *spheroid* ($a = b$) and a *sphere* ($a = b = c$) follow from the above expression.

To account for the creeping and travelling wave contribution in the resonance region, one applies correction terms to the above echoing areas. It is due to these contributions that the echoing area of a sphere, which is isotropic in a monostatic case, exhibits variations depending on the ratio of the sphere's radius and the used wavelength. At frequencies at which the above ratio exceeds 10, the echoing area becomes equal to the optical cross-section πa^2.

Ellipoids are used for modelling aircraft fuselages, engine nacelles, wing tanks and wing tips.

The *ogive* is a geometrical body obtained by rotating an arc of a circle of radius R_1 about a chord of length L located at a distance $(R_1 - a)$ from the centre of the circle.

The general expression for the echoing area of this body for $0° \leqslant \theta \leqslant (90° - \alpha)$, according to Crispin and Siegel (1968), is

$$\sigma(\theta) = \frac{\lambda^2 \tan^4\alpha}{16\pi \cos^6\theta\,(1 - \tan^2\alpha \tan^2\theta)^3}, \tag{7.6}$$

where the aspect angle θ is the angle between the chord L and the directional vector pointing towards the radar.

Finite circular *cones* are often used for missile modelling, while truncated elliptic cones are used for the modelling of wing surfaces and portions of fuselage surfaces.

The nose – on echoing are of a cylindrical cone, according to Crispin and Maffett (1965*a*), is

$$\sigma_{\text{tip}} = \frac{\lambda^2}{16\pi} \tan^4\alpha, \tag{7.7}$$

where α is the vertex half angle of the cone.

The expression

$$\sigma_{\perp} = \frac{8\pi \sin\alpha}{9\lambda \cos^4\alpha}(L_2^{3/2} - L_1^{3/2})^2 \tag{7.8}$$

yields the echoing area of a truncated circular cone at normal incidence. L_1 and L_2 are the distances of the truncating planes from the vertex.

For other than normal incidence

$$\sigma = \frac{\lambda L \tan\alpha}{8\pi \sin\theta} \tan^2(\theta - \alpha), \tag{7.9}$$

where L takes the value of either L_1 or L_2 depending on whether the main contributor is the small or the large end of the cone.

Cones may be terminated by other surfaces than planes normal to the cone axis. A number of these terminations gives sharp discontinuities. Formulae for the computation of their echoing area are available in the quoted source and in Crispin and Maffett (1965*a*).

Cylinders may be used for modelling aircraft fuselages, wing tanks and engine nacelles. Very thin circular cylinders having radii much smaller than the wavelength have been used for modelling of the sharp edges of some wing surfaces. If the wavelength is small in comparison with both the length L, and the radius of the cylinder, the echoing area of an elliptic cylinder at normal incidence, $\theta = 90°$, is

$$\sigma_\perp = \frac{2\pi L^2 a^2 b^2}{\lambda(a^2 \cos^2\psi + b^2 \sin^2\psi)^{3/2}}, \tag{7.10}$$

where a and b are the semi-major and semi-minor axes respectively, while the angle ψ is measured in the x–y plane of a spherical coordinate system with reference to the x-coordinate.

The echoing are of a perfectly conducting *wire* of length L, many wavelengths λ long the radius a only a fraction of the wavelength is, according to Chu (1947),

$$\sigma = \frac{\pi L^2 \sin^2\theta \left[\dfrac{\sin \dfrac{2\pi L\cos\theta}{\lambda}}{\dfrac{2\pi L\cos\theta}{\lambda}}\right] \cos^4\psi}{\left(\dfrac{\pi}{2}\right)^2 + \left(\ln \dfrac{\lambda}{\gamma\pi a \sin\theta}\right)^2}. \tag{7.11}$$

Here $\gamma = 1{\cdot}781\,072$ is Euler's constant, θ is the angle between the wire and the direction of incidence and ψ is the angle between the polarization direction and the plane formed by the wire and the direction of incidence.

Aircraft and missile engine ducts and exhausts are cavities which are often bordered by a metallic loop – a *torus* described by

$$(\rho - a)^2 + Z^2 = b^2$$

with $(a - b)$ the inner and $(a + b)$ the outer radius of the torus in the x–y plane.

Based on geometrical optics, at normal incidence when $\theta = 0^\circ$, for the short wavelength case the echoing area, according to Crispin and Siegel (1968), is

$$\sigma = \frac{8\pi^3 ba^2}{\lambda}. \tag{7.12}$$

The echoing area of an *open-ended tube* of diameter d, in the nose-on aspect in the high frequency region, is given by the semi-empirical formula

$$\sigma = 0{\cdot}05\lambda^2 (kd)^3 \tag{7.13}$$

due to Crispin and Siegel (1968). It offers reasonable estimates in cases for which $kd \geqslant 10$. The constant $k = 2\pi/\lambda$.

In a spherical coordinate system the echoing area of a *circular flat plate* in the $X-Y$ plane is

$$\sigma = \frac{\pi a^2}{\tan^2\theta}\left[J_1\left(\frac{4\pi a \sin\theta}{\lambda}\right)\right]^2 \tag{7.14}$$

(Crispin and Siegel, 1968), with a denoting the radius of the plate and $J_1(X)$ the first order Bessel function of argument X. At $\theta = 0^\circ$, the echoing area becomes

$$\sigma = 4\pi A^2/\lambda^2 \tag{7.15}$$

with $A = \pi a^2$, the area of the plate.

The echoing area of an *infinite thin wedge*, with ψ denoting the angle of incidence and γ the wedge angle, is

$$\sigma(\psi, \gamma) = \frac{\pi L^2}{(2\pi - \gamma)^2}\left[\frac{\cos\beta_1}{1 + \sin\beta_1} \mp \frac{\cos\beta_1}{\sin\beta_1 + \cos\beta_2}\right]^2 \tag{7.16}$$

with

$$\beta_1 = \frac{\gamma/2}{2 - \gamma/\pi} \quad \text{and} \quad \beta_2 = \frac{2\psi}{2 - \gamma/\pi}.$$

The upper sign applies to fields having the electric vector parallel to the edge of the wedge; the lower sign applies to the fields with the magnetic vector in that direction.

7.5 Multiple reflection

Although with modern aircraft, most of which have their wings swept back, multiple reflections are rarely considered, they must be considered when determing the echoing area of corner reflectors which are often used as beacons and standards for the experimental determination of the radar echoing area of targets.

Multiple reflections were studied by Crispin and Siegel (1968), who give formulae for the computation of echoing areas of interest.

7.6 Dielectric bodies

Dielectric cylinders and spheroids are used for cockpit covers and radomes.

Detailed instructions for the computation of the echoing area of long, thin, not perfectly conducting dielectric bodies by the mode theory, together with graphs of values of various coefficients required for the computation, are contained in Crispin and Siegel (1968).

7.7 The reduction and enhancement of the echoing area

The reduction of the target echoing area is of interest to designers of military equipment, e.g. aircraft and missiles, in order to reduce their detection probability by enemy radars. On the other hand, it may sometimes be required to enlarge the echoing area beyond that offered by the geometry of the device, e.g. in beacons and decoys.

A few words on methods for achieving of either effect follow.

7.7.1 The reduction of the echoing area

The reduction of the echoing area by coating the surface of targets by layers of absorbing material had already been intensively studied during the Second World War. Its theory is based on the principles of physical optics and its aim is the reduction of the horizontal and vertical reflection coefficients over a range of frequencies and angles of incidence.

Weston (1963) has found that, assuming a plane wave incident upon a body along an axis about which a 90° rotation does not alter the shape of the body and the material medium, the far-zone backscattered field is identically zero if the complex relative permeability of the body equals its complex relative permittivity. In general, the only materials satisfying this condition are ferrites.

If such an absorbing material is applied to targets which are large compared with the wavelength and in which the main contributors to the echoing area are specular points, the performance of the absorber may be predicted on the basis of its flat plate reflection coefficient. Further, the backscatter echoing area will be zero if the above condition of equality of the relative permeability and permittivity is satisfied and if the body is made of materials which make the tangential field components satisfy the surface boundary condition

$$\boldsymbol{E} - (\boldsymbol{n} \cdot E)\boldsymbol{n} = \eta_s(\boldsymbol{n} \times \boldsymbol{H}) \tag{7.17}$$

where $\boldsymbol{n}$ is the unit outward normal to the surface and η_s is an equivalent surface impedance for the material. The above conditions will be satisfied if the variation of the permeability and permittivity, in a tangential direction to the surface, is small and if the index of refraction $(\mu\epsilon)^{1/2}$ of the material is large and has a large imaginary part; further, the fields on the body's surface are to vary slowly in comparison with the wavelength in the material and the radii of curvature of the surface are to be large in terms of the wavelength in the material.

In order to predict the performance of absorbing materials on a specific body in the high-frequency and upper resonance region, the effect of the absorber on the various analytic components of the echoing area must be estimated. If the condition given by eqn. (7.17) is satisfied, the influence of the absorbing material on the creeping wave field may be estimated. Weston has shown that a perfect flat-plate absorber reduces the creeping wave contribution to the echoing area of a sphere by about 20 dB. For axial incidence on a body of revolution for which the above conditions are satisfied, the creeping wave contribution, due to the symmetry of the body, is zero. However, there is a definite creeping wave contribution to the echoing area for other than axial incidence.

The influence of absorbing materials on the travelling-wave contribution is difficult to assess. In general, a lossless dielectric layer on a body tends to enhance this contribution.

The effect of absorbing materials in the middle- and low-frequency regions of the resonance region is difficult to assess (Ruck *et al.*, 1970) and very little work has been done so far on the analytical solution of this problem. However, experimental data show that the performance of a specific absorber begins to deteriorate as the resonance region is approached and continues to decline until in the Rayleigh region it is no longer an effective absorber.

It can be shown (Ruck *et al.*, 1970) that an echoing area reduction in the Rayleigh region may be achieved by coating the body by a magnetic material for which the relative permeability is greater than the relative permittivity.

7.7.2 Echoing area enhancement

Large echoing areas are advantageous for tracking, ship buoys, runway demarcation, etc. Analytical techniques for enhancing the echoing area have concentrated on two main configurations, the corner reflector, and the biconical reflector.

7.7.2.1 The corner reflector: The corner reflector is a multiple scatterer which consists of three mutually orthogonal planes.

The scattered energy is concentrated in a number of beams, each of which is centred about a direction of specular reflection (Crispin and Siegel, 1968). The maximum value of the echoing area is approximately

$$\sigma = 4\pi \frac{A^2}{\lambda^2}, \tag{7.18}$$

where the area of the aperture A, expressed in terms of the directional cosines of the direction to the transmitter, satisfying the condition $1 \leqslant m \leqslant n$, is for a square corner reflector

$$\begin{aligned} A &= 4\frac{lmb^2}{n}, \qquad m \leqslant n/2, \\ A &= l(4 - n/m)b^2, \quad m \geqslant n/2, \end{aligned} \tag{7.19}$$

and for a triangular corner reflector

$$A = 4\frac{lm}{l+m+n}b^2, \qquad (l+m \leqslant n),$$
$$A = \left(l+m+n-\frac{2}{l+m+n}\right)b^2, \qquad (l+m \geqslant n), \tag{7.20}$$

where b is the side length of the corner reflector.

The average echoing area of a trihedral square corner reflector having sides of length l is, according to Ruck *et al.* (1970),

$$\bar{\sigma} = 0{\cdot}7l^4/\lambda^2;$$

the average echoing area of a trihedral triangular corner reflector is

$$\bar{\sigma} = 0{\cdot}17l^4/\lambda^2,$$

while that of a circular one is

$$\bar{\sigma} = 0{\cdot}47l^4/\lambda^2.$$

The angle between the beam direction and the direction in which the echoing area has fallen to half is,

$$\psi = 7{\cdot}5^\circ\ \lambda/B, \tag{7.21}$$

where B is the radius of gyration of the aperture taken about an axis through the aperture's centre of gravity and perpendicular to the plane in which the deviation from the centre of the beam is taken. The direction which makes equal angles with the three axes of the reflector is an axis of symmetry for both the square and triangular corner reflector. The diffracting aperture in this direction may be considered to be symmetrical yielding a bistatic echoing area

$$\sigma = \frac{108\pi b^4}{\lambda^2\tau^4}\,[\sin\tau\,\sin\tau/3]^2 \tag{7.22}$$

and

$$\tau = (\pi b/\lambda)(\sin\beta/\sqrt{2} - \cos\beta).$$

7.7.2.2 The bi-conical reflector: The bi-conical reflector consists of two vertex-to-vertex truncated circular cones. The radius of the intersection of the two cones is denoted by a, while the maximum radius of the cones is denoted by b. The incident radiation is reflected from one of the cones to the other one from which it is scattered.

The radar echoing area for this condition

$$\sigma = \frac{32\pi}{9\lambda}\,(b\sqrt{2b-a} - a^{3/2})^2 \tag{7.23}$$

has been found to be in excellent agreement with experimental results.

7.8 The echoing area of complex shapes

The theoretical method for calculating the echoing area of complex shapes has been reduced to three steps (Crispin and Maffett, 1965*b*). In the first step the complex shape is considered to be an ensemble of simple shapes each of which approximates the echoing area of the component feature it replaces. This first step is then the geometrical modelling of the configuration. In the second step one computes the echoing areas of the simple shapes as outlined in Section 7.4 above. The third step consists of the appropriate combination of the echoing areas of the simple shapes in order to yield the echoing area of the complex body.

Most missiles consist of a conical or ogival nose section with a tip in the shape of a segment of a sphere. The conical section is joined onto a cylindrical section which may be terminated with a segment of a sphere, spheroid or a truncated cone. Booster sections of the missile look like open-ended cylinders or truncated cones with sharp ring-like edges. Fins may be approximated by flat plates or thin wedges. The aerials are thin wires or slots cut into the skin.

The modelling of aircraft follows the pattern indicated for missiles except that the number of simple shapes will be larger to account for the greater complexity of the aircraft's shape. Discontinuities of the surface of the target are usually modelled as wedges, rings, wires, etc. In many cases travelling and also creeping waves must be considered. Multiple scattering may occur in complex shapes, e.g. aircraft, but are rarely present in missiles.

In the case of a twin-engined aircraft, e.g. at near nose-on aspect, one would approximate the two outer wing edges, the two engines, the inner wing edges and the nose with possibly some corner reflectors for the join between the wings and the fuselage. The individual sources must be appropriately identified with co-ordinates and all simple shape approximations must be for the sought polarization.

While the approximation of a complex shape may, occasionally, be simple, the determination of its echoing area can be rather involved.

7.9 Computer programmes

Since the computation of the radar echoing area pattern for varying aspects consists of many repetitive calculations for slightly differing values of the same parameters, it is well suited for processing by digital computers. Echoing area computer programs are particularly useful in the design stage of missiles and aircraft where it is important to determine how some alteration of the configuration of the vehicle will affect the radar echoing area. No doubt, many programes have been written and are being used, but few have been published so far. Crispin and Siegel (1968) mention two programs used by the Conductron Corporation and the special issue of the *Proceedings of the I.E.E.E.* on radar reflectivity (Vol. 53, No. 8, August, 1965) contains a few papers dealing with the solution of scattering problems by digital computers.

Generally, programs will require as input a specification of the simple shapes to be used for the approximation of the features of the real targets, their echoing area and dependence on the aspect and some data on the shadowing of some components by others at particular aspect angles.

There seem to be two trends of thought on the preferred form of programs. One advocates the use of a number of subroutines which can be easily selected or discarded, modified and replaced, and offer a rapid means of computation. The other one prefers a main program which accepts the description of the target as a series of sections, each of which is evaluated separately.

The choice of program configuration is best left to the programmer.

7.10 Target simulation by physical modelling

Target simulation here means the scaled modelling of radar targets which has been used extensively since the early days of radar. It has the advantage of being cheap, compared with full scale target trials, and repeatable.

There are two currently used methods of scaled target modelling, though a third one, namely modelling in a water tank, finds limited application. The earliest method used optical simulation which was limited by the incoherence and very short wavelength of light permitting only unquantified identification of the sources of specular echoes. Laser beams used nowadays have removed this limitation. Ultrasonic waves, due to the coherent nature of their source, have replaced light waves before the advent of the laser. However, since water, the only practicable transmission medium, has an inherent inability to support shear waves, the modelling of polarization and polarization dependent effects is not possible with ultrasonic waves. These limitations are acceptable for some investigations. The second and most often used method of scaled simulation is radio modelling. This method, if carried out properly, provides a true model. It permits measurements to be made in a well controlled environment so that the measurements are repeatable.

For the sake of completeness of this survey a brief description of the two most often used methods follows.

7.10.1 Optical modelling

This method consists of illuminating a target model, which has a highly reflecting surface finish, by a beam of light and measuring the light reflected by the target by a photometer. The experimental set-up is calibrated by substituting a standard sphere or cylinder of a known echoing area for the target model and subtracting the stray background illumination at the photometer. The photometer current is a direct measure of the mean echoing area in the direction of the photometer.

The advent of the laser, which offers coherent high intensity light sources, may lead to the resurrection of this, now rarely used, method of target echoing area determination.

7.10.2 Radio modelling (Ruck *et al.*, 1970; MacDonald, 1972)

Two types of radio modelling may be performed in practice, viz. full-scale modelling, involving the flying of targets to be modelled, or scaled modelling usually performed in an anechoic chamber. Full-scale modelling is expensive and it does not offer easy repetition of measurements under identical conditions. For these reasons scaled modelling, which is cheaper and permits repeating measurements under controlled conditions, is preferable.

The possibility of using scaled-down models of real targets is based on the linearity of Maxwell's equations, which allow some latitude in the choice of model parameters. The model may be specified in a linear medium by four scaling factors, i.e. length, time, the electric and the magnetic field. Using these four factors, it is theoretically possible to model an electromagnetic system and to relate the measured quantities to the corresponding parameters of the full-scale system. The measurements yield absolute values of the model system parameters. However, in practice the choice of scaling factors is limited by the values of conductivity, permittivity and permeability obtainable in model materials. For most measurements, the permeability and permittivity of the model material will necessarily be those of the full-scale target. Usually only the length and either the electric or the magnetic scaling factor can be chosen. When only the length is chosen, the model becomes a geometrical model which does not permit the measurement of absolute power in the two systems. The measured relative powers in the model system may be converted to the absolute value by calibration based on the comparison of model returned signals with those returned from a target of known echoing area. This applies generally since the values of the electric and magnetic field strength are seldom known. A sphere for which a full theoretical analysis of the echoing area is available, is a convenient calibration source.

As it has been mentioned earlier, scaled modelling is usually performed in an enclosed area padded with material which absorbs electromagnetic radiation at the frequencies in use and thus reduces the strength and the number of unwanted interfering echoes. The scale of target dimension reduction is dictated by the size of the available anechoic chamber. The use of large scaling factors is attractive since this increases the effective size of the anechoic chamber. However, the availability of stable radio-frequency sources dictates the upper limit of the scaling factor choice.

The models have usually the same reflective properties as the real target, but they must be light in weight to permit their suspension by a few thin nylon threads. Cellulose sprayed wood, rubbed down and silver coated to at least five times skin effect depth is the currently most often used model material. A surface finish to about $10^{-4}\,\lambda$ is desirable in order to reduce its effect on deep minima of the polar diagram to about 1 dB. More complex target components, e.g. radar installations and cockpit contents, are made in metal and added later and so are the dielectric radomes and cockpit canopies.

Larger targets, up to full size, may be modelled on outdoor ranges which, however, must be greater than $2l^2/\lambda$ where l is the largest dimension of the target, in order to satisfy the far-field requirement (see Section 7.2.4). This criterion permits a 45° phase variation across the target. A relatively uniform amplitude distribution near the target model may be obtained by placing the model at a height which phases-in the ground-reflected and direct ray. This height depends on the source height and frequency and on the distance between source and target model. The range surface is to be such as to provide specular reflection at the frequencies used.

7.10.3 Data recording and interpretation

Model ranges are, as a rule, equipped with recording facilities to produce plots of reflected signal level, hence echoing area, against aspect angle. In addition, especially to make the time-consuming analysis of records easier, signals are recorded in digital form and used as input to computer programmes.

Since the recorded data are, generally, too irregular for further study, some form of smoothing is usually applied. Smoothing out oscillations caused by variations of one parameter permits the study of the effects of other parameters. Smoothing is usually performed by averaging the measured data over an arbitrary angular range or over the phases by which contributions of individual scatterers are combined. Smoothing with respect to other parameters is also possible.

Averaging over an angular range is used very often and is particularly suitable for the comparison of two sets of data, particularly if two aspect angles are involved, e.g. in the determination of the echoing area of thin wires (Van Vleck, Bloch and Hamermesh, 1947), when care must be exercised to make all experimental measurements in the same plane with respect to the target (Crispin and Maffett, 1965*b*).

Great care must be exercised both in the analysis and interpretation of experimental data and in their comparison with theoretical computations. When comparing theoretical and experimental results, account must be taken of accuracy limitations in the experiments which may be introduced by, e.g. background noise, model construction and suspension, smoothing introduced by the recording equipment, etc. These tend to reduce the accuracy requirements on theoretical methods so that agreement of 2 to 5 dB between theory and experiment is considered to be satisfactory.

Occasionally, an experimentally obtained echoing area is to be used in radar performance computations. In such cases a mathematical expression approximating the obtained smoothed echoing area curve needs to be found. Methods of curve fitting are available in many textbooks on mathematics. Alternatively, an approximation based on expressions used for other physical phenomena may sometimes be used.

Figures 7.2 and 7.3 show the measured dependence (dashed line) of the relative echoing area of a medium size and an interceptor aircraft respectively on the aspect angle together with the same dependences computed (full line) by algorithms given below.

Since the curves for vertical polarization are reminiscent of band-pass filter characteristics, expressions similar to these, with the aspect angle replacing frequencies, were applied in deriving the algorithms given below. Tuned circuit impedance curve expressions, with judiciously chosen centre frequencies and Q-factors were used for reducing deviations of the computed curves from the measured ones.

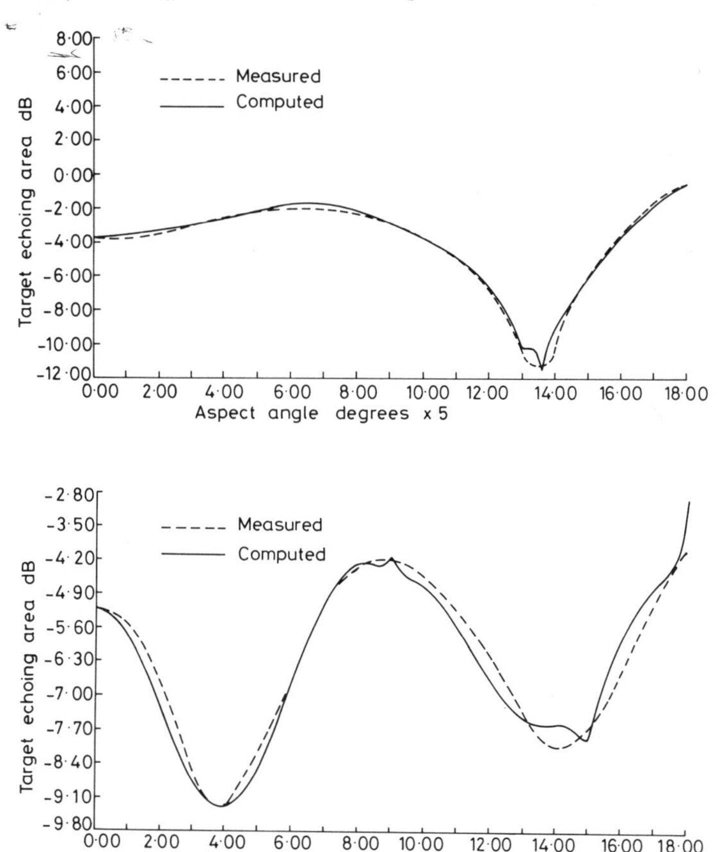

Fig. 7.2 *The relative radar echoing area of a medium size aircraft at vertical (top) and horizontal (bottom) polarization*

The algorithm of the $\sigma = \sigma(\psi)$ dependence of the upper curve of Fig. 7.2, where ψ is the aspect angle, is

$$\sigma_{dB} = -8{\cdot}44 + \cosh^{-1}[1 + |50{\cdot}9(2{\cdot}162 \times 10^{-4}\psi^2 - 1)|] + \frac{2{\cdot}8}{\sqrt{1 + 2\left(\dfrac{\psi^2 - 1225}{35\psi}\right)^2}} - \frac{5{\cdot}67}{\sqrt{1 + 20\left(\dfrac{\psi^2 - 4624}{68\psi}\right)^2}}$$

$$+\frac{4{\cdot}83}{\sqrt{1+20\left(\dfrac{\psi^2-8100}{90\psi}\right)^2}}-\frac{1{\cdot}7}{\sqrt{1+1000\left(\dfrac{\psi^2-4225}{65\psi}\right)^2}}.$$

The second term of this algorithm is similar to the expression for the attenuation characteristics of Campbell's band-pass filter

$$\alpha = \cosh^{-1}\left[1-2\frac{f'^2-f_2^2}{f_2^2-f_1^2}\left(\frac{f^2}{f'^2}-1\right)\right]$$

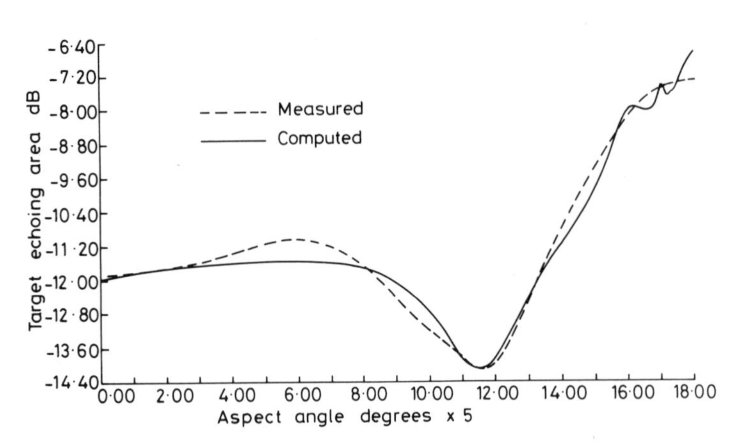

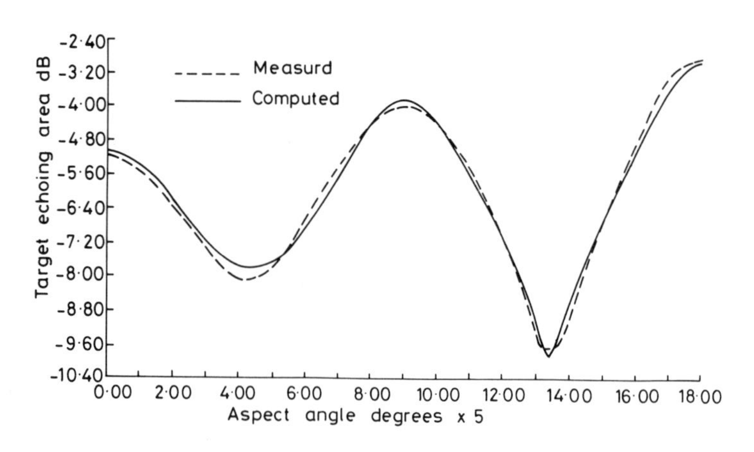

Fig. 7.3 *The relative radar echoing area of an interceptor aircraft at vertical (top) and horizontal (bottom) polarization*

with f' the frequency of maximum attenuation and f_1 and f_2 the pass-band limiting frequencies.

The subsequent terms represent the impedance Z of parallel L, C, R circuits.

$$|Z| = \frac{R}{\sqrt{1 + Q^2 \left(\frac{f^2 - f_0^2}{f f_0}\right)^2}}$$

with f_0 the centre frequency and $Q = \omega_0 CR$.

The algorithm of the lower curve is

$$\sigma_{\mathrm{dB}} = 0{\cdot}0233\,|\psi| - 7{\cdot}5 + 2{\cdot}25 \cos 9\psi + 2{\cdot}25 \sin |\alpha|$$

$$- \frac{1{\cdot}6}{\sqrt{1 + 500\left(\frac{\psi^2 - 5625}{75\psi}\right)^2}} + \frac{1{\cdot}7}{\sqrt{1 + 10^4\left(\frac{\psi^2 - 8100}{90\psi}\right)^2}}$$

$$+ \frac{0{\cdot}8}{\sqrt{1 + 10^3\left(\frac{\psi^2 - 2025}{45\psi}\right)^2}}$$

with $\alpha = 8{\cdot}57\,(|\psi| - 45)$ for $\alpha > 0$ only.

Similarly, the upper curve of Fig. 7.3 may be expressed by

$$\sigma_{\mathrm{dB}} = 0{\cdot}03\,|\psi| - 11{\cdot}9$$

$$- \frac{4{\cdot}4}{\sqrt{1 + 20\left(\frac{\psi^2 - 3364}{58\psi}\right)^2}} + \frac{2{\cdot}2}{\sqrt{1 + 200\left(\frac{\psi^2 - 6400}{80\psi}\right)^2}}$$

$$+ \frac{0{\cdot}75}{\sqrt{1 + 10^4\left(\frac{\psi^2 - 7225}{85\psi}\right)^2}} + \frac{2{\cdot}9}{\sqrt{1 + 300\left(\frac{\psi^2 - 8100}{90\psi}\right)^2}}$$

while the lower one by

$$\sigma_{\mathrm{dB}} = 1{\cdot}5 \cos 8\,|\psi| - 6{\cdot}6 + \frac{1{\cdot}2}{\sqrt{1 + 10\left(\frac{\psi^2 - 2025}{45\psi}\right)^2}}$$

$$- \frac{3{\cdot}0}{\sqrt{1 + 200\left(\frac{\psi^2 - 4489}{67\psi}\right)^2}} + \frac{2{\cdot}3}{\sqrt{1 + 20\left(\frac{\psi^2 - 8100}{90\psi}\right)^2}}.$$

In deriving algorithms by approximating given curves by characteristics of electrical networks, it should be noted that it is not necessary for the network to be physically realizable so that there is no restriction on the choice of circuit constants and Q's.

All the above algorithms are easily programmable. The goodness of the fit is perceivable from the figures.

The method provides a generally applicable method of curve fitting which, with some practice, may lead to close fits in a reasonably short time.

Algorithms like those given above may be incorporated into a subroutine of the RDPRO model for the computation of the aspect angle ξ_n' based on a geometry intimated in Section 4.4 and a factor modifying the radar cross-section SIGMA introduced as one of the input data.

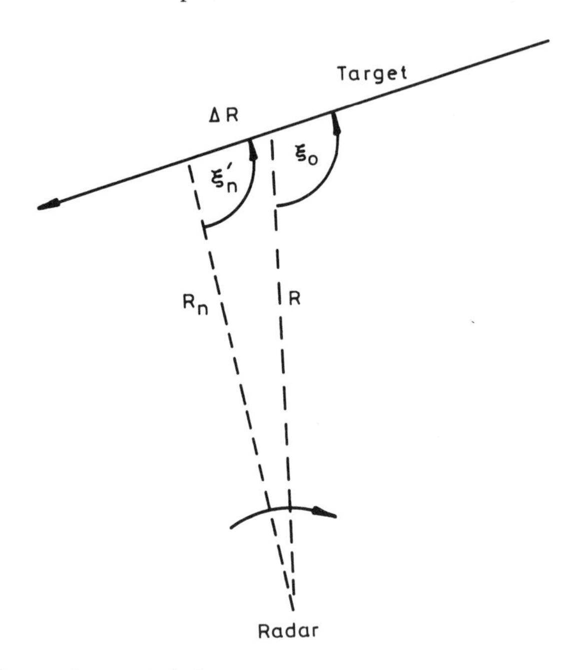

Fig. 7.4 *A general target trajectory*

Assuming that the target trajectory at the instant of interception at a range R, e.g. the radar horizon, is at an angle ξ_0, as shown by Fig. 7.4, the range on the nth aerial sweep is

$$R_h = \sqrt{R^2 + [(n-1)\Delta R']^2 + 2R(n-1)\Delta R' \cos \xi_0}$$

and the aspect angle is

$$\xi_n' = \sin^{-1} \frac{R \sin \xi_0}{R_1}$$

with $\Delta R' = (60/\text{r.p.m.}) \times$ target velocity.

Table 7.1 *Radar echoing area of some targets of interest*

Target designation	Target name	Echoing area m^2
Aircraft		
Aero 500	Aero Commander	6.5
B-29	Super Fortess	100
B-47	Stratojet	16
B-52	Strato Fortress	125
B-57B	Canberra	10
707	Boeing 707	32
720 and 727		25
Britannia	Bristol Britannia	25
C-121	Constellation	100
C-130	Hercules	80
	Caravelle	16
	Comet	40
Cessna 180		1.5
310B		4
Convair 240, 340, 440	Metropolitan	40
DC-3	Dakota	20
DC-8		32
DC-9		25
F-1/FJ	Sabre	5
F-4	Phantom	10
F-9F	Cougar	12.5
F-27	Fokker Friendship	25
F-86	Sabre	5
F-104	Starfighter	5
IL-28	Beagle	8
	Javelin	8
	Lamps	25
MIG-21	Fishbed	4
P-3A	Orion	80
P-3B	Orion	95
TU-16	Badger	25
TU-20	Bison	40
TU-95	Bear	125
	Viscount	16
Ships		
Small cargo ship		150
Large cargo ship		16 000
Submarine surfaced		530
Submarine schnorkel		5
Aircraft carrier		3000

Table 7.1 (*continued*)

Target designation	Target name	Echoing area m^2
Destroyer		1400
Patrol Boat		85
Trawler		15
Missile nose-on		1

7.11 Radar echoing area of some targets of interest

Since the echoing area of complex targets, e.g. aircraft and ships, may change to a very large extent even for small changes in aspect angle, it is impossible to give exact figures for this parameter.

The echoing areas in Table 7.1 are to be taken as representative averages only. Those given for aircraft are for a nose-on, while those for ships are for a bow-on aspect.

An empirical formula for the echoing area of ships in m^2

$$\sigma = 52\sqrt{fD^3},$$

where f is the radar's operating frequency in MHz and D the ship's full load displacement in kilotons, is given by Skolnik (1974). Other empirical formulae claim the echoing area of wooden crafts in a bow-on aspect to be equal to one third of the bow-on projected area, while that of metal hulled ones to the bow-on projected area.

7.12 Conclusions

In general, the approach using theoretical calculations is limited only by the time, effort and cost. An expensive computer program makes the computation of the echoing area of practically any object possible. While such an approach is often time consuming and expensive, it is the recommended procedure to adopt if high precision is required. In most cases, however, high precision is not warranted and the echoing area computation may be performed fairly cheaply and rapidly. Experience has shown that often it is adequate to know the echoing area to within 2 to 4 dB so that approximate methods of calculation are acceptable.

Most echoing areas are studied both theoretically and experimentally. Here, it must be remembered that both studies should be performed under 'identical' conditions if a reasonable agreement between the two methods is to be achieved. The results obtained from experimental studies will be only as good as the quality and accuracy of the model. Differences in modelling will result in differences in echoing area values obtained.

Chapter 8

Search radar – purpose, accuracy, resolution and required performance

In order to fulfil the requirements specified by the radar user, the radar designer has to choose the various equipment parameters so as to satisfy the set specifications. This does not apply to the range performance only but also to such characteristics as measurement accuracy and resolution dealt with in this chapter.

8.1 Purpose

There are various types of radar depending on their purpose which, in turn, determines those characteristics and performance requirements which are necessary or most suited for meeting the allocated task. A search or surveillance radar, as its name implies, searches an allotted space for targets of interest. Detection of targets at long ranges then provides adequate time for alerting the necessary services for the target's arrival. Search radars have therefore many civilian and military applications. The most common civilian use of surveillance radars is that at airports where they are used for air-traffic control mainly of further removed targets since secondary radars are used at most well-equipped airports for the surveillance of closer ones.

Some of the constraints imposed on the choice of some radar parameters have been mentioned earlier (Chapter 4). A cursory look at the requirements which search radars are expected to satisfy, follows.

One of the basic requirements is that of search, a proposal for the assessment of which is in Chapter 2 of this treatise. It is mentioned there that detection of a target is the only measure of a search's success. The searched volume of space is a function of the aerial's coverage diagram which is to be examined closely later.

Detection from a theoretical point of view is treated in Chapter 3, where the necessity of specifying the probability of detection and the probability of false alarm is brought out. Detection probability is a function of the signal-to-noise ratio and this, besides various other parameters, depends on the target's echoing area, behaviour and range. The interrelation between search volume, target characteristics and detection is thus established and the importance of specifying besides the radar's parameters, the required probability of detection and tolerable probability of false alarm as well as the parameters of the aerial and target, whenever a performance assessment is to be made, is demonstrated.

8.2 Accuracy and resolution

Search radars pinpoint targets in space by determining two coordinates which, except in the case of height finding radars, are the azimuth angle and either range or velocity. A height finding radar determines the target's altitude at the target's coordinates measured by the search radar.

8.2.1 Range accuracy

Range is measured by the time interval between the transmission of a pulse and the reception of the corresponding datum of its echo pulse, e.g. the time interval between the leading edge's crossing of a threshold. Noise, which is always present in a receiver, distorts the received pulse by widening and raising it and thus introducing inaccuracies into the range measurements. The distorted pulse edge has largely the same slope as that of the original one so that one may determine the range error from the time difference Δt in their threshold crossing by

$$\Delta R = \frac{c\,\Delta t}{2}\,.$$

The time interval Δt expressed on the basis of the similarity of the triangles formed on the one hand by this quantity, the noise amplitude $n(t)$ near the threshold crossing and the leading edge of the distorted pulse and on the other hand correspondingly by the rise time t_r, amplitude A and leading edge of the undistorted pulse at the output of the video filter as,

$$\Delta t = t_r \frac{n(t)}{A}$$

and inserted into the expression for ΔR yields

$$\Delta R = \frac{ct_r}{2}\frac{n(t)}{A}\,.$$

The last term of this relation may be expressed in terms of the customary signal-to-noise power ratio SN obtained earlier in Chapter 3; besides that, since for a band-limited case in which the pulse-width–bandwidth product is much larger than unity, the pulse rise time is inversely proportional to the i.f. bandwidth B. The range error, using the above modifications, becomes

$$\Delta R = \frac{c}{2B}\frac{1}{\sqrt{SN}} \qquad (8.1)$$

showing that higher accuracy, i.e. a reduction of the range ambiguity ΔR, is obtained by increasing the bandwidth and/or the signal-to-noise ratio.

While the derived expression for ΔR gives an acceptable measure of the accuracy of the range measurement, it is not strictly correct since the bandwidth B, the i.f. bandwidth, was taken to be proportional to the reciprocal of the pulse rise time when, as it was pointed out earlier, experimental and analytical evidence shows that the best bandwidth–pulse duration product $B\tau$, from the detection point of view, is approximately equal to unity and hence B is proportional to the reciprocal of the pulse width. Most radars use bandwidths satisfying this condition.

The mentioned time difference of threshold crossing of the pure and distorted pulses is clearly due to phase shift caused by noise. Noise being a random quantity is appropriately treated statistically. The problem of the accuracy of range determination could be considered then on the basis of the probability distribution of the phase.

A radar pulse is a burst of sinusoidal r.f. energy embedded in noise assumed to have a Gaussian distribution. One has to find an expression for the probability density of the phase of such a signal.

The joint probability distribution of the envelope of signal plus noise and phase is given by

$$q(\rho, \phi) = \frac{\rho}{2\pi\sigma^2} \exp\left[-\left(\frac{\rho^2 + A^2 - 2A\rho\cos\phi}{2\sigma}\right)\right],$$

where ρ is the envelope, ϕ the phase of signal plus noise and σ^2 is the variance of the orthogonal projections of ρ.

The probability density of the phase alone is of interest at present and this may be shown to be

$$q(\phi) = \int_0^\infty q(\rho, \theta)\, d\rho = \frac{1}{2\pi}\exp\left(-\frac{A^2}{2\sigma^2}\right)$$
$$+ \frac{A\cos\phi}{2\sigma\sqrt{2\pi}}\left[1 + \operatorname{erf}\left(\frac{A\cos\phi}{\sqrt{2}\sigma}\right)\right] \exp\left(-\frac{A^2\sin^2\phi}{2\sigma^2}\right),$$

where A is the signal's amplitude and σ is the r.m.s. value of the noise so that A^2/σ^2 is the signal-to-noise power ratio.

It was shown by eqn. (8.1) that the range ambiguity is reduced by a large signal-to-noise ratio. Assuming then that this ratio is large and the absolute value of the phase ϕ is of such a small value that the approximations $\sin\phi \doteq \phi$, $\cos\phi \doteq 1$ and $\operatorname{erf}(A\cos\phi/\sqrt{2}\sigma) \doteq 1$ are acceptable, the above expression for the probability density becomes

$$q(\phi) \doteq \frac{A}{\sigma\sqrt{2\pi}} \exp\left(-\frac{A^2\phi^2}{2\sigma^2}\right),$$

which is a normal distribution with a zero mean value and variance σ^2/A^2 the width of which is inversely dependent on the voltage signal-to-noise ratio and approaching an impulse function for a signal-to-noise ratio tending to infinity.

The definite integral of $q(\phi)$ within chosen limits yields the probability that the phase shift is within the range of the limits, i.e.

$$P(|\phi| < |\phi_0|) = \int_{-\phi_0}^{\phi_0} q(\phi)\,d\phi = 2\frac{A}{\sigma\sqrt{2\pi}} \int_0^{\phi_0} \exp\left(-\frac{A^2\phi^2}{2\sigma^2}\right) d\phi$$

bearing in mind that this is valid only for $A^2/\sigma^2 \gg 1$ and ϕ_0 small enough to make the mentioned approximations acceptable. Recalling that the customary signal-to-noise ratio as used in radar practice is $SN = A^2/\sigma^2$ the above probability expressed in terms of the usual S/N ratio SN reads

$$P(|\phi| < |\phi_0|) = \frac{2}{\sqrt{\pi}}\sqrt{\frac{SN}{2}} \int_0^{\phi_0} \exp\left(-\phi^2\frac{SN}{2}\right) d\phi$$

and the probability of the phase being within the chosen limits becomes

$$P(|\phi| < |\phi_0|) \doteq \operatorname{erf}\left(\phi_0\sqrt{\frac{SN}{2}}\right) \tag{8.2}$$

which shows that the probability that ϕ will be within the specified limits increases with an increasing singal-to-noise ratio.

A phase-shift of π occurs in τ seconds, where τ is the pulse width and corresponds to the range of $c\tau/2$. A phase-shift of ϕ radians, expressed in units of length, is $c\tau\phi/2\pi$ so that the probability that the range error due to noise will be within the specified limits of $\pm\,\Delta R_0$ is

$$P(|R| < |\Delta R_0|) \doteq \operatorname{erf}\left(\frac{2\pi}{300}\,\frac{\Delta R_0}{\tau}\sqrt{\frac{SN}{2}}\right) \tag{8.3}$$

with the pulse duration τ in μs.

The accuracy of the target's range measurement increases with an increasing signal-to-noise ratio. This result satisfies also the intuitive expectation of the target's position being more accurately known when the echo is strong to overcome disturbing noise.

A gating signal is sometimes fed to a mixer having the received signal containing the sought time delay information corrupted by noise as its other input. The product of the two signals is integrated over the significant modulation interval. Mallinckrodt and Sollenberger (1954) show that the Fourier transform of the optimum form of the gating signal for minimizing the pulse position measurement error is

$$F_g = \frac{j\omega F_s}{|F_n|^2},$$

where F_s and F_n are the Fourier transforms of the input signal $s(t - T)$ and the input noise $n(t)$ respectively and $j\omega$ denotes differentiation so that the optimum gating waveform appears as the time derivative of the received waveform provided that the noise spectrum is constant.

If the above mixer is preceded by a matched filter, the optimum gating waveform has the Fourier transform

$$F_g(\omega) = j\omega$$

when the amplitude response of the matched filter is

$$G(\omega) = F_{si}^*(\omega),$$

where F_{si} is the Fourier transform of the receiver input signal at a point where the noise is not yet band-limited. The two transforms are related by

$$F_s = GF_{si} = |F_{si}|^2$$

and similarly, based on the assumption that $|F_{ni}|^2$ is constant or white,

$$|F_n|^2 = |G|^2 |F_{ni}|^2 = |G|^2.$$

The mean time error which is proportional to the mean square variation of noise with respect to the sensitivity, i.e. the slope of the discrimination function at cross-over, as given by Mallinckrodt and Sollenberger for optimum gating, is

$$\overline{\Delta t^2} = \frac{1}{(4\pi)^2} \frac{1}{\int_0^\infty f^2 \frac{|F_s|^2}{|F_n|^2} df},$$

which, when referring the spectral densities F_s and F_n to the receiver input and introducing the signal energy

$$E = 2 \int_0^\infty |F_{si}|^2 df,$$

becomes

$$\overline{\Delta t^2} = \frac{1}{8\pi^2} \frac{\eta}{E} \frac{\int_0^\infty |F_{si}|^2 df}{\int_0^\infty f^2 |F_{si}|^2} = \frac{1}{2\beta^2} \frac{\eta}{E}$$

with

$$\beta^2 = \frac{\int_0^\infty (2\pi f)^2 |F_{si}|^2 df}{\int_0^\infty |F_{si}|^2 df}$$

the effective bandwidth and η the input noise spectral density.

Reverting to the leading edge detection accuracy determination and using the relations

$$B\tau \approx 1,$$

$$A = \frac{E}{\tau},$$

$$n(t) = \frac{\eta B}{2},$$

$$SN = \frac{2E}{\eta B\tau} = \frac{2E}{\eta},$$

where E is the energy content of a pulse, the time component of eqn. (8.1) is

$$\Delta t = \frac{1}{B\sqrt{SN}} = \frac{1}{B}\sqrt{\frac{\eta}{2E}}$$

while that based on the optimum gating is

$$\Delta t = \frac{1}{\beta}\sqrt{\frac{\eta}{2E}} = \frac{1}{\beta\sqrt{SN}} \tag{8.4}$$

and then

$$\Delta R = \frac{c}{2}\frac{1}{\beta\sqrt{SN}},$$

showing that the difference is due only to the bandwidths considered in the two cases. The case using the conventional bandwidth, though slightly inferior to the optimum one, is generally used in radar accuracy calculations.

8.2.2 Accuracy of angle measurement

One would assume intuitively that the accuracy of target position determination in azimuth and elevation, in case of a height-finding radar, depends, besides the signal-to-noise ratio of the received echo, on the beamwidth of the aerial in the appropriate plane of interest. Since this study is concerned with search radars measuring range and azimuth angle only, no further reference to height finders or velocity measuring radars will be made in future.

The signal strength of a signal fed into a receiver by a scanning aerial receiving an echo of a stationary target, considered here to be a point source, is a function of the coverage diagram of the aerial which, when taken to be time dependent due to its scanning, leads to an expression identical with that of eqn. (8.4) except that the bandwidth β in this case is replaced by a ratio of the Fourier transforms of the amplitude distributions $A(x)$ across the apertures.

The angular accuracy is then given by

$$\Delta\theta = \frac{1}{\gamma\sqrt{SN}}, \tag{8.5}$$

where γ (Skolnik, 1962) is

$$\frac{\int_{-\infty}^{\infty} (2\pi x)^2\,|A(x)|^2\,dx}{\int_{-\infty}^{\infty} |A(x)|^2\,dx}$$

with $A(x)$ the previously mentioned amplitude distribution across the aerial's aperture which is a function of the distance x measured in the plane of interest from the axis of symmetry of the aerial.

The theoretical angular error for optimum angular accuracy, based on a one-way voltage gain of the aerial and assuming a unity bandwidth-pulse duration product is, according to Manasse (1960),

$$\Delta\theta = \frac{\lambda\sqrt{1+(SN/2)}}{d_0(SN/2)}$$

which for large values of the signal-to-noise ratio reduces to

$$\Delta\theta = \frac{\lambda}{d_0\sqrt{SN}}$$

where, in case of an aerial with a rectangular aperture of length D, the coefficient

$$d_0 = \frac{\pi D}{\sqrt{3}}.$$

The half-power beamwidth θ of a rectangular distribution is $0{\cdot}88\lambda/D$ which when inserted into the expression for the angular error yields

$$\Delta\theta = \frac{0{\cdot}627\theta}{\sqrt{SN}}. \tag{8.6}$$

This result is similar to eqn. (8.1) for the range error.

In case of a circular aperture of diameter D the coefficient $d_0 = \pi D/2$ so that, with a half power beamwidth of $\theta = 1{\cdot}22\lambda/D$, the angular error becomes

$$\Delta\theta = \frac{0{\cdot}53\theta}{\sqrt{SN}}.$$

The theoretical angular error at high signal-to-noise ratios is directly proportional to the half-power beamwidth of the aerial and inversely proportional to the signal-to-noise ratio.

The limits imposed by receiver noise on the accuracy of a target's angular position determination by a pulsed radar was studied by Swerling (1956).

A target paints a short arc of a length proportional to the beamwidth on a PPI (plan position indicator) display and the operator, by assuming the target to be at an azimuth angle corresponding to the centre of the arc, can estimate the target's coordinate for large signal-to-noise ratios with an accuracy considerably better than the beamwidth. However, noise, be it noise generated within the radar receiver or external noise, diminishes the accuracy of such an estimate.

Swerling assumes the scanned angular sector to be large compared with the aerial's two-way beamwidth and the target to be stationary but having a rapidly fluctuating echoing area described by the Rayleigh probability density function during a single scan. He derives then the lower bound of the r.m.s. error of the regular unbiased angle estimate, i.e.

$$\Delta(\theta) \geqslant \frac{\theta/2}{SN\sqrt{k}\sqrt{\dfrac{1}{2}\displaystyle\int_{-\infty}^{\infty}\frac{f'^2(\theta)}{[1+SNf(\theta)]^2}\,d\theta}}, \tag{8.8}$$

where $f(\theta) = \exp(-\theta^2)$ is a Gaussian approximation of the beam shape and SN the signal-to-noise power ratio at the input to the second detector for a pulse emitted when the 'nose' of the beam is pointed directly at the target. The signal-to-noise is averaged over the target fluctuations. The term denoted by k is the number of hits per scan.

Evaluating eqn. (8.8) Swerling arrived at the expressions

$$\Delta\theta_{\min} = \frac{0{\cdot}58\theta}{SN_c\sqrt{k}} \quad \text{for } SN_c \ll 1 \tag{8.9}$$

and

$$\Delta\theta_{\min} = \frac{0{\cdot}49\theta}{\sqrt{k\, SN_c}} \quad \text{for } SN_c \gg 1$$

when using the hitherto used one-way half-power beamwidth with SN_c the signal-to-noise ratio at the beam centre.

The relative accuracy of azimuth angle measurement, i.e. the accuracy referred to the half-power beamwidth, is then $\Delta\theta_{\min}/\theta$.

The difference in the results obtained by the two methods of angular error determination may be attributed to Swerling's use of the two-way aerial radiation pattern as against the one-way one used in the former approach to the problem.

The angular accuracy expressed in units of length is

$$\Delta L = R\,\Delta\theta.$$

8.2.3 *Resolution*

Besides accuracy, i.e. the ability of determining a target's coordinates without appreciable error, a radar must satisfy certain requirements of resolution which may be described as the ability of separating two or more closely spaced similar targets and rejecting clutter. Clutter will be dealt with later in this treatise so it is only the separation of targets that is of interest here.

Resolution is usually specified in terms of the minimum distance in range or angular difference in azimuth at which two closely spaced similar targets are discernible. The similarity of echoes of targets of interest makes resolution more difficult.

Denoting by

$$y_1(t) = s_1(t - t_1) + n(t)$$

the echo of one of the targets and by

$$y_2(t) = s_2(t - t_2) + n(t)$$

that of the other, one is interested in maximizing their mean-square difference, a measure of their dissimilarity. Evaluating

$$\int_{-\infty}^{\infty} [y_1(t) - y_2(t)]^2\, dt,$$

one then arrives at the correlation function

$$R_{s_1 s_2}(t) = 2 \int_{-\infty}^{\infty} s_1(t-t_1)s_2(t-t_2)dt, \tag{8.10}$$

since the correlation functions of the other terms of the expansion either cancel out, are zero or constant. Assuming that the signals from the two targets are identical functions differing only in time, the above cross-correlation function changes into the autocorrelation function

$$R_s(t_1, t_2) = 2 \int_{-\infty}^{\infty} s(t-t_1)s(t-t_2)dt, \tag{8.11}$$

which should be zero for maximum range resolution at all values of $t_1 \neq t_2$ and equal to the received signal energy when the two signals coincide at $t_1 = t_2$, i.e. for stationary targets at the same range.

An expression for angular resolution may be obtained by similar reasoning.

A heuristic approach to the problem would suggest that two identical targets must differ in range by

$$\Delta R = m\frac{c\tau}{2} \tag{8.12}$$

in order to be resolvable. The coefficient m is a safety factor with a value $1 < m < 2$. Identically, the angular separation of two resolvable identical targets at the same range is at least

$$\Delta\theta = m\theta \tag{8.13}$$

with θ the half-power beamwidth of the aerial. However, since the edges of the aerial beam are not sharply defined, the angular resolution may be affected by receiver gain. The length of the arc painted on the PPI display depends on the beamwidth of the aerial and is proportional to range.

The pulse width-bandwidth product of approximately unity generally used in radar receivers suggests the resolvable range difference

$$\Delta R \doteq m\frac{c}{2B}$$

to be inversely proportional to the bandwidth. A narrower pulse, i.e. an appropriately wider bandwidth or energy spectrum, will reduce ΔR and hence increase range resolution. A narrow bandwidth tends to spread out the amplified pulses and thus reduce resolution.

The cathode-ray-spot size may also affect resolution.

8.3 Required performance

Performance requirements of a system or appliance are dictated by its purpose and application. As it was mentioned earlier, this treatise is concerned with surveillance radars only and the performance requirements outlined below apply only to these types of radar.

Air traffic nowadays has reached such density and, with the introduction of jet aircraft into the civilian air travel industry, such velocities that the increasing demands on air-traffic control necessitate the extension of air safety facilities beyond the air space in the immediate vicinity of airfields. An extensive radar network with adequate overlap of coverage to permit a continuous observation of the air space is necessary for safety in the air. For this reason search or surveillance radars should possess an extended detection range and height coverage capability. They should be equipped with clutter suppression facilities and be quickly adaptable to changing operational requirements, e.g. switchable linear to circular polarization, compatibility with other navigational systems, switchable from moving target indicator (MTI) to normal displays, fast frequency change, etc.

Modern radar receivers, fed by signals from high-gain aerials, can intercept targets of interest almost at the radio horizon which, for an aerial at a height h and a target at a height H, is at a range given by the length of a straight line, tangent to the earth, connecting these, or

$$R = \sqrt{h(2r+h)} + \sqrt{H(2r+H)}$$

with r denoting the corrected radius of the earth.

Since $h^2 \ll 2rh$ and $H^2 \ll 2rH$, an approximate expression for this range in km, when using the height dimensions in metres and assuming a 'standard' atmosphere, reads

$$R \approx 4{\cdot}12(\sqrt{h} + \sqrt{H}). \tag{8.14}$$

The maximum intercept range for a fixed aerial height then depends on the target height. Since most civilian air transport is channelled into corridors of about 20 km widths and heights of up to 10 km, the maximum intercept range of radars catering for the above approach geometry becomes

$$R \approx 412{\cdot}19 + 4{\cdot}12\sqrt{h} \quad \text{km},$$

or, for an aerial height of, e.g. 19·8 m, $R \approx 430$ km. This, for an aircraft velocity of say 500 km/h, would provide ample time for ground preparations for landing. However, due to multipath effects the decrease of the field strength at low radiation angles, of up to about $\frac{1}{2}°$, follows an inverse square and that of the echo an R^{-8} dependence on range so that the detection ranges will, generally, be shorter than the computed maximum intercept ranges. Elevating the aerial improves the capability of detecting low-flying targets.

The required vertical coverage is determined on the one hand by the above indicated expected maximum altitude of targets to be observed and on the other by the requirement that the intensity of the target paint from the maximum displayed range (see Chapter 4) down to the nearest point of interest on the target's trajectory should remain approximately constant.

If the elevation angle of a ray from the aerial to a target at a slant range R and altitude H is ϵ, then, using orthogonal coordinates, one may write

$$R = \frac{H}{\sin \epsilon} = H \operatorname{cosec} \epsilon.$$

The intensity of the target paint on the display depends on the echo signal power which in turn is a function of the field strength

$$\bar{E} = \frac{\text{const.}}{R} f(\epsilon) = \frac{\text{const.}}{H \operatorname{cosec} \epsilon} f(\epsilon),$$

where $f(\epsilon)$ expresses the dependence of the aerial gain on the ray elevation angle which, in order to yield a constant field intensity and hence a constant echo signal power and display paint, must be

$$f(\epsilon) = \operatorname{cosec} \epsilon.$$

In terms of energy the aerial coverage diagram has a cosec^2 characteristic, a term generally used to denote aerials satisfying more or less the above conditions.

In practice, the vertical coverage diagram deviates from the above character at higher ray elevation angles to counteract the decrease in signal strength due to the often used automatic receiver overload control.

The maximum ray elevation angle is limited by the maximum tolerable cone of silence, i.e. un-illuminated space, above the aerial and the maximum tolerable difference between the diaplayed slant range and the ground range of interest

$$\Delta R = \frac{1 - \cos \epsilon}{\cos \epsilon}.$$

The customary maximum elevation angle of 45° causes a 41% range difference.

Some of the constraints imposed on other search radar parameters were dealt with in Chapter 4.

Chapter 9

Radar siting

9.1 Introduction

The influence of terrain features on the propagation of electromagnetic waves at very high and ultra-high frequencies was described elsewhere (Rohan, 1948). Small portable communication sets operating at these frequencies use either omni-directional whip aerials or simple directional aerials having considerable side and back lobe radiation. A judicious siting of such aerials may exploit terrain features either to guide the radiation or to reflect it in such a manner as to reinforce it in the desired direction.

Surveillance radars, except for the simple portable types, use more efficient aerials which have (for most purposes) negligible side and back lobe radiation so that the philosophy of siting surveillance radar aerials differs from that used for the site selection for simple communication equipment.

While there is a proliferation of publications on many radar topics, there is a scarcity of papers dealing with radar site selection; the following chapter endeavours to give guide-lines for this important topic, based on available information and the extrapolation of knowledge of the phenomena affecting the propagation of electromagnetic waves. Some of these, though dealt with earlier, will be repeated here in order to provide a self-contained treatise, independent on what preceded.

9.2 The effect of ground reflections

The most often quoted radar equation, in its simplest form, determines the free-space detection range based on the known radar and target-echoing-area parameters and the required signal-to-noise ratio for the probability of detection and false alarm of interest. The envelope of the ranges calculated from this equation, taking into account the vertical polar diagram of the aerial, is the 'aerial coverage pattern' given in the manufacturer's sales literature and equipment handbooks. This radiation pattern applies, however, only to the hypothetical case of a radar in free space with all reflecting surfaces absent. It does not give, therefore, a true picture of the radar's target detection capabilities.

The surface of the earth or of the sea modifies the aerial coverage diagram since radar aerials, in order to permit the detection of low flying targets, inevitably radiate part of the transmitted energy towards the ground or sea. This energy is partly dispersed by the unevenness of the surface and partly absorbed by it. Part of the reflected energy propagates towards the target where it combines with that part which propagates directly between the aerial and the target and, depending on the relative phase of the direct and reflected rays, they either reinforce or weaken one another. In general, maxima and minima of radiation are interspersed and the resulting vertical polar diagram consists of lobes separated by wedges of space where the field strength is so low that a target might elude detection.

The simple radar equation has to be modified therefore to take this phenomenon into account and yield a truer picture of the radar's target detection capabilities. Mathematical expressions for the modifying factor have been derived and used in computer programes for modelling radars, as shown, e.g. by Chapters 4, 5, 6 and 10.

The modifying factor takes into consideration the reflection coefficient of the reflecting surface, the phase shift on reflection, the phase difference due to the difference in propagation path lengths of the direct and reflected ray, the aerial polar diagram, and the roughness of the surface to account for dispersion.

It may be shown that the number of lobes n in the vertical polar diagram is a function of the aerial height h and of the wavelength of the transmission λ,

$$n = 2h/\lambda. \tag{9.1}$$

It is directly proportional to the first and inversely proportional to the second. On the basis of this, since the wavelength is determined by the designer, the only way of increasing the number of lobes and thus making the areas of low detection probability small, is to elevate the aerial or to erect it on the edge of a cliff facing in the direction of the most likely target approach. However, with radars operating at wavelengths of a few cm, the aerial height above the reflecting surface need only be of the order of 3 m (or greater) for the lobe structure to become very fine and therefore negligible to a first approximation.

The above will illustrate the importance of ensuring adequate height of the radar aerial above the surrounding terrain. However, careful consideration of the contours about the radar site will also be worthwhile, as the following example will show. (It is sometimes even possible to obtain a focusing effect from the ground contours.) Also very careful attention needs to be paid to the angle of elevation at which the lowest edge of the coverage diagram (which is dependent partly on the terrain) cuts off. The terrain reflections, together with atmospheric refractivity variations, are the main determinants of coverage on low altitude targets which is now increasingly important operationally.

Besides elevating the aerial Norton and Omberg (1947) recommend siting the aerial on a slope (if available), as shown in Fig. 9.1, towards improving low angle cover. The height of the aerial above the slope is to be

$$h_1 = n\frac{\lambda}{4\sin\theta'}, \tag{9.2}$$

where n is an odd integer and θ' is the ray elevation angle as shown in the figure. The height h_1', at which the aerial is to be placed above the bottom of the slope is

$$h_1' = m\frac{\lambda}{4\sin\theta}, \tag{9.3}$$

where m is also an odd integer. Such sites are of importance for coastal radar particularly.

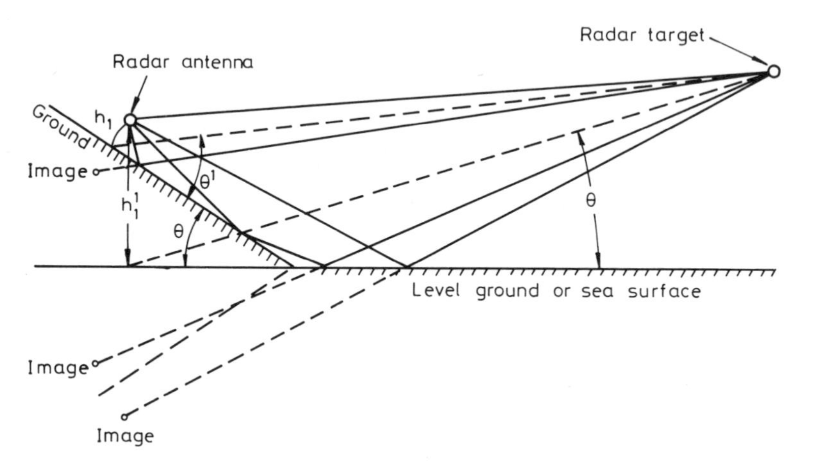

Fig. 9.1 *Increasing the number of ground reflection (After Norton and Omberg, 1947) (Copyright, 1947, IRE now IEEE)*

The number of ground reflected waves may be tripled in this manner and, if all combine in phase, the energy received from a distant target may be increased almost 16 times or by 12 dB. The maximum range for targets at certain angles, at which in-phase combination occurs, may often thus be increased to nearly twice that obtainable without the slope. However, due to multiple phase shift on reflection, the lobing pattern will, in general, differ from that obtained with a single reflection.

Since the footprint of the radiation is not a point, as is often assumed for theoretical studies, but larger in area, the slope must be sufficiently extensive to support at least one complete Fresnel zone (see Chapter 6). The size of this is a function of the aerial height and wavelength. Further, in order to prevent the occurrence of large path length differences causing unacceptable phase shifts between the energy components reflected from the area of the Fresnel zone, the permissible surface height deviation for an aerial height h and a well developed kth lobe is

$$\Delta h = \frac{h}{4(2k-1)}. \tag{9.4}$$

9.3 The calculation of ground clutter effects

9.3.1 Land clutter effects

At the frequencies at which most search radars operate, it is imperative that the line of sight from the aerial to the target be not interrupted by terrain features. These obstruct the visibility to the targets behind them and cause clutter on the display. It is important therefore to survey the terrain all around the radar site and choose this so as to permit detection at the greatest range, preferably out to the maximum unambiguous range, given by

$$R_u = \frac{c}{2\ \mathrm{PRF}}. \tag{9.5}$$

Here, c is the velocity of propagation of electromagnetic waves and PRF is the pulse repetition frequency.

It is further important to remember that, due to the variation of the index of atmospheric refraction with height, the 'rays' do not propagate in straight lines but, even in standard atmospheric conditions, are bent downwards. It is possible therefore to obtain echoes from terrain features obstructed from direct view and very flat sites are subject to this disadvantage. To make a first correction to this, when drawing terrain profiles in any radial direction from the aerial, the terrain feature heights read from maps are to be adjusted by using the corrected earth radius appropriate to the most frequent atmospheric refraction at the area of the radar's operation.

The corrected earth radius may be expressed by (see eqn. (5.11))

$$a = \frac{r}{1 + r\dfrac{dn}{dh}} \tag{9.6}$$

where r is the true radius of the earth and dn/dh the gradient of the index of refraction with respect to height h above the earth. The corrected earth radius for a standard atmosphere is 4/3 of the true radius.

Wenisch (1971) suggests a method for the advance determination of ground echoes from terrain features without vegetation or with immobile vegetation. If the vegetation is agitated by wind, the echo will consist of a fixed part and an echo with a varying amplitude superposed onto it. The amplitude distribution of the resultant fluctuating echo will depend on the relative values of the fixed echo and of the fluctuating one.

The instantaneous value of a signal received at a time t after the start of a transmitted pulse is the sum of the elementary echoes due to those parts of the terrain illuminated by the beam between the distances $ct/2$ and $c(t+\tau)/2$, where τ is the pulse width and c the velocity of propagation. These two distances define on the terrain a segment of an annulus centred at the radar and containing the terrain features which contribute to the echo; Wenisch calls this the annulus of confusion.

It is to be borne in mind, that only those terrain features which are illuminated by the radar will contribute to clutter. Their contribution will be proportional to the square of the aerial gain in their direction. While it is customary to consider only those features which lie within the half-power beamwidth of the aerial, one must not neglect the possibility of echoes from strongly reflecting surfaces which may be illuminated by sidelobes or which may be at the edges of the main lobe and contribute as strong an echo, if not stronger, as the terrain features directly illuminated by the main lobe. The contribution of terrain features illuminated only by the energy diffracted by features which obstruct their direct illumination is negligible. (However, refraction may be important.)

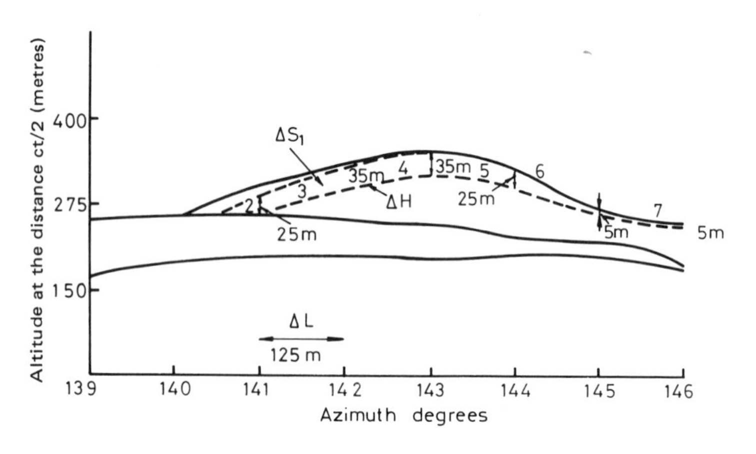

Fig. 9.2 *Panoramic view of the terrain (After Wenisch, 1971)*

The incident energy received from the radar by a terrain cell of surface area S is directly proportional to its projection S_1 as seen by the radar (Fig. 9.2 and 9.3). The energy backscattered by this surface equals the above energy multiplied by the coefficient of reflection of the terrain ρ, which has a value of less than 1 and takes into consideration absorption losses. ρS_1 is then the radar cross-section of the terrain. The ratio of this radar cross-section and the actual terrain surface cross-section, known as clutter reflectivity, is to be denoted by

$$\sigma_0 = \frac{\rho S_1}{S}; \tag{9.7}$$

it depends on the angle ψ at which the radar rays illuminate the terrain and on the exact definition of the surface S.

If S is the true surface and not its horizontal projection, one may write

$$\sigma_0 = \rho \sin \psi. \tag{9.8}$$

The energy reflected by a surface S_1 is the vectorial sum of energies reflected by the elementary surfaces ΔS_1 of the surface S_1. One common model assumes that these have random phases with a Rayleigh probability distribution. In reality, for a given small area it is often true that the elementary physical reflectors are too regular or too small with respect to the wavelength to obey Rayleigh's law, but since these elementary reflectors appear in such quantity in the observed cell, Rayleigh's law usually applies to their ensemble.

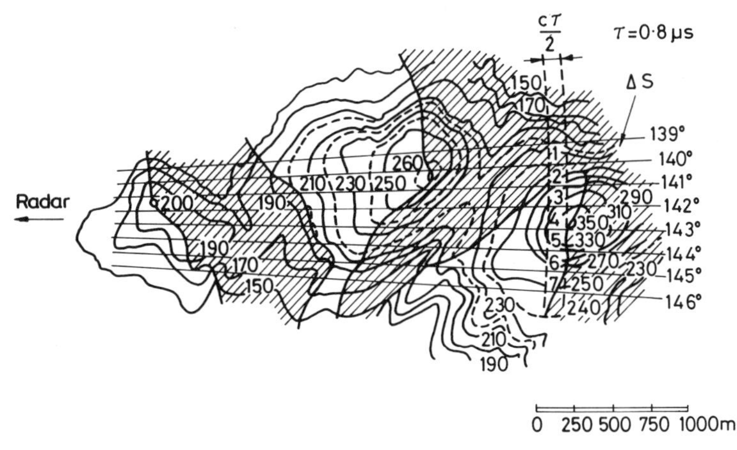

Fig. 9.3 *Map of the terrain of Fig. 9.2 (After Wenisch, 1971)*

In some cases, one of the elementary reflectors has an appreciable strength with respect to the other members of the ensemble, or the configuration of the elementary reflectors is regular rather than random. In this case, Rayleigh's law does not apply and one speaks of 'specular' echoes. These may originate on an artificial feature, e.g. roads, buildings, etc., or in a regular terrain feature such as a steep shore, beach or snow field.

Rayleigh's law also determines the probability that the amplitude of the resultant echo will be greater or smaller than a certain value referred to the square of the sum echo of the elementary echoes. The Rayleigh probability distribution is shown in Fig. 9.4 from which one may determine, e.g. that there is a 0·1 probability that it will be 3 dB above the sum squared echo of the elementary echoes ($m = 0$). The 0 dB level in this figure corresponds to the echo from an apparent surface to which echoes from elementary surfaces contribute. If s', one of the elementary surfaces, is more prominent than the others denoted by s_1, the probability curves are determined by the non-zero values of the parameter $m = s'/\Sigma s_1$. The value of $m = 0$ applies for a diffuse echo.

The likely values of the echo in a given direction may be obtained by multiplying the elementary radar cross-sections by the squared aerial gain in this direction and their position relative to the gain maximum $(G/G_{max})^2$, by zero or one depending on whether the surface is hidden or illuminated and by $\rho \sin \psi$ depending on the nature of the terrain slope and the angle of elevation of the radar beam.

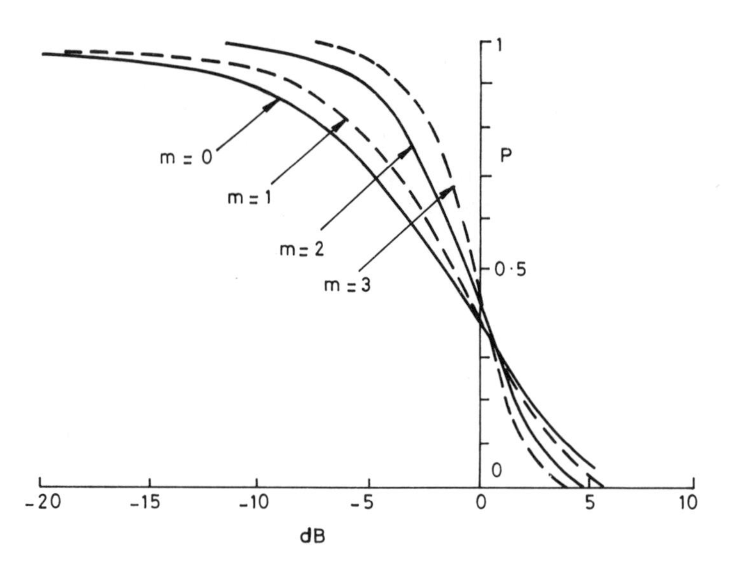

Fig. 9.4 *Cumulative probability distribution of a ground clutter (After Wenisch, 1971)*

The probable difference between the predicted and real values of the echo given by Rayleigh's law is shown by curve $m = 0$ in Fig. 9.4. This applies strictly to a motionless terrain yielding a truly fixed echo. If one could measure real echo values of a featureless terrain zone, one would find an amplitude distribution in agreement with Rayleigh's law.

Due to the influence of the wind, some reflectors will oscillate with respect to their mean position, while others will remain fixed. The resultant echo from a given point will fluctuate in time and its amplitude probability distribution will not, in general, be Rayleigh. It will depend on the relative value of the fixed and fluctuating parts and is also illustrated by the curves marked $m = 1, 2, 3$ in Fig. 9.4.

The fixed part of the same terrain may have a strong or weak value according to the probabilities of Rayleigh's law, while the mean square value of the fluctuating part remains the same. The resultant distribution laws differ therefore according to the value taken by the fixed part.

The following method of ground echo prediction was used by Wenisch (1971) for the prediction of echoes shown in Figs. 9.6 and 9.7.

(*a*) One determines first the visible and obscurred parts of the terrain as viewed from the radar's aerial by using topographic maps of the environment and the usual graphical methods.

One should use the diagram of the earth's curvature corrected for the radar rays if justified by the distance of the studied zones. The exact aerial height must be kept in mind and the terrain height should be corrected for the height of vegetation (e.g. forrests) and also for the crests of masking, and shaded zones. The illuminated and hidden zones are thus delineated on the map (Fig. 9.3).

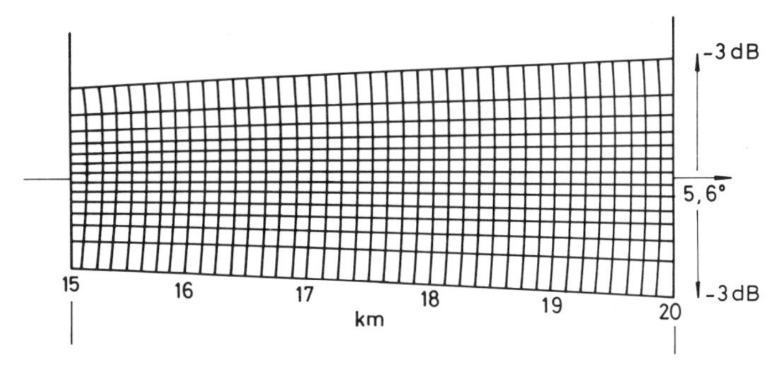

Fig. 9.5 *Protractor for $\tau = 0{\cdot}8\ \mu s$ and a $5{\cdot}6^\circ$ half-power beamwidth (After Wenisch, 1971)*

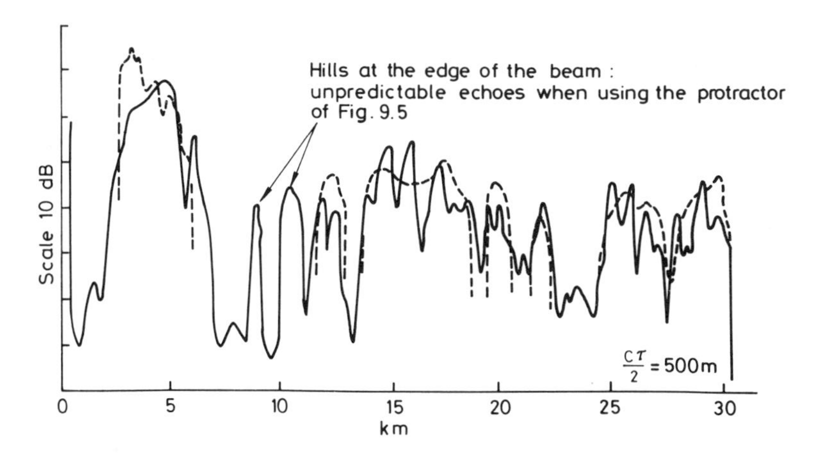

Fig. 9.6 *Amplitude plot of echoes received on a PPI screen (full lines) and of the amplitude predicted on the basis of a topographic map (dashed lines) in a given direction (After Wenisch, 1971)*

(*b*) One traces on the map of the illuminated and hidden parts the $c\tau/2$ wide annulus of confusion at the sought distance and one notes the aerial gain at each point of the annulus.

This may be done with the aid of a protractor (Fig. 9.5) marked in the map scale. The protractor is limited in width by the half-power beamwidth and in length by the ranges of 15 and 20 km, in this example. It is divided into strips of a width corresponding to the pulse width ($\tau = 0{\cdot}8\,\mu s$ in the illustration).

(*c*) One divides the zone of confusion into rectangular elements sufficiently small to permit the assumption that they are homogeneous as far as aerial gain, terrain slope and vegetation is concerned (Figs. 9.2 and 9.3). Since the radar ray is nearly enough horizontal, one may determine the frontal surface S_1 of each square element by multiplying the width ΔL of the element between the right and left side by the difference in height ΔH between the near and far side as seen by the radar. The heights may be read from the map height-contours. If a surface element is hidden from the aerial, a zero value is assumed for the appropriate ΔS_1.

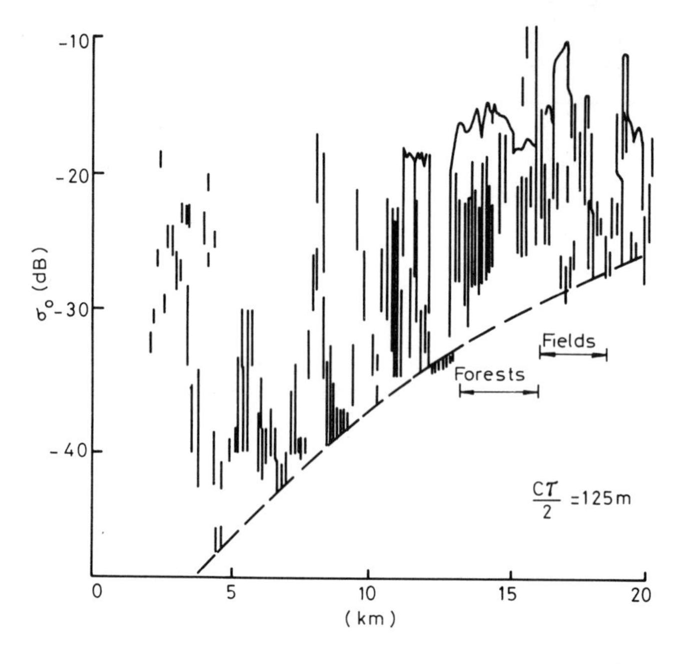

Fig. 9.7 *Plot of echo amplitude variation (vertical lines) reduced to σ_0 in each 125 m cell in a given direction. The curve of the σ_0 predicted on the basis of the map (plain curve) with an assumed $\rho = 1$ (After Wenisch, 1971)*

(*d*) The so-determined ΔS_1 value is then multiplied by the reflection coefficient ρ of the terrain (obtainable from standard texts) and by the square of the relative aerial gain $(G/G_{max})^2$ in the considered direction.

The sum of the surfaces $\Delta S_1 \rho (G/G_{max})^2$ over the annulus of confusion yields the apparent radar cross-section which permits the calculation of the echo signal strength from the radar equation. The probabilities of the real echoes with respect to those predicted, are given by Rayleigh's law (Fig. 9.4, $m = 0$) for featureless country.

A good agreement between the measured and predicted echo amplitudes has been achieved by this method in case of clutter due to wooded hills, as shown by Figs. 9.6 and 9.7.

9.3.2 Sea clutter effects

So far, the treatise has dealt with the site selection from the point of view of land clutter only. Although, clearly, sea clutter characteristics cannot affect site selection except perhaps in the determination of the height of the radar above sea level, estimates of sea clutter strength are nevertheless important in evaluating likely radar performance at a site looking over the sea.

Sea clutter, i.e. backscattering from the surface of the sea, often limits the ability of a radar to detect targets, especially targets near or on the sea surface. Its strength and extent depends on a number of parameters, e.g. wind velocity and aspect angle, wave height and the wavelength and angle of incidence of the electromagnetic radiation.

Many investigations have been performed in an effort to arrive at acceptable formulae for clutter reflectivity σ^0. The empirical expressions derived on the basis of a great number of measurements and statistical data are of the general form (Boring *et al.*, 1957)

$$\sigma^0 = CW^a \psi^b \lambda^i 10^{(k \cos\theta + mK)}$$

where C, a, b, i, k and m are constants dependent on the frequency and polarization of the radiation, W is the wind velocity, ψ the grazing angle, θ the wind aspect, K the sea wave height and λ the wavelength of the electromagnetic radiation.

The dependence of clutter reflectivity on wind speed for vertical polarization for various angles of incidence, according to Grant and Yaplee (1957), is shown in Fig. 9.8. (Note that the angle of incidence here is the angle made by the ray with the vertical.)

Reversal of polarization, e.g. from vertical on transmission to horizontal on reception, or from right to left circular polarization, reduces the relative back-scatter; however, since the target's echoing area is similarly affected, little is gained by the application of this technique in radars aiming at detecting low-flying or surface targets.

Clutter reflectivity is a dimensionless quantity which must be multiplied by the area intercepted by the radar's radiation, the clutter cell, in order to yield the clutter cross-section.

It is a constant only for a given grazing angle, radar frequency, polarization, pulse length and wind conditions. Tables of sea clutter reflectivity for vertical and horizontal polarization and various grazing angles and frequencies are available in Nathanson (1969) together with tables of land clutter reflectivity for various terrain types.

The beamwidth-limited clutter cell, used for the calculation of the clutter echoing area, for ranges for which

$$\tan\psi > \frac{\tan\phi_2 R}{c\tau/2} \tag{9.9}$$

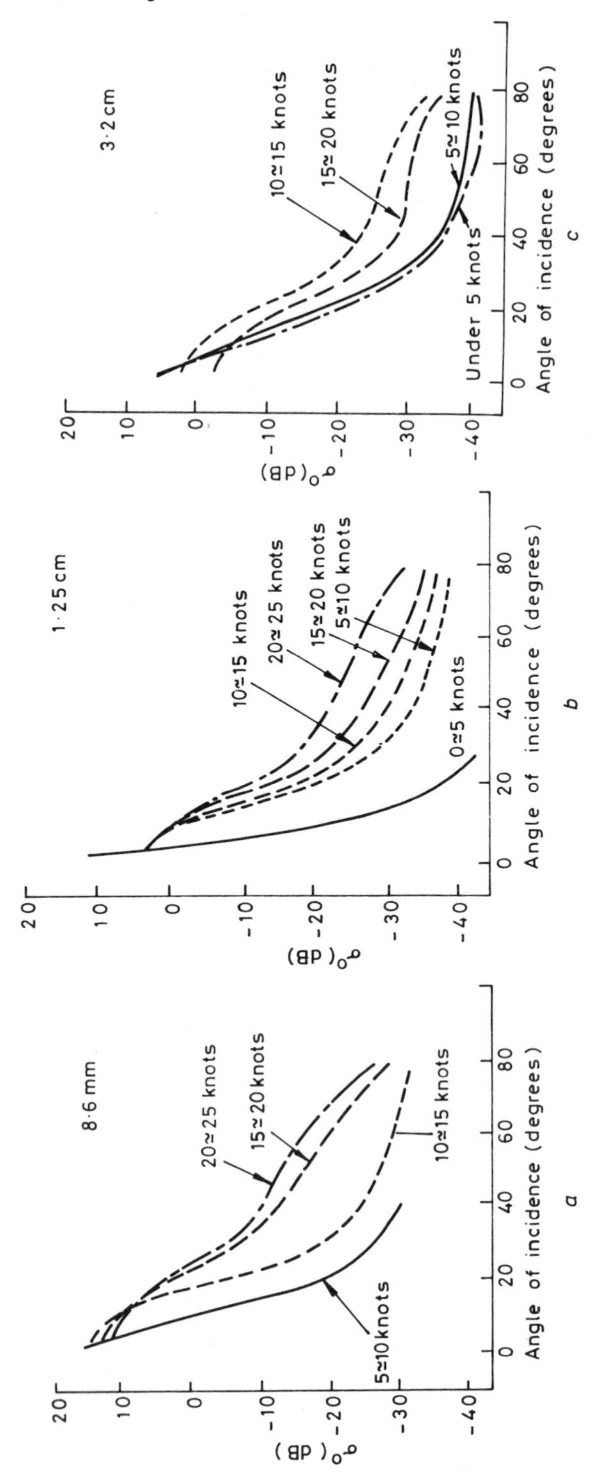

Fig. 9.8 *σ^0 as a function of wind speed for vertical polarization (After Grant and Yaplee, 1957) (Copyright, 1957, IRE now IEEE)*
(*a*) $\lambda = 8{\cdot}6$ mm
(*b*) $\lambda = 1{\cdot}25$ cm
(*c*) $\lambda = 3{\cdot}2$ cm

is, according to Nathanson (1969), given by

$$A_c = \pi R^2 \tan\frac{\phi_1}{2} \tan\frac{\phi_2}{2} \operatorname{cosec}\psi, \tag{9.10}$$

while the pulse-length-limited clutter cell, used when

$$\tan\psi < \frac{\tan\phi_2 R}{c\tau/2}, \tag{9.11}$$

is

$$A_c = 2R\frac{c\tau}{2} \tan\frac{\phi_1}{2} \sec\psi. \tag{9.12}$$

The meaning of the symbols in the above expressions is as follows: ψ is the grazing angle of the electromagnetic radiation at the surface of the sea, ϕ_1, ϕ_2 are the 3 dB horizontal and vertical beamwidth, respectively, R is the range, c is the velocity of propagation of electromagnetic waves, and τ is the pulse width in seconds.

The most often used formula for clutter reflectivity is due to Barton (1969); it represents the approximate average for all polarizations and wind directions. Above a so-called 'critical angle', which Katzin (in Skolnik, 1970) defines as

$$\psi_c = \sin^{-1}\frac{\lambda}{2K_w}, \tag{9.13}$$

where K_w is the sea wave height exceeded by 10 per cent of the waves, the clutter reflectivity is

$$\sigma_{DB} = -64 + 6K_B + 10\log\sin\psi - 10\log\lambda. \tag{9.14}$$

Below the 'critical angle' this expression must be decreased by

$$40\log\frac{\psi_c}{\psi}, \tag{9.15}$$

where K_B is the Beaufort wind scale.

Since sea clutter value, in general, is more a function of weather conditions than site, choice of the latter will not be strongly dependent on sea clutter calculations. However, it is clear that observed clutter at a given radar will, in general, be dependent, for a given range, on the height of the radar above sea level, and therefore, in some cases, sea clutter estimates as a function of height may be worth making, towards the choice of the optimum height above sea level for the radar aerial.

Statistical data on prevailing winds and sea currents may assist in predicting sea clutter.

9.4 Meteorological effects

Consideration of meteorological effects can usually have only a secondary effect on site selection. It would obviously be undesirable to locate a microwave radar in an area where rainfall was exceptionally heavy due to a localized geographic situation, if other sites outside the area were equally acceptable. However, in most situations the meteorological environment will be broadly unchanging over a comparatively large region. In assessing the future performance of a radar on a selected site it is necessary to take into account the local meteorology, since this will have very significant effects on coverage and sensitivity of a search radar, particularly for low elevation targets.

The effects which are of interest are rain, hail, fog and effects which affect the atmospheric refractivity and may cause super-refraction or sub-refraction (see Chapter 5).

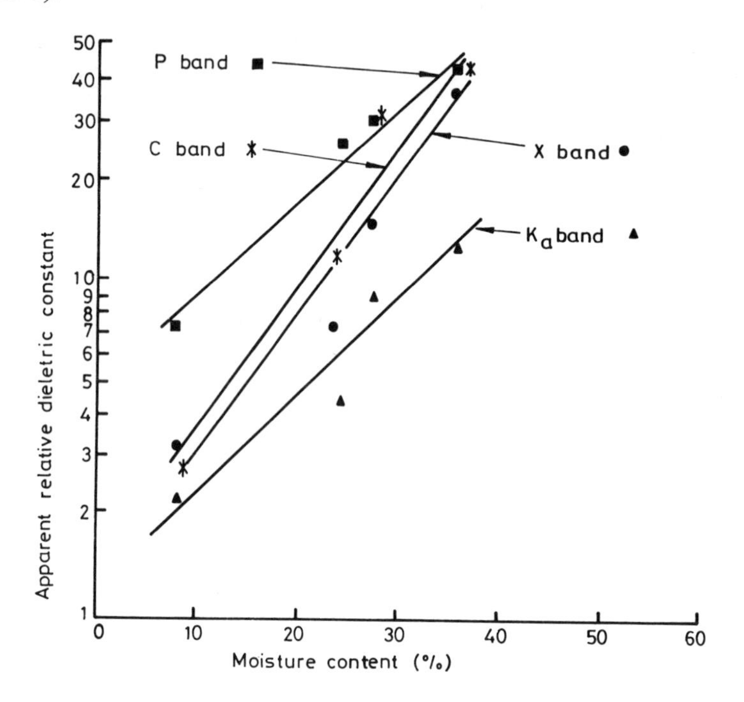

Fig. 9.9 *Apparent relative dielectric constant vs. moisture content of silt loam (After Lundien, 1966, as shown by Skolnik, 1970)*

9.4.1 Rain, hail and fog

It has been mentioned earlier that lobe formation in the vertical polar diagram of a radar aerial is due to the reflection of part of the radiated energy from the ground. The intensity of the field due to the reflected energy is a function of the complex dielectric constant

$$\epsilon_R = \epsilon - j60\sigma\lambda \tag{9.16}$$

of the reflecting medium and of the angle of incidence (Chapter 6). In this expression ϵ is the dielectric constant and σ the conductivity.

Since both of these will change with moisture, it is obvious that a change in the field distribution will occur after a drenching rain. Figure 9.9 shows the variation of the apparent relative dielectric constant of silt loam with moisture content as given by Lundien (1966) and shown by Skolnik (1970). The effect of such changes is taken into account in the computations described in Section 9.2.

However, this is not the only influence rain has on radar performance.

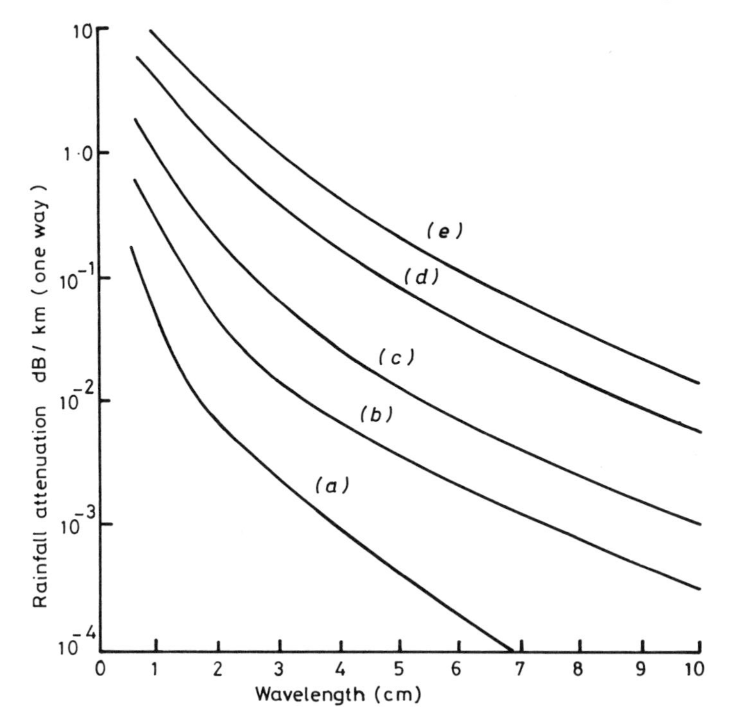

Fig. 9.10 *One-way attenuation (dB/km) in rain at a temperature of 18°C (After Skolnik, 1962)*
(*a*) drizzle — 0·25 mm/h
(*b*) light rain — 1 mm/h
(*c*) moderate rain — 4 mm/h
(*d*) heavy rain — 16 mm/h
(*e*) excessive rain — 40 mm/h

Attenuation due to absorption by water vapour and attenuation and scattering by raindrops will reduce the received signal power. The ability of a radar to detect targets in inclement weather is then decreased due to attenuation of the signal in the path between the radar and the target and due to clutter, which obscures wanted signals, from water particles in the vicinity of a target. Figure 9.10 shows one way attenuation by rain (Skolnik, 1962), while Figure 9.11 for example shows, for frequencies of about 35 GHz, the relative back-scatter power as a function of rainfall rate (Nathanson, 1969).

While figures read from the graphs are roughly indicative of the attenuation one may expect from rain, a possible inaccuracy has to be borne in mind, since our knowledge of drop-size distribution in rains of varying rates of fall under differing climatic and weather conditions is very limited. However, studies of this problem seem to indicate that a certain most probable drop-size distribution can be attached to rain of a given rate of fall.

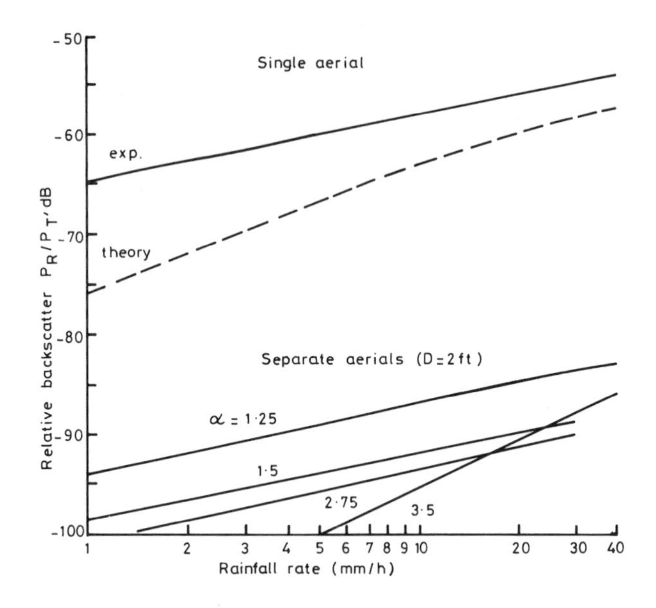

Fig. 9.11 *Relative backscatter power as a function of rainfall rate for a K_a-band CW radar in uniform rain; single and separate aerials (After Nathanson, 1969)*

A further complication in the determination of attenuation by rain is introduced by the dependence of the dielectric properties of water on temperature which will affect the total attenuation cross-section. Tables of correction factors for rainfall attenuation at various temperatures may be found in Skolnik (1970).

The volume reflectivity of precipitation per metre, η_V, may be computed, after Barton and Ward (1969), from

$$\eta_V = 6 \times 10^{-14} r^{1 \cdot 6} \lambda^{-4}, \tag{9.17}$$

where r is the rainfall rate in mm/h and λ is the wavelength of the radiation in m.

The clutter volume V_m is given by (Skolnik, 1962) as

$$V_m = \frac{\pi}{4} R^2 \phi_1 \phi_2 \frac{c\tau}{2}, \tag{9.18}$$

where the symbols have the same meaning as those used in eqns. (9.9) to (9.12).

Rain clutter will fluctuate because of the changing position of raindrops. For rain above a rough sea, the echo will have fluctuations also because of changes in the phase and amplitude of the various polarized components of a forward scattered electromagnetic wave. A rigorous solution of this problem, due to its complexity, is rather difficult, especially since not enough data are available on the polarization of the scattered electromagnetic waves.

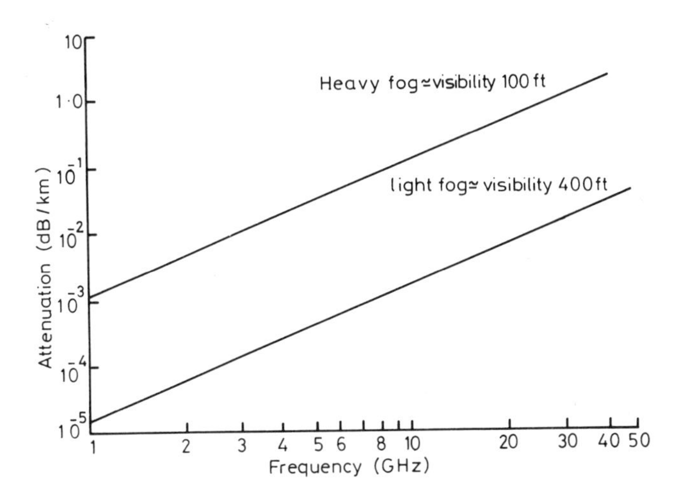

Fig. 9.12 *One-way attenuation (dB/km) of electromagnetic waves in fog at a temperature of 18°C (After Skolnik, 1962)*

The attenuation caused by hail is about one-hundredth of the attenuation caused by rain (Skolnik, 1970). Ice crystal clouds and snow, even at the rate of fall of 125 mm per hour, cause very little attenuation. However, the scattering by spheres surrounded by a concentric film of a different dielectric constant, e.g. melting ice spheres, may reach up to 90 per cent of the value of an all-water drop.

Since clouds and fog consist of water droplets, they also influence the detection capability of radars. Figure 9.12 is the one-way attenuation of electromagnetic waves by fog at a temperature of 18°C (Skolnik, 1962).

9.4.2 Atmospheric refractivity

Theoretical studies of electromagnetic wave propagation avail themselves often of the methods of geometrical optics. This applies particularly to studies of ray path investigations.

The path followed by a ray in the atmosphere depends on the gradient of the refractive index along the path. The horizontal gradient is usually negligibly small, so that it is mainly the vertical gradient which affects the ray. The numerical value of the vertical gradient of the refractive index depends on the vertical distribution of atmospheric temperature, humidity and pressure. Normally, in a turbulent atmosphere, temperature and humidity decrease with height. However, there are periods of time when the air is fairly calm and temperature inversions can be widespread in area and persist over a relatively long period of time. They may exercise a stabilizing influence on air motion such that turbulence is suppressed and strong humidity gradients may develop.

Super-refraction, due to the mentioned effects, creates ducts which may deflect, guide or trap electromagnetic waves and thus influence the detection range to a great extent. Ducts may start at ground level or at some greater height; there is a great variability in the thickness of the duct (Chapter 5).

Duct formation is frequent on sea shores where dry air from above a warm land surface may flow out over a colder sea with moist air above so that air with a different refractive index is brought into the propagation path.

Temperature inversions which may occur when the hot ground cools faster than the upper air during the night so that an atmospheric stratification is formed above the ground, may cause trapping of electromagnetic radiation near the ground. These inversions may be due also to the subsidence of air in the wake of a high-pressure system since the sir subsiding from a high level in the atmosphere is dry, and may also be warm, while that at lower levels, especially over the sea, is cooler and more humid.

Other meteorological conditions, e.g. fog formation, reduce the possibility of duct formation and may lead to sub-refraction.

Super-refraction may extend the detection range of a radar well beyond the radio horizon and cause clutter by echoes from land masses several hundred km away (Booker, 1946).

Effects due to super-refraction and sub-refraction are generally called anomalous propagation. Detailed consideration of the implications for radar siting of the prevalence of super-refraction in the area in which a site is to be selected is outside the scope of this treatise. For further details the quoted references are recommended. The influence of meteorological conditions on low-angle cover is, however, considerable, and serious consideration of such conditions for proposed sites is necessary if low-angle cover is of operational importance.

9.5 Other clutter effects ('Angels')

A further type of clutter, usually referred to under the name of 'angels' may also appear on the PPI display (Ottersen, 1970). Two types of 'angels' are known. These are the 'layer angels' and 'dot angels'. The first type are persistent, diffuse, layer type echoes due to backscattering from clean-air refractive index perturbations associated with free convection, or with turbulent mixing induced by wind shear in zones of enhanced static stability. 'Dot angels' are short duration, coherent echoes, from apparent point targets, usually attributed to birds or insects. This latter type of angel is particularly troublesome for ground radars looking over the sea and ship-borne radars operating near the coast. Although the radar cross-section of a single bird is small compared with that of an aircraft, bird echoes may be relatively strong at shorter ranges because of the inverse fourth-power variation with range.

Insects, though small, may be detected by radar. Extremely heavy angel activity may be produced by insect concentrations scarcely detectable visually.

Dot angels are more likely to be observed at lower altitudes at dawn and twilight.

In general, in present conditions, the importance of clutter due to radar angels is secondary to that due to ground and rain effects, and the possibility of angel clutter is usually neglected in radar siting.

9.6 Résumé

In an effort to give guide-lines and background information for radar site selection, the effect of terrain features and other phenomena affecting the performance of radars have been briefly summarized. Of these only terrain features in the area of a possible future radar site can be fully known. Long term observations may provide statistical data on the other, particularly meteorological effects which, if available, should also be considered. A detailed map of the intended site topography is an important and necessary prerequisite. The technical factors affecting site selection given above imply a logical sequence of investigations to be taken if successful operation of the radar is to be assured. These may be summarised as follows:

(*a*) Construct a protractor (Fig. 9.5) for the radar of interest.

(*b*) Obtain a detailed topographic map of the terrain surrounding the intended site.

(*c*) According to Section 9.3.1 determine and mark on the map those parts of the terrain which are illuminated by and hidden from the radar ray originating at the proposed aerial height (see Fig. 9.3).

(*d*) With the aid of the protractor trace on the map the $c\tau/2$ wide annulus of confusion at the range of interest and note the aerial gain at each point of the annulus.

(*e*) Divide the zone of confusion into rectangular elements (Figs. 9.2 and 9.3) small enough to permit the assumption of homogeneity in aerial gain, terrain slope and cover by vegetation.

(*f*) Determine the frontal surfaces ΔS_1 by multiplying the width of the individual elements by the difference in height between the near and far sides of the elements, as read from the map contours.

(*g*) Multiply the above surfaces by the coefficient of reflection (computed for constants obtained for the type of the illuminated terrain, from standard text – or handbooks) and by the square of the relative aerial gain. The sum of these products over the annulus of confusion, the apparent radar cross-section, when inserted into the radar equation, permits the calculation of the land clutter signal strength.

(*h*) On the basis of statistical data, if available, estimate the influence of sea clutter (Section 9.3.2) and of meteoreological phenomena (Section 9.4) on the radar performance.

(*i*) Check whether there is a more suitable site in the vicinity.

Alternatively, feed the terrain data obtained from the map into the MAP model (Section 10.2) and assess the suitability of the proposed site by mathematical modelling.

Chapter 10

Mathematical models for radar performance assessment

Three mathematical models developed for radar performance assessment have been used for obtaining the results presented in the following chapter.

This chapter is devoted to the description of the models.

The model which computes and plots the vertical coverage diagram will be referred to as COVER, while that producing map contours will be called MAP and the third one which computes and plots the dependence of the probability of detection on range will be known as RDPRO for short. All three models use the radar–ground–target geometry described in Chapter 5 together with relations given in Chapter 6 for the determination of quantities connected with electromagnetic wave propagation. Formulae of Chapter 3 are used for signal-to-noise ratio and detection probability calculations. While COVER and MAP compute the instantaneous signal level for a given geometry and detection probability, model RDPRO computes the signal strength and probability of detection. This is the most comprehensive of the three models as it will be shown in the following.

Besides computing and plotting the vertical coverage diagram COVER may be used for the derivation of algorithms to describe the free-space coverage diagram of an aerial. MAP has also two applications. It plots contours delineating the ranges at which targets at specified heights may be detected with a chosen probability and it aids in site selection when supplied with topographical data of the terrain surrounding the location of interest. The third model serves as a tool for the thorough examination of a radar's performance against a given target moving along a specified trajectory.

10.1 The COVER model

As it was mentioned earlier this model serves either as a tool for the derivation of algorithms approximating the gain versus ray elevation angle dependence of aerials or for computing and plotting the vertical coverage diagram of radars.

A number of input data is required for the computation and plotting. A glossary of their code-names in order of their appearance is in Table 10.1. Figure 10.1 shows the flow-chart of the model, while Appendix A provides a listing of the model together with that of the mentioned algorithms of the Adelaide airport's radar aerial as derived by Rohan.

10.1.1 Input data

A few words about the input data follow.

Chapter 3 of this book contains, amongst others, Fig. 3.4 taken from DiFranco and Rubin (1968) which supplied data V and W used as a look-up table in subroutine PHIPFA the listing of which is shown in Appendix A by Fig. A.2. It is used for the computation of the detection threshold for a given probability of false alarm. The values of the variables V and W, listed in Figs. A.6(*a*) and (*b*) represent P_{fa}, the probability of false alarm, and ϕ^{-1} (P_{fa}), the inverse of eqn. (3.94), respectively. The format statements for these are given in the listing, Fig. A.1, by lines 16 resp. 14.

Table 10.1 *Glossary of COVER input data code names*

Code name	Description
V, W	Quantities read from Fig. 3.4
L	$L = 1$ for free space coverage diagram, $L = 2$ for the lobing pattern
GRAD	Gradient of the atmospheric refractive index (NBS constant C_e)
REF	Surface refractivity (NBS constant $N_s \times 10^{-6}$)
PERM	Relative permittivity of the ground
RHO	Conducitvity of the ground, mho/m
ATM	Atmospheric attenuation, dB/km (for frequencies below 1 and above 10 GHz only, see text)
UNDUL	Terrain irregularities in height, m
SHORE	Distance to the sea shore, km
AERSEA	Height of aerial base above sea level, m
SEA	Sea wave height, m
IPOL	= 1 for vertical polarization = 2 for horizontal polarization = 3 for circular polarization
AERHIT	Aerial height, m
HBMWTH	Horizontal beamwidth of the aerial, degrees
VBMWTH	Vertical beamwidth of the aerial, degrees
TILT	Aerial tilt, degrees
ASR	Angular scan rate (see comment in listing)
RPM	Aerial scan rate, r.p.m.
SCA	Aerial scan angle, degrees (see comment in listing)
PKW	Transmitter peak power, kW
FREQ	Frequency, MHz
FDB	Receiver noise figure, dB
BMHZ	Receiver bandwidth, MHz; the expanded bandwidth when pulse-compression is used,

Table 10.1 *(Cont.)*

ELOSS	Plumbing and transmission line losses, dB
FIF	Intermediate frequency of the receiver, MHz
PRF	Pulse repetition frequency, pps
AEGDB	Aerial gain, dB
PFA	Probability of false alarm (e.g. 1.0E–06)
PRODET	Probability of detection, %
SIGMA	Target echoing area, m^2
SLST	Distance to the beginning of the terrain slope, km
TERHTS	Height above sea level of the above location, m
SLF	Distance to the end of the terrain slope, km
TEHF	Height above sea level of the above location, m
K	Target type selector $K = 1$ for a Swerling I target $K = 2$ for a Swerling II target $K = 3$ for a Swerling III target $K = 4$ for a Swerling IV target. $K = 5$ for a non-fluctuating target
NOPER	Integration improvement factor selector (see comment in listing)
SIT	Scan integration time, s
PCR	Numerator of the pulse compression ratio (1 when no pulse compression is used)
ANGRD	Gradient of the atmospheric refractive index for anomalous propagation conditions
ANREF	Surface refractivity for anomalous propagation conditions
HD1, HD2	height of duct near the radar resp. at a remote location, m
DMS	Distance between the above two locations, km
JAM	Jammer type selector (see comment in listing)
PJ	Jammer peak power, kW
GJ	Jammer aerial gain, dB
BJ	Jammer bandwidth, MHz
DJ	Jammer losses, dB
RJ	Distance of jammer, km
ANTJAM	Height of jammer aerial, m
DECLJ	Direction of jammer, angle between the radar-jammer and radar-target lines, degrees
ICIR	Circular polarization sense selector, ICIR = 0 for opposite } sense circular polarization ICIR = 1 for same } sense circular polarization
TITLE	The modelled radar's designation

The fixed point constant L selects either the computation and plotting of the free-space radiation pattern or the lobing pattern depending on its being chosen to be equal to 1 or 2. Its format is that given by line 18 of the listing in Fig. A.1.

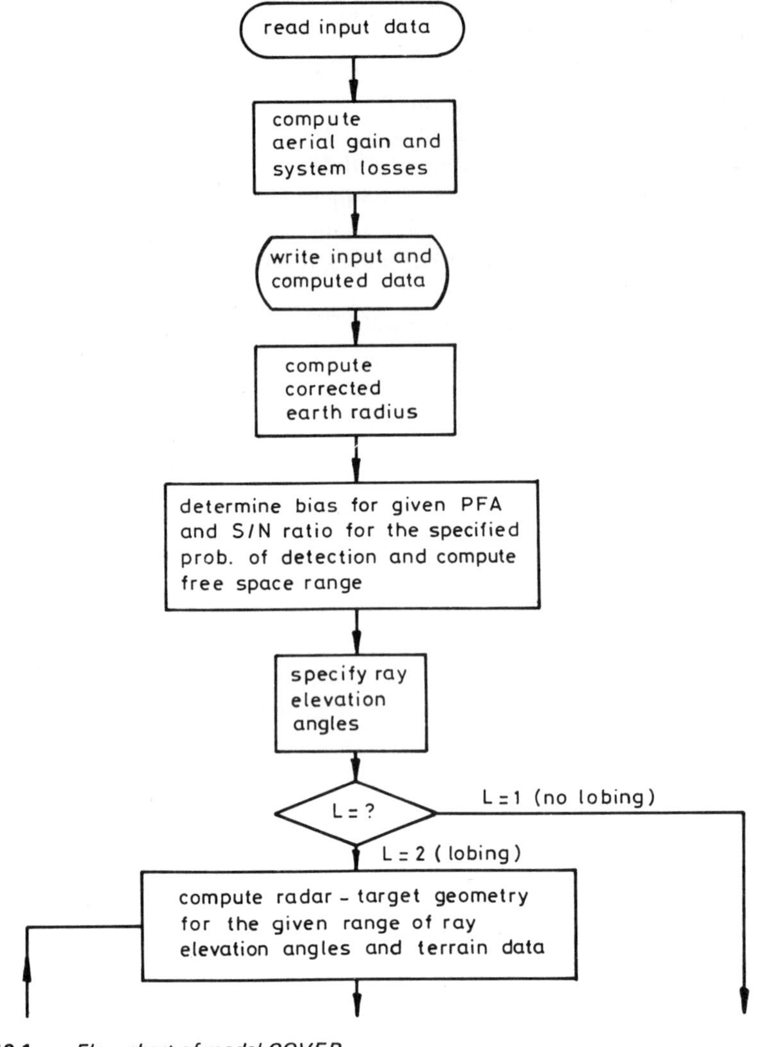

Fig. 10.1 *Flow-chart of model COVER*

GRAD is the gradient of the refractive index of the atmosphere, as given by the NBS constant c_e (Bean and Dutton, 1966). It was taken to be equal to $13{\cdot}575\,887\,40 \times 10^{-2}$/km for a 'standard' atmosphere.

REF is the surface refractivity, equivalent to NBS's constant $N_s \times 10^{-6}$ which, as used by Rohan, is $2{\cdot}890\,362\,74 \times 10^{-4}$ for a 'standard' atmosphere.

The formats of these two quantities are shown by lines 21 and 24 of Fig. A.1 respectively.

PERM, the relative permittivity and RHO, the conductivity, of the ground are used for the computation of the magnitude and phase of the reflection coefficient (Chapter 6) and their values are obtained from data published for various types of ground in, e.g. Kerr (1951). The format statements of these two input data are those given by lines 26 resp. 28 of Fig. A.1.

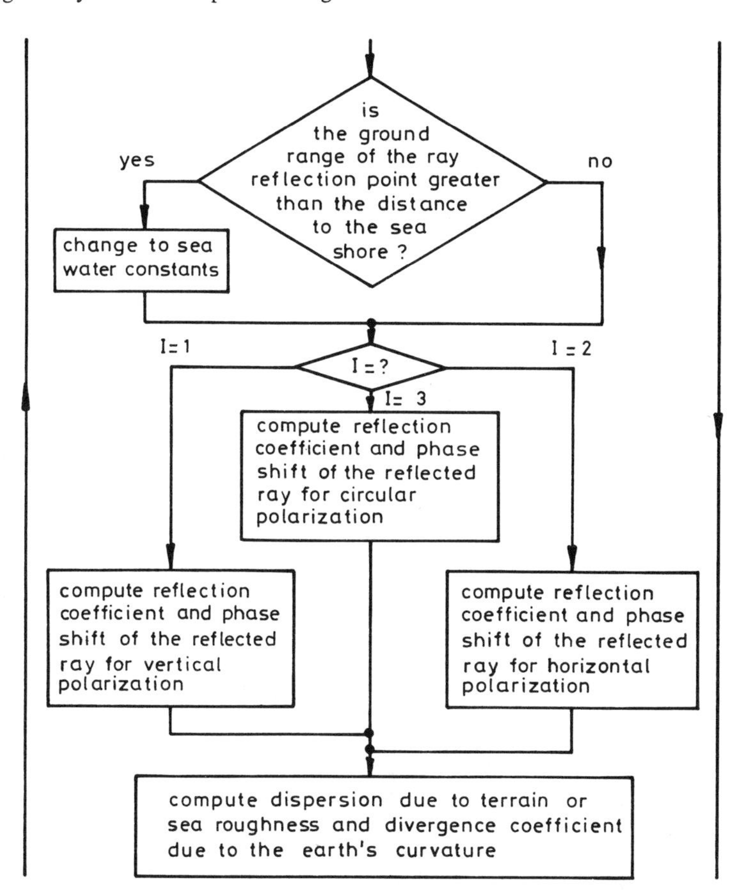

Fig. 10.1 *Continued – page 2*

The value of the atmospheric attenuation ATM should be obtained from Fig. 4.5 if the radar operates at a frequency below 1 and above 10 GHz, however, the model computes the appropriate attenuation for frequencies within the above limits by using eqns. (4.22) to (4.25). Line 30 of Fig. A.1 shows the used format. Since only the mentioned applications are envisaged for this model, no provision was made for separate inputs for attenuation by rain, fog or hail. Should it be required to obtain the coverage diagram in inclement weather, it is recommended to include the

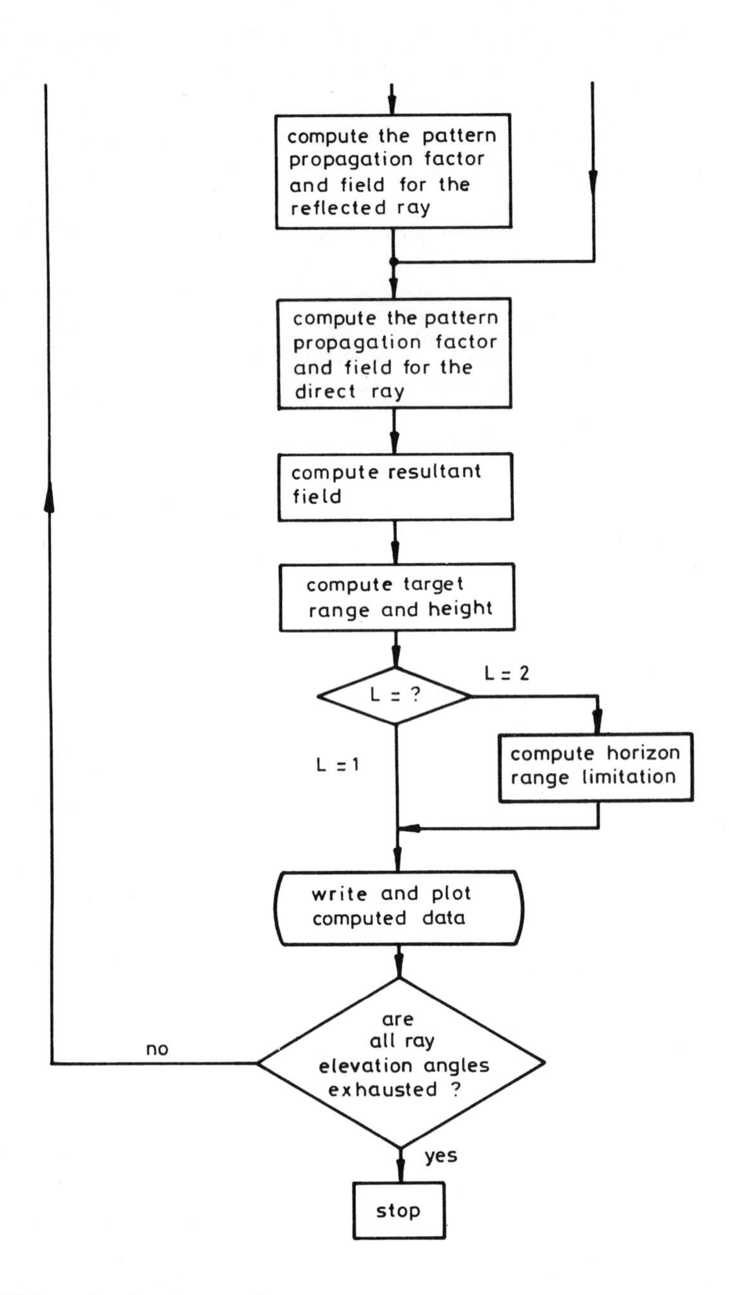

Fig. 10.1 *Continued – page 3*

appropriate attenuation constants into ATM. Only model RDPRO has the facility of using these entered separately.

UNDUL, having the format given by line 32, expresses the unevenness of the surface from which reflection occurs. It determines dispersion (Chapter 6) which influences the field of the reflected wave.

SHORE determines the distance at which the permittivity and conductivity of the ground are replaced by those of the sea, computed by statements 287 to 296 of the model. The formulae for their computation were obtained by linear approximation to data for these quantities given for a number of frequencies by Kerr (1951). The statement in line 33 defines the used format.

The following data, namely AERSEA, SEA, IPOL, AERHIT, HBMWTH, VBMWTH and TILT are adequately defined by the glossary of code-names in Table 10.1. Their format statements appear in lines 36, 38, 42, 44, 46, 48 resp. 50. TILT permits the consideration of the change of the aerial gain in a given direction when the aerial is tilted.

Since the aerial scan rate, ASR, is occasionally specified by the number of azimuth degrees scanned in a given interval, this may be used as input. Otherwise, when the canning rate, RPM, in revolutions per minute, is available this should be used and the value of ASR should be taken to be equal to zero. Lines 55 resp. 60 show the used format statements for these two input data.

SCA, the scan angle of a sector scan, as used in all three models, is adequately commented on in the listing of the programme. It is defined by the format in line 64.

The following eleven input data refer to equipment parameters and the required probabilities for which computations are to be made. They do not require elucidation, except that the widened bandwidth is to be used when the radar uses pulse compression. The format statements of these data follow the appropriate READ statements.

The four input data SLST, TERHTS, SLF, TEHF refer to environmental topographic data concerning the slope of the terrain from which reflection occurs. This influences the magnitude of the grazing angle or the angle of incidence and hence the magnitude and phase of the complex reflection coefficient (Chapter 6). The relevant statements are those of lines 336 and 516 to 528. The format statements are in lines 88 and 90 of Fig. A.1.

The meaning of K is adequately explained in the provided glossary.

The value of NOPER depends on the type of pulse integration used by the radar. A zero value is appropriate to post-detection integration. Unity is used for operator–cathode-ray-tube integration and two for coherent or pre-detection integration. The format of this input datum is shown by line 96. Lines 213 to 228 of the model listing compute the integration improvement factor. The integration improvement of post-detection integration is k^{β} where k is the number of integrated pulses. The determination of the exponent β for post-detection integration, as used here, was obtained by least square exponential fitting to Fehlner's (1962) data by Tenne-Sens (1971). Figure 10.2 shows the computed value of β for three values of the probability of false alarm and probability of detection and numbers of integrated pulses from 1 to 10000.

SIT is the scan integration time applicable when integration is extended over an interval exceeding that of a single scan. The format of this input is given by line 98 and the use of the quantity by lines 195 ro 197 of the listing.

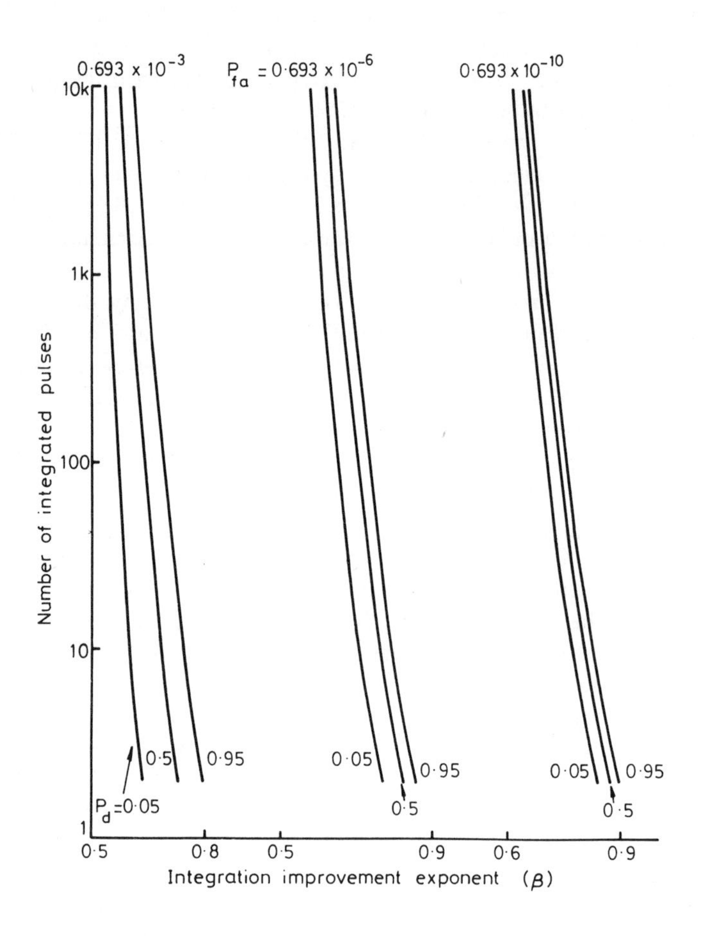

Fig. 10.2 *Post-detection integration improvement exponent β*

PCR is the numerator of the pulse compression ratio with a format shown by line 100.

ANGRD and ANREF are the equivalents of GRAD resp. REF when anomalous propagation conditions prevail. Their values may be obtained by meteorological measurements. Although the model is not applicable to coverage diagram determination when ducting prevails, it uses these data for the computation of the appropriate earth radius correction factor and adjusts all relevant distances accordingly. The format statements are identical with those given for GRAD resp. REF.

HD1 and HD2 are the 'track widths' (Chapter 5) of a duct at the radar and at a distance DMS from it. The input formats of these quantities are given by lines 104, 106 and 110 respectively.

The application of these five parameters is implemented by the statements in lines 463 to 483 of the mentioned listing.

Datum JAM selects either a target-borne or a stand-off type noise jammer by using either zero or unity in a format given by the statement in line 113.

The following seven input data stand for jammer parameters and need no elaboration.

ICIR determines the sense of circular polarization as specified in the glossary of Table 10.1, and TITLE is the name of the modelled radar. Its format statement, line 131, caters for a 30 character long designation.

10.1.2 The flow-chart and listing

Having defined the required input data one may look at the flow-chart in Fig. 10.1 and identify its steps with corresponding groups of statements in the listing given by Fig. A.1.

The first step calling for the reading of the input data extends from line 13 to line 131 of the listing and does not need further explanation.

The computer is instructed then to write the input data starting with the radar's name, the type of target determined by the selected value of K, to convert the aerial gain from the input format in dB to one expressed as a multiplier (statement in line 188), to compute the number of hits per scan in lines 198 to 203 and the beam shape loss (eqn. (4.11) and lines 204 to 208), integration improvement factor (eqn. (4.13) and lines 213 to 236), collapsing loss (eqn. (4.18) and lines 237 to 238) and operator loss by eqn. (4.20) in lines 240 and 241. The sum of all losses is then computed and the computer continues writing the various computed and supplied data when, in lines 263 to 277, it computes the coefficient of atmospheric attenuation from eqns. (4.22) to (4.25) for frequencies within the range of 1 to 10 GHz. For operating frequencies outside this range the value fed-in as the ATM datum will be written and used in later computations. Writing continues until, in lines 288 to 297, the relative permittivity and conductivity of sea water for the appropriate operating frequency is computed. Writing is resumed until line 318 is reached.

The corrected earth radius is then computed (eqn. (5.11)) in lines 322 to 325 for both, the coefficients of the 'standard' atmosphere and those supplied as input for anomalous propagation conditions.

After converting, in lines 329 to 340, some of the input data to the format used further in the model and reducing, in lines 341 to 343, the target cross-section when circular polarization is used, one proceeds to the computation of the quantity M, used in the computation of the signal-to-noise ratio of certain types of target, as given by eqn. (3.116) and the determination of the detection threshold by calling, in lines 347 to 348, subroutine PHIPFA mentioned earlier.

The signal-to-noise ratio for the specified probability of detection and false alarm and target type is computed by statements in lines 351 to 399 using formulae given for the various target types in Chapter 3 together with their correction factors.

Computing the noise power in line 405 one proceeds, if jamming is present, to convert the supplied jammer data to the required format and using then the radar equation (eqn. (4.26)), in lines 417 to 423, one determines the free space range RFS for the given and computed parameters.

The ray elevation angles are specified in 0·01° steps in the range of 0° to 17°, or roughly three times the vertical beamwidth, and in 0·1° steps up to 45°, the maximum radiation angle of the used aerial as given by the manufacturer of the Adelaide airport's radar.

The programme then splits into two branches; one branch for free-space coverage diagram computation and plotting chosen by L equal to unity, and the other branch, with L equal to two, calling for the computation and plotting of the lobing pattern. This division is called by lines 461 to 462. When only free-space propagation is of interest all computations between lines 463 and 621 are by-passed.

If the lobing pattern is to be determined, the program proceeds, in lines 468 to 516, to determine the quantities required for completing the radar–ground–target geometry, to then consider the terrain slope in the statements given by lines 517–527 in order to modify, in line 528, the angle of incidence GAM used later for the computation of the magnitude and phase of the complex reflection coefficient as given in Chapter 6.

Lines 529 to 533, depending on whether the distance to the point of reflection is greater or smaller than the distance to the sea shore, choose either the relative permittivity and conductivity of the sea or those of the ground to proceed to line 537 selecting the computation of the appropriate reflection coefficient. RHOVER, RHOHOR and RHOCIR are the magnitudes, while AMUVER, AMUHOR and AMUCIR the phase shifts for the three polarizations catered for by the model. The statement in line 589 determines whether the quantities for same or opposite circular polarization are used depending on whether ICIR was specified to be equal to 1 or 0. The statements in lines 540 to 560 apply to vertical, lines 564 and 578 to horizontal and those from 582 to 596 to circular polarization.

Divergence due to the convexity of the reflecting surface (eqn. (6.30)) is taken into account by lines 599–602.

Lines 603 to 609 determine whether ground or sea surface irregularities cause dispersion. Line 607 applies particularly to the westerly direction considered in computing the coverage diagram of the Adelaide airport's radar shown in Fig. 11.7. The constant 2000 is the distance (in metres) between the radar and the boundary beyond which the irregularities of the terrain surface increase due to housing development as will be mentioned later. It may be replaced by another suitable distance expressed in metres or the statements in lines 607 to 609 may be omitted if the above limitation is not applicable.

Dispersion due to irregularities of the reflecting surface is computed by Ament's formula (eqn. (6.16)) in line 610.

PADI, the path length difference between the direct and reflected ray is computed by the statement in line 613 after which subroutine AERPLD is called to compute the reflected ray's pattern propagation factor, i.e. the aerial gain referred to the gain in the direction of maximum radiation.

Statements 615 to 620 determine the relative magnitude of the field due to the reflected ray and chosen polarization.

The previously mentioned second branch used for the determination of the free space coverage diagram joins at the statement in line 621 the path followed so far. The pattern propagation factor and field contribution of the direct ray is then computed by statements in lines 622 to 627 and subroutine AERPOD, while the statements in lines 628 and 629 yield the resultant relative field of both contributions.

After computing the exponential term determining the range reduction due to atmospheric attenuation one proceeds, in line 633, to calculating the range.

The influence of jamming is taken into account by the statements in lines 634 to 663, while the height for the given ray elevation angle and computed range is obtained by lines 667 and 668 and obstruction by terrain features by those of 671 to 685.

The SDLB (side lobe) subroutine called by the statement in line 656 and listed in Fig. A.3 computes the radar aerial's sidelobe propagation factor when it is required to consider the effect of a stand-off jammer on the radar's performance. It assumes a rectangular aerial aperature in the horizontal plane.

The radar's horizon ZB is given by line 686, while that of a target at the computed height, denoted by ZC, by line 687. Their sum ZD is the radio horizon which, determined by the statements in lines 689 to 693, may limit the radar's coverage at low ray elevation angles.

The ground range GR is given by the statement in line 695.

Converting then the computed quantities to convenient dimensions the programme proceeds to write them, together with the specified ray elevation angles, as directed by the statements in lines 700 and 701.

The statements in lines 703 to 705 change the limits of computations and their intervals to those appropriate for elevation angles above 17° given earlier.

Instructions for producing a plot and for its annotation are contained in the remaining statements of the main programme.

10.1.3 Applications

It was pointed out earlier that the COVER model has two applications, it serves either as a vehicle for the development of algorithms for aerial coverage diagrams, or it may be used for the computation and plotting of either the free-space coverage or the lobing pattern of a radar's aerial.

The following section shows the uses of COVER for the derivation of coverage diagram algorithms.

10.1.3.1 Coverage diagram algorithms: Expressions for the ray elevation angle dependence of the gain of a number of aerial types is available in the relevant literature. Some of these will be given here for the sake of completeness; an example of obtaining the algorithm for the coverage diagram of the Adelaide airport's radar is demonstrated.

The shape of the coverage diagram of an aerial depends to a great extent on the shape of the aerial's aperture and the distribution of its illumination.

(*a*) *Circular aperture:* Some radars use aerials with parabolic reflectors having circular apertures illuminated by a horn radiator placed so that the distribution of the illumination across the aperture follows either a uniform parabolic law given by a $1 - r^2$ or a parabolic squared law expressed as $(1 - r^2)^2$ where r is the ratio of the radius at any point of the aperture to that at its edge and the radiation pattern is given accordingly by the equation

$$\pi\frac{D^2}{4}\,\frac{2^{n+1}n!J_{n+1}(x)}{x^{n+1}},$$

where n may assume values of 0, 1 and 2, and $J_{n+1}(x)$ is a Bessel function of the first kind and order $n+1$ (Jasik, 1964). The argument

$$x = \pi\frac{D}{\lambda}\sin\epsilon$$

with D the diameter of the aperture in terms of the wavelength.

The computation of the Bessel function may be made by the author's program (1973) the listing of which is given by Fig. A.7. Figure 10.3 shows the free-space coverage diagrams of aerials with the above three aperture illumination laws.

Considering then, e.g. the $J_1(x)/x$ coverage diagram, one may compute the angles of the minima of radiation by determining the angles for which the Bessel function is zero. The first zero of $J_1(x)$ occurs for $x = 3{\cdot}832$ (Jahnke and Emde, 1945), or

$$\sin\epsilon = \frac{3{\cdot}832\lambda}{\pi D},$$

i.e.

$$\epsilon = \sin^{-1} 1{\cdot}22\frac{\lambda}{D}. \tag{10.3}$$

However, since at the frequencies where high gain parabolic reflectors with circular apertures are practicable the wavelength is much smaller than the diameter, one may write, with acceptable accuracy,

$$\epsilon = 1{\cdot}22\frac{\lambda}{D}$$

for the ray elevation angle of the first zero radiation direction.

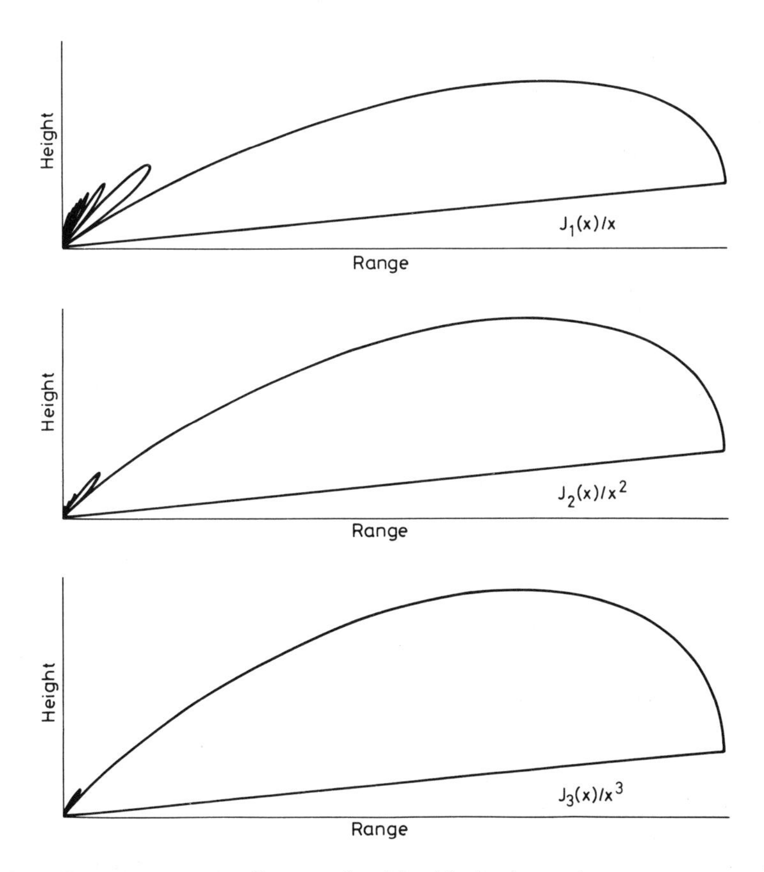

Fig. 10.3 *Free-space coverage diagrams of aerials with circular aperture*

The complete expression for the computation of the field pattern is then

$$\frac{J_1\left(\dfrac{3{\cdot}832 \sin \epsilon}{\sin 1{\cdot}22\lambda/D}\right)}{\dfrac{3{\cdot}832 \sin \epsilon}{\sin 1{\cdot}22\lambda/D}} \tag{10.4}$$

which is zero for $\epsilon = 1{\cdot}22\lambda/D$.

(*b*) *Rectangular aperture:* The field of a uniformly illuminated rectangular aperture, Fig. 10.4, is of the

$$\frac{\sin x}{x} \tag{10.5}$$

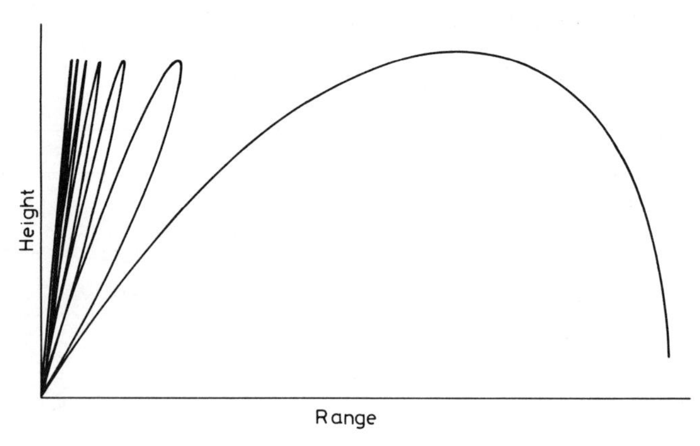

Fig. 10.4 *Free-space coverage diagram of an aerial with a rectangular aperture*

form (Skolnik, 1962), where the value of the argument x depends besides on the ray elevation angle ϵ, on the linear dimension of the aperture, as compared with the used wavelength. One may then write the above expression as

$$\frac{\sin (a \sin \epsilon)}{a \sin \epsilon}$$

and, since at a ray elevation angle equal to one half of the beamwidth BW in the considered plane the field must fall to $1/\sqrt{2}$ of the field in the direction of maximum radiation, one may determine the value of a as

$$a = \frac{\pi D}{\lambda} = \frac{1{\cdot}391\,578\,9}{\sin BW/2} \tag{10.6}$$

since (10.5) equals $1/\sqrt{2}$ when $x = 1{\cdot}391\,578\,9$ (Jahnke and Emde, 1945) so that the field factor of this type of aerial in the considered plane becomes

$$\frac{\sin \dfrac{1{\cdot}391\,578\,9 \sin \epsilon}{\sin BW/2}}{\dfrac{1{\cdot}391\,578\,9 \sin \epsilon}{\sin BW/2}} \tag{10.7}$$

Figure 10.5 shows the lobing pattern of an aerial having a rectangular aperture. The upper plot applies to horizontal, while the lower one to vertical polarization. Since the magnitude and phase of the vertical polarization's reflection coefficient is generally smaller than that of the horizontal one (see Chapter 6), the reflected ray is less developed when vertical polarization is used and the minima are therefore less deep. The variation of the above quantities at vertical polarization also influences the amplitude of the sidelobes as is perceivable from the comparison of the two plots.

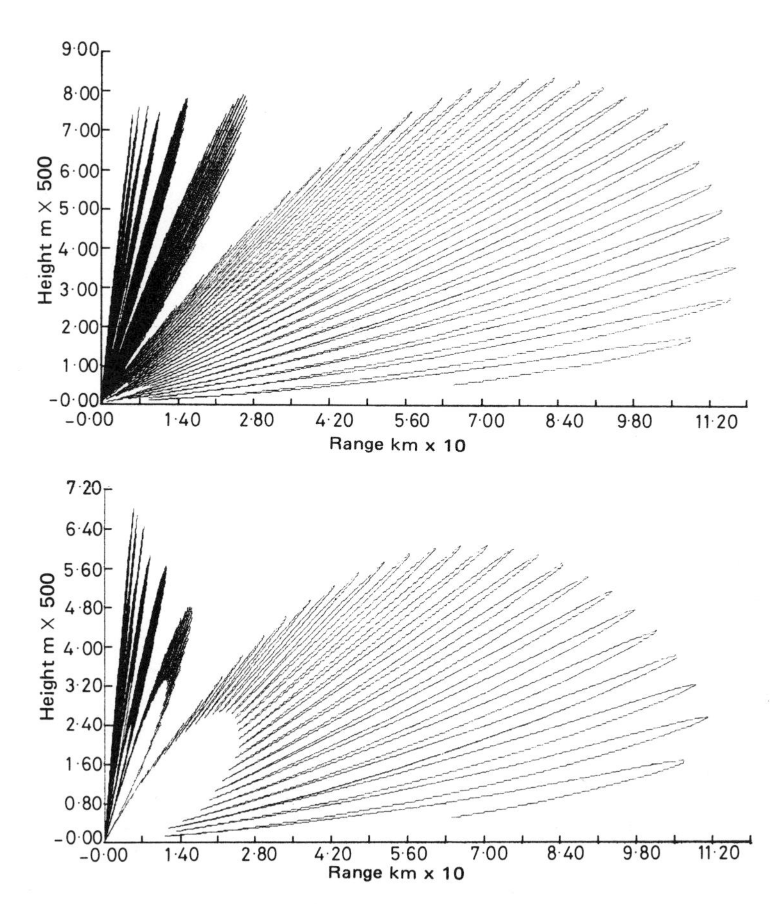

Fig. 10.5 *The lobing pattern of a rectangular aperture aerial*
PD = 50%, PFA = 10^{-6}, RCS = $1{\cdot}00\ m^2$, aerial height = 15·0 m
(*a*) Horizontal polarization
(*b*) Vertical polarization

(*c*) *Derivation of algorithms for other types of aerials:* More modern radars use aerials with radiation patterns shaped so as to yield for targets of a given radar cross-section, more or less constant paint intensity on the PPI display over the whole displayed range.

An idealized radiation pattern of this kind is shown by the quadrangle OABH in Fig. 10.6 where one may express the range R as

$$R = \frac{H}{\sin \epsilon} = H \operatorname{cosec} \epsilon \tag{10.8}$$

with H the expected maximum height of targets and ϵ the ray elevation angle (Bartholomä, 1961).

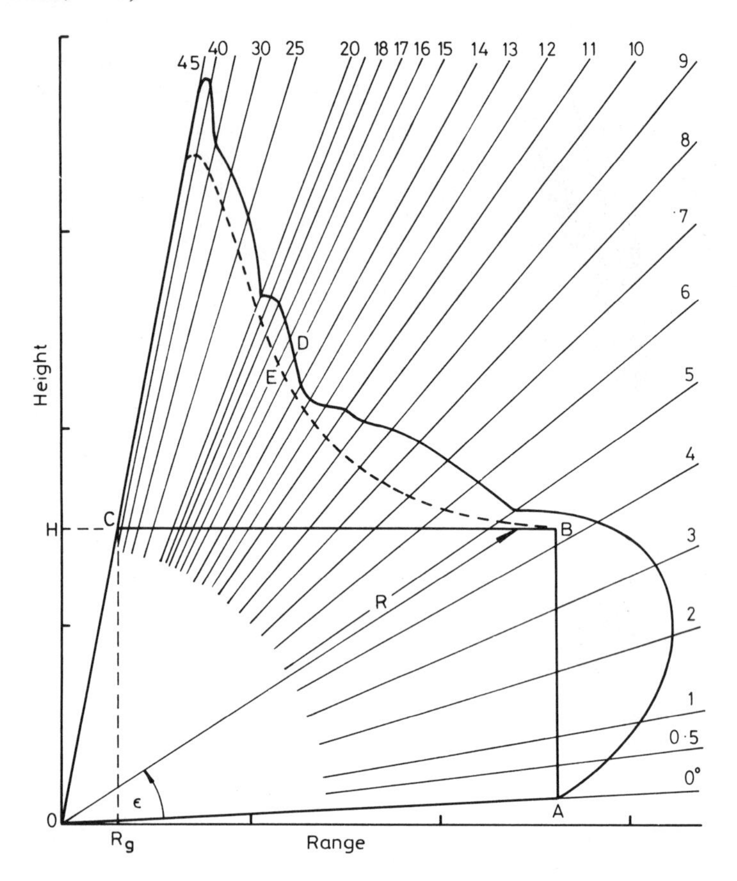

Fig. 10.6 *A "cosec²" aerial's free-space coverage diagram*

The field at a range R is

$$\bar{E} = \frac{c}{R} f(\epsilon) = \frac{c}{H \operatorname{cosec} \epsilon} f(\epsilon), \tag{10.9}$$

where c is a constant containing the radar parameters and $f(\epsilon)$ implies the aerial gain dependence on the ray elevation angle ϵ. The paint intensity depends on the field strength at the target and for this to be constant the above expression must remain constant for all angles ϵ of interest, or the function $f(\epsilon)$ must be

$$f(\epsilon) = \operatorname{cosec} \epsilon$$

so that

$$\bar{E} = \frac{c}{H} = \text{const.} \tag{10.10}$$

Since with most radars a common aerial is used for transmission and reception the above characteristic applies for both functions so that the received signal power is proportional to $\operatorname{cosec}^2 \epsilon$, hence the designation of these aerials by the collective name of 'cosec-square'.

In reality the radiation pattern is made to deviate from this ideal pattern in order to counteract the sensitivity time control's action used to prevent the overloading of the radar receiver by strong signals from near targets. The radiation pattern is made to follow approximately the contour E. Contour D shows the radiation pattern of an aerial of this type as given by a manufacturer in his sales literature. The position of point C depends on the tolerable extent of the cone of silence above the radar and the acceptable range error

$$\Delta R = \frac{1 - \cos \epsilon}{\cos \epsilon}. \tag{10.11}$$

The required radiation pattern is achieved by appropriately shaping the aerial's reflector (Woodward, 1946).

The designation adopted for these aerials is not specific enough. More than 30 years ago Silver (1949) already dealt with four different cosec^2 aerials. It is impossible therefore to adopt a universal algorithm for modelling the coverage diagram of these aerials. One has to resort to deriving individual algorithms based on patterns obtained, hopefully, by measurements and provided by manufacturers in equipment handbooks.

A demonstration of coverage diagram algorithm derivation, using the diagram shown in Fig. 11.2, follows.

Some preparatory work is necessary before one may commence with the algorithm derivation.

Ranges and heights at various ray elevation angles are read first from the available coverage diagram. If this represents free-space coverage, it is replotted directly, together with rays emanating from the aerial at the specified height at elevation angles of interest, on transparent paper in the scale of the expected computer plots, as shown by Fig. 10.7. If there is doubt about the given coverage diagram's being that for free-space, one computes the maximum free-space range for the given parameters and adjusts all the ranges obtained from the given diagram accordingly. Some error is thus introduced since a constant contribution of the reflected ray at all elevation angles is assumed. However, since the greatest error occurs at higher ray elevation angles at which the search radar coverage is of lesser interest than that at lower ones, the distortion of the coverage diagram is acceptable in most cases.

Further, the program is modified so as to print, besides the values asked for by the WRITE statement in line 700 of the listing, also the value of FORM computed by the AERPOD subroutine. The format of this variable is chosen to be F11.9.

The dashed curve in Fig. 11.3 represents the coverage diagram redrawn from that given by Fig. 11.2 for a 4·5 dB noise figure, as is also the one shown by Fig. 10.7.

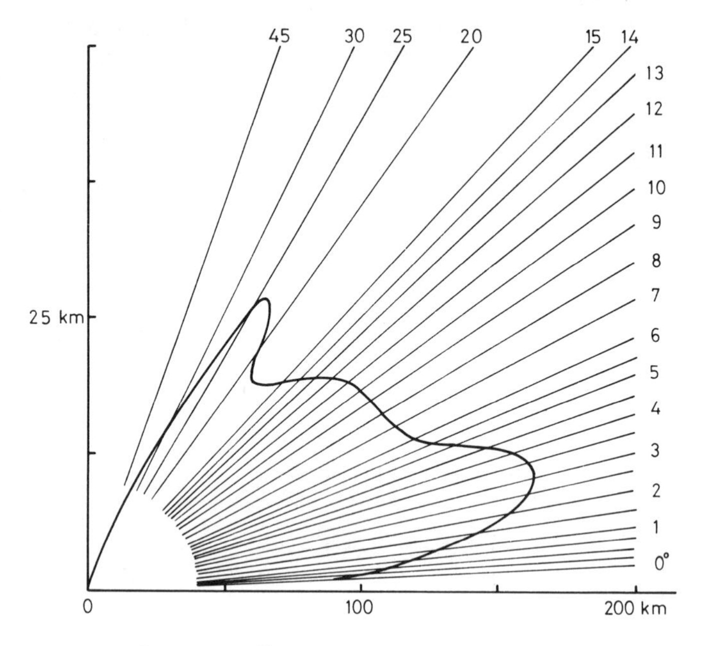

Fig. 10.7 *Coverage diagram stencil*

Figure 10.8 shows the steps of the piece-wise approximation used to obtain the coverage diagram given by the full line in Fig. 11.3.

The bottom part of the diagram may be approximated by an exponential expression of the form

$$E(\epsilon) = \exp\left(-XK(\epsilon - BW)^2\right), \tag{10.12}$$

where ϵ is the ray elevation angle in radians, BW is one half of the vertical 3 dB beamwidth of the modelled aerial and the coefficient XK is determined from the requirement that the above expression should be equal to $1/\sqrt{2}$ at a zero ray elevation angle or

$$XK = \frac{\ln 2}{2BW^2} \tag{10.13}$$

Step 1 of Fig. 10.8 shows by the full line the computer plot of this function, while the dashed line represents the given coverage diagram.

The next step corrects the deviation of the computed from the given curve below the maximum range ray elevation angle.

A coefficient XK such as to reduce at $\epsilon = 0°$ the computed maximum range of 162·19 km, read from the computer print-out for $\epsilon = BW$ to the given range of

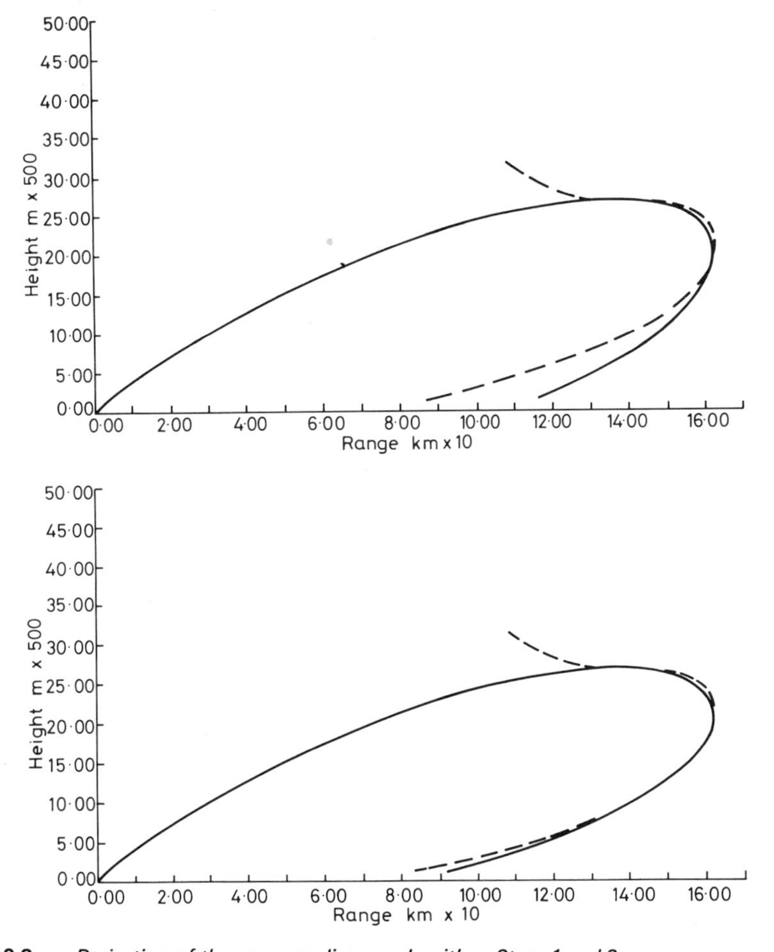

Fig. 10.8 *Derivation of the coverage diagram algorithm. Steps 1 and 2*
Free space coverage PD = 80%, PFA = 10^{-6}, RCS = 2·00 m², aerial height = 19·8 m, tilt = 0·00°, non-fluctuating target

about 90 km is required, or

$$\frac{90}{162 \cdot 19} = \exp(-XK\, BW^2)$$

which, for $BW = 3/57 \cdot 3$ radians, yields $XK = 214 \cdot 826\,414\,3$. The result of this correction is obvious from step 2 of Fig. 10.8.

Placing the transparency over the last plot, one then determines that this starts deviating from that of the transparency at a ray elevation angle of 5·3° and that it may be approximated by a smooth curve of an approximately uniform slope up to about 6°. Since the coverage diagram from that ray elevation angle onwards resembles a cosec function, the next task is to find the exponent of a cosec function to make it merge, as far as possible, with the given curve.

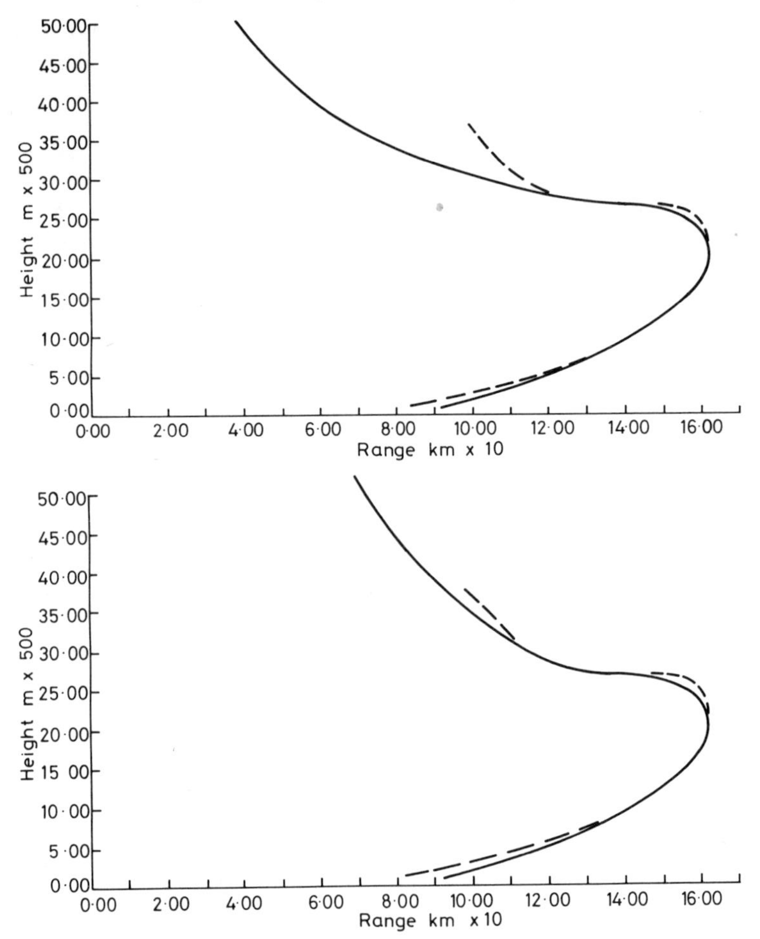

Fig. 10.8 *Continued – Steps 3 and 4*

Reading the height H_1 at the ray elevation angle of 5·3°, i.e. the point where the deviation of the two curves starts, from the computed curve, and height H_2 at angle of 6° from the given curve, one may write that the range R_1 at the first one of the above mentioned points referred to the maximum range R_0 is

$$\frac{R_1}{R_0} = \frac{XA}{\sin^{YA} \epsilon_1},$$

while that at the second point is

$$\frac{R_2}{R_0} = \frac{XA}{\sin^{YA} \epsilon_2},$$

where XA is a coefficient to be determined.

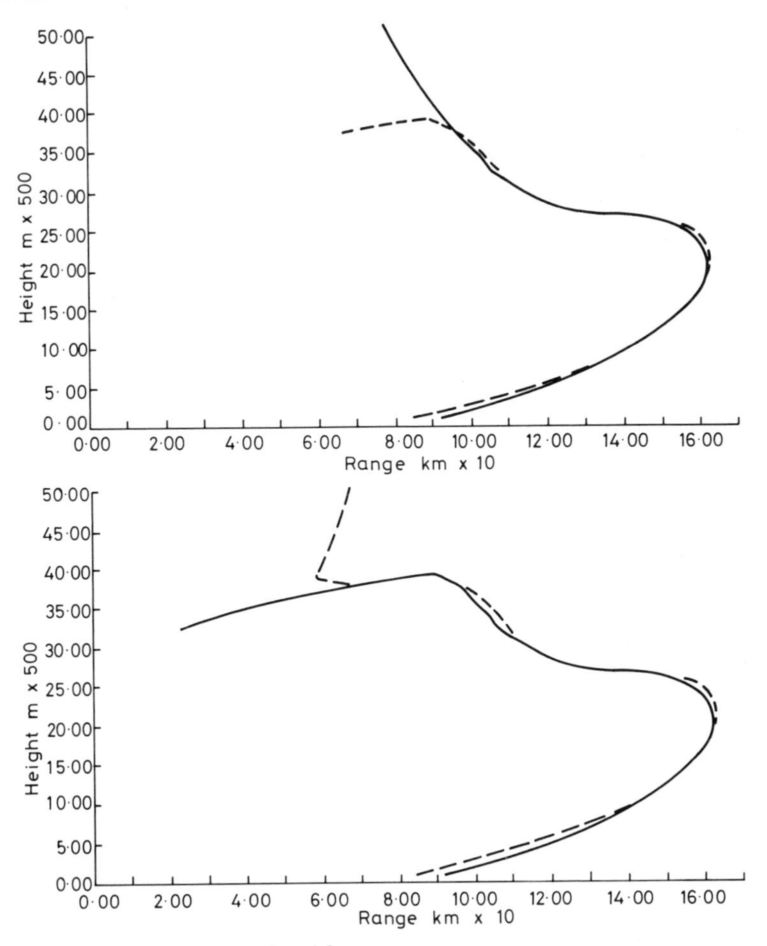

Fig. 10.8 *Continued – Steps 5 and 6*

In the used orthogonal system of coordinates one may write further that

$$R_1 = \frac{H_1}{\tan \epsilon_1} \quad \text{and} \quad R_2 = \frac{H_2}{\tan \epsilon_2}$$

or

$$\frac{H_1}{R_0 \tan \epsilon_1} = \frac{XA}{\sin^{YA} \epsilon_1},$$

$$\frac{H_2}{R_0 \tan \epsilon_2} = \frac{XA}{\sin^{YA} \epsilon_2},$$

from where

$$XA = \frac{H_1 \cos \epsilon_1 \sin^{YA-1} \epsilon_1}{R_0} = \frac{H_2 \cos \epsilon_2 \sin^{YA-1} \epsilon_2}{R_0}$$

and then

$$\frac{H_1 \cos \epsilon_1}{H_2 \cos \epsilon_2} = \left[\frac{\sin \epsilon_2}{\sin \epsilon_1}\right]^{YA-1}$$

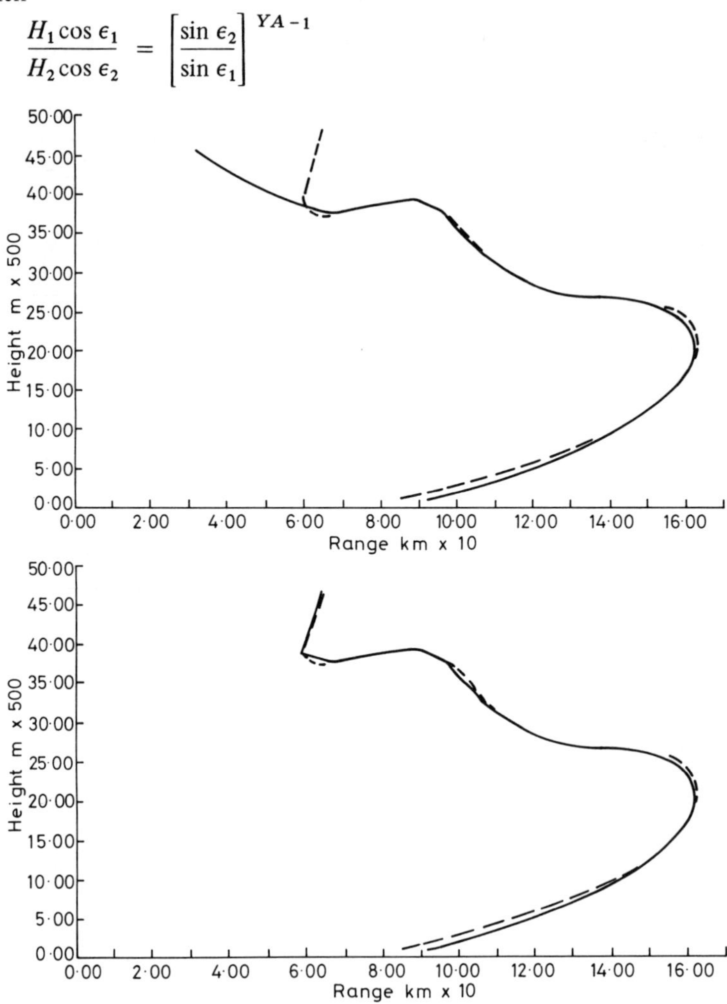

Fig. 10.8 *Continued – Steps 7 and 8*

Taking logarithms on both sides one arrives at

$$YA = 1 + \frac{\log \dfrac{H_1 \cos \epsilon_1}{H_2 \cos \epsilon_2}}{\log \dfrac{\sin \epsilon_2}{\sin \epsilon_1}} \qquad (10.14)$$

as the sought exponent of the cosec function.

Using the figures obtained for the assessed radar's coverage diagram, namely

$$\epsilon_1 = 5{\cdot}3^\circ \qquad H_1 = 13\,394{\cdot}19\,\text{m},$$

$$\epsilon_2 = 6{\cdot}0^\circ \qquad H_2 = 14\,000{\cdot}00\,\text{m},$$

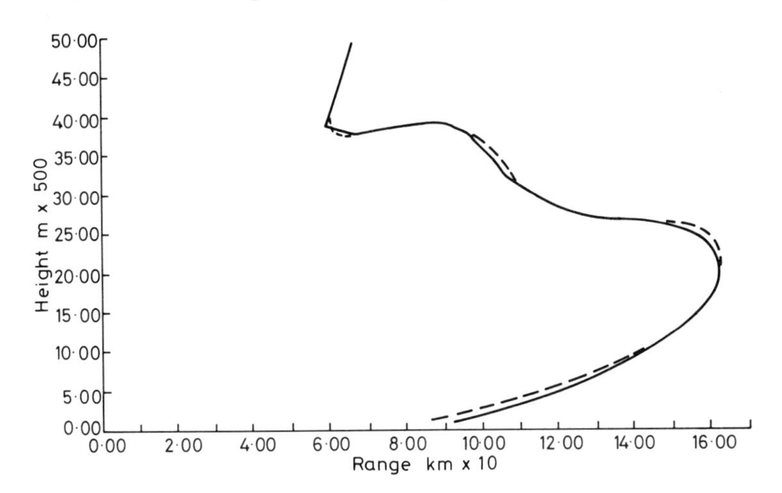

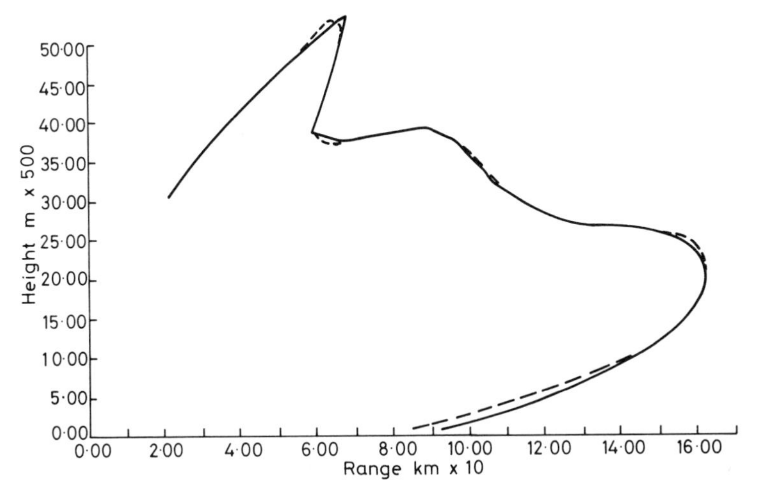

Fig. 10.8 *Continued – Steps 9 and 10*

one computes $YA = 0{\cdot}652\,023\,264$, as in line 50 of the AERPOD subroutine listing in Fig. A.4.

The previously mentioned coefficient XA is determined from the requirement that the ratio

$$\frac{XA}{\sin^{YA} 5{\cdot}3^\circ}$$

must equal the value of FORM computed at this ray elevation angle for the preceding approximation. In the used case it is

$$\text{FORM} = 0{\cdot}815\,700\,313$$

so that

$$XA = 0{\cdot}172\,597\,515$$

as shown by the statement in line 49 of the above listing.

The closeness of the fit is perceivable from the illustration of step 2 in Fig. 10.8.

Continuing in the outlined fashion one then puts the transparent overlay onto the obtained plot and determines that a curve appropriately fitted could proceed up to a ray elevation angle of 8·5° without a sudden change in its slope.

Determining the heights at the two ray elevation angles, i.e. 6° and 8·5°, one proceeds determining the values of the exponent *YA* of the cosec function and the constant *XA*.

The final result of the approximation is shown by step 10 of Fig. 10.8 and by Fig. 11.3.

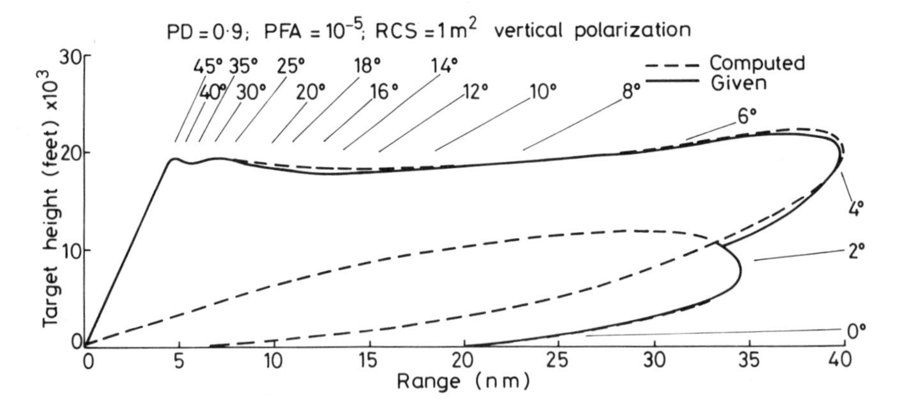

Fig. 10.9 *A composite aerial free-space coverage diagram*

Figure 10.9 shows another coverage diagram approximation derived by the shown method for a radar using two back-to-back aerials, one for high angle and one for low angle coverage.

The described method seems to be tedious at first sight. However, with some practice one may arrive at acceptable results fairly quickly and the time spent is amply repaid by the higher accuracy of the results of radar modelling.

The obtained algorithm constitutes the AERPOD subroutine.

Since it is only the bottom part of the aerial gain versus ray elevation angle dependence which affects the reflected ray, the AERPLD subroutine contains only those statements of the AERPOD subroute which are relevant to that part of the derived dependence as shown by the listing in Fig. A.5.

10.2 The MAP model

The MAP model is proposed for use either for the assessment of the suitability of a site for the required radar operation or for site selection thus obviating the necessity of an expensive on-the-spot survey.

Basically this model is very similar to the earlier described COVER model and, since Appendix B provides the complete listing of the main MAP model as used for the assessment of the Adelaide airport's site, the following lines will touch only on those of its parts which differ from those of the listing in Appendix A.

Some of the initial dimensional statements of the MAP model, namely those referring to the terrain data, i.e.

SHO – the diatance to the sea shore in km,
SLO – the distance to the beginning of the terrain slope in km,
TER – the height above sea level of the above location in m,
SF – the distance to the end of the terrain slope in km, and
TEF – the height above sea level of the above location in metres,

differ from those in COVER since each of these requires, depending on the chosen azimuth intervals, many more input data than the corresponding inputs in COVER. The shown statements cater for 361 of each of the above data since data renewals for each degree of a 360° azimuth were used. The 361st and the first datum of each set is identical in order to return the plotter to the initially plotted point on each contour.

The model proceeds with computing some auxiliary quantities and writing them, just as it was described for COVER, up to the statement in line 415 on page 6 of Fig. B.1 which calls for four sets of consecutive computations for targets at the four heights specified by the input value of TARGHT the lowest target height in line 124 and the statement in line 416.

Lines 418 to 425 determine the ray elevation angles of the direction of the maxima of the intercepting lobes and the computation continues as in COVER up to line 589 where the delineation of the boundary, as used for the sample site, commences. Lines 589 to 623 determine the distance from the radar to the airport's boundary (Fig. 11.1) over which a surface smooth at the used frequency is assumed.

Both of the hitherto mentioned models, i.e. COVER and MAP, are identical then up to line 676 of the MAP listing.

In lines 677 to 681 the ray elevation angle to the target, while in lines 683 to 690 that to the terrain peak at the considered azimuth angle, is computed. The statements in the following six lines determine whether the radar's view, when looking at the target, is obstructed by the terrain. If the first of the above angles is the greater of the two, no obstruction occurs. On the other hand, if it is smaller, the model computes the range to a target at the specified height intercepted by a ray touching the terrain peak, thus determining the greatest range at which a target approaching the radar at the given height from the considered direction may be detected with the specified probability at the given probability of false alarm.

The remaining statements compute the ground range of the target from the radar and call for writing and plotting of the results.

The subroutines used by this model are identical with those used by COVER.

Figure 10.10 shows the flow-chart of the MAP model. Examples of its output are shown by Figs. 11.9 and 11.10.

10.3 The RDPRO model

This, as it has been mentioned earlier, is the most comprehensive of the three models developed and used in the above radar performance predictions and assessments. While the previous two models use, for the determination of the signal-to-noise ratio for specified probabilities of detection and false alarm, the approximate formulae corrected by their appropriate coefficients, all of which are given in Chapter 3, this model uses, for the computation of the probability of detection of a given target at a specified range and height, the 'exact' formulae given by Fehlner (1962) and shown also in Chapter 3 of this treatise.

Assuming that the radar is capable of intercepting a target as soon as it appears on the horizon, the target's range and probability of detection is computed on each aerial scan. The number of scans is determined from the radar's and target's horizon, the aerial scanning rate and the target's velocity.

Figure 10.11 shows the simplified flow-chart of the model; Fig. C.1 lists the main RDPRO program.

The majority of input data is identical with those used in the earlier described models; besides these there are a few additional ones to be specified below in order of their appearance in the listing. They are:

RAIN	rain attenuation in dB/km,
RR	extent of rain in km,
FOG	attenuation by fog in dB/km,
RF	extent of the fog in km,
BKB	Beaufort wind scale,
STATE	sea state,
RRT	rainfall rate in mm/h,
AGEX	frequency agile excursion in MHz,
SCV	sub-clutter visibility in dB,
FASTEP	number of frequency agile steps,
TAU	pulse width in μs,
XAD	a scaling factor for the x-axis of the plots,
VELOC	the target's velocity in km/h,
DESRT	the target's descent rate in m/km,
TARGHT	the target's height at the horizon in m,
MTI	see comment in listing

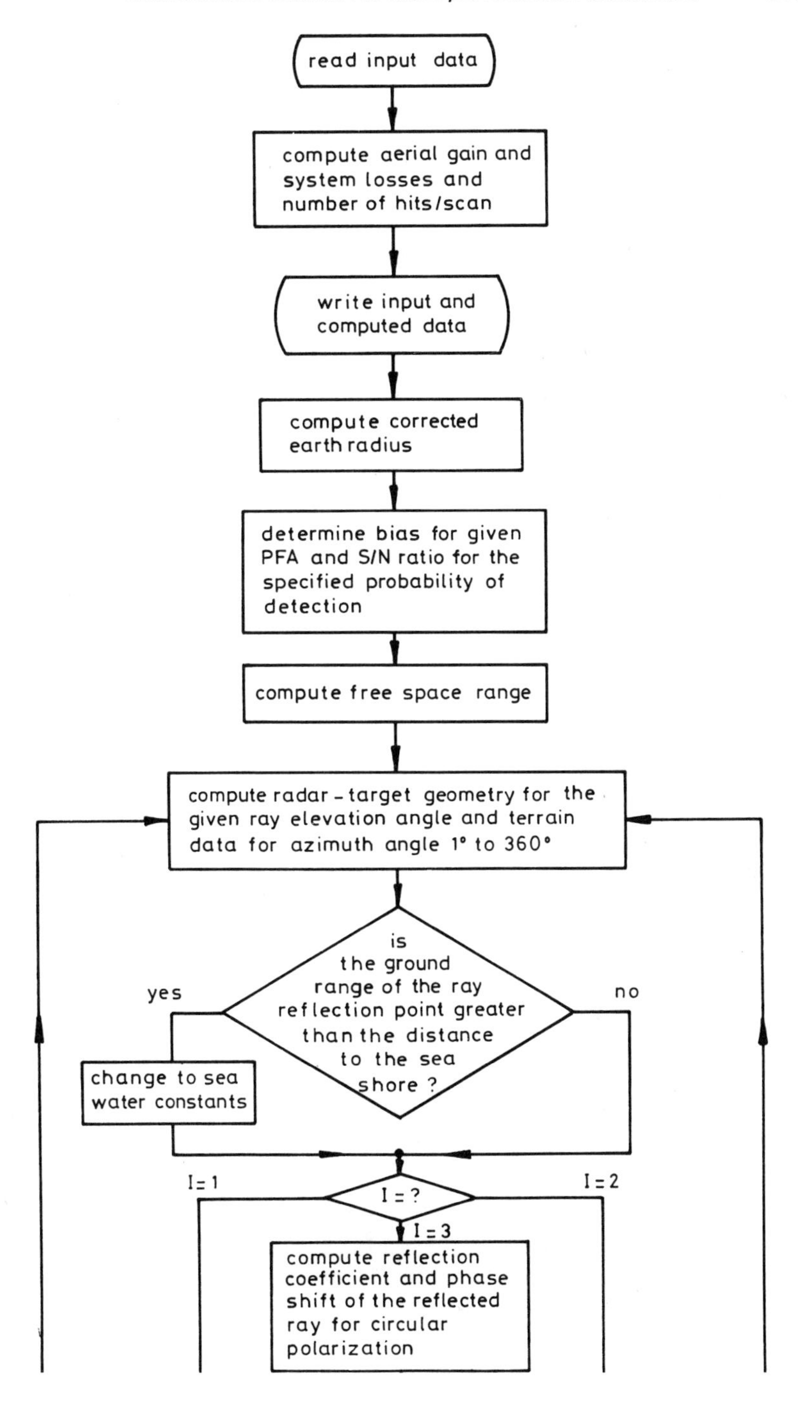

Fig. 10.10 *Flow-chart of model MAP*

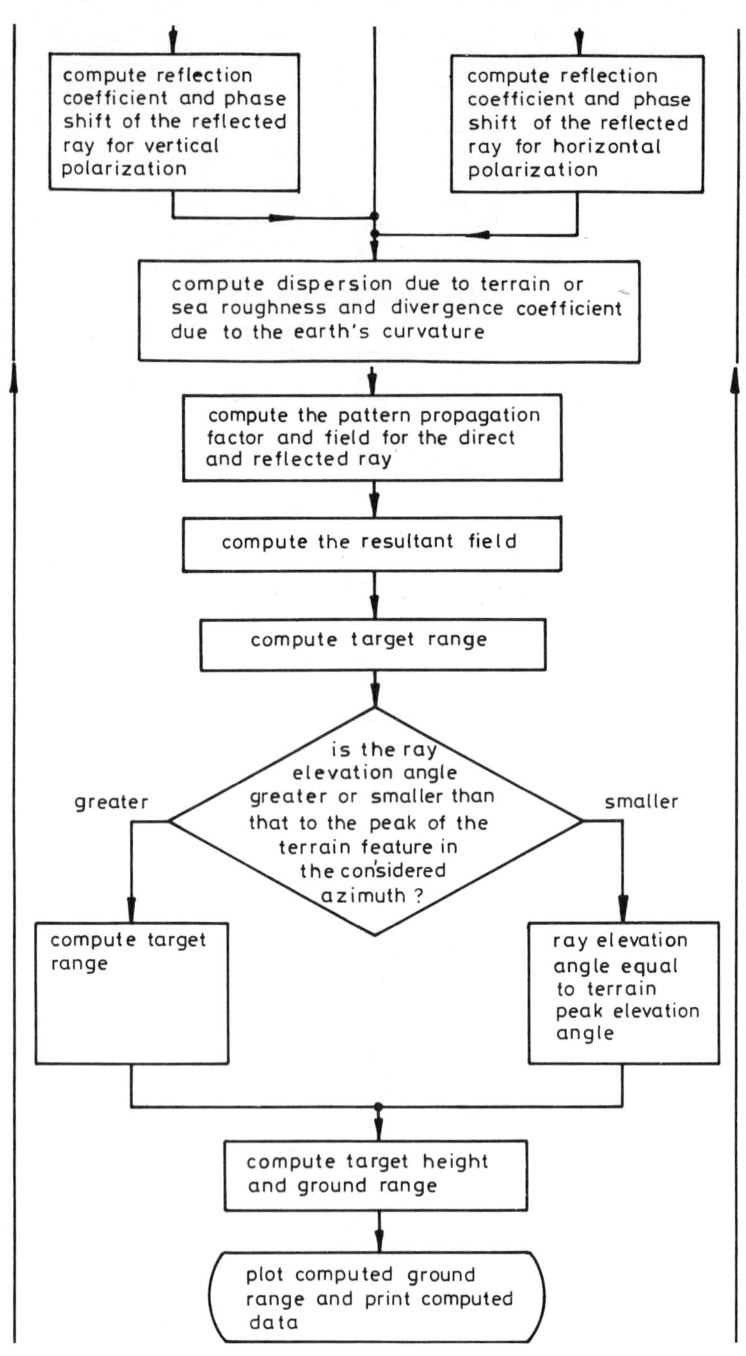

Fig. 10.10 *Continued – page 2*

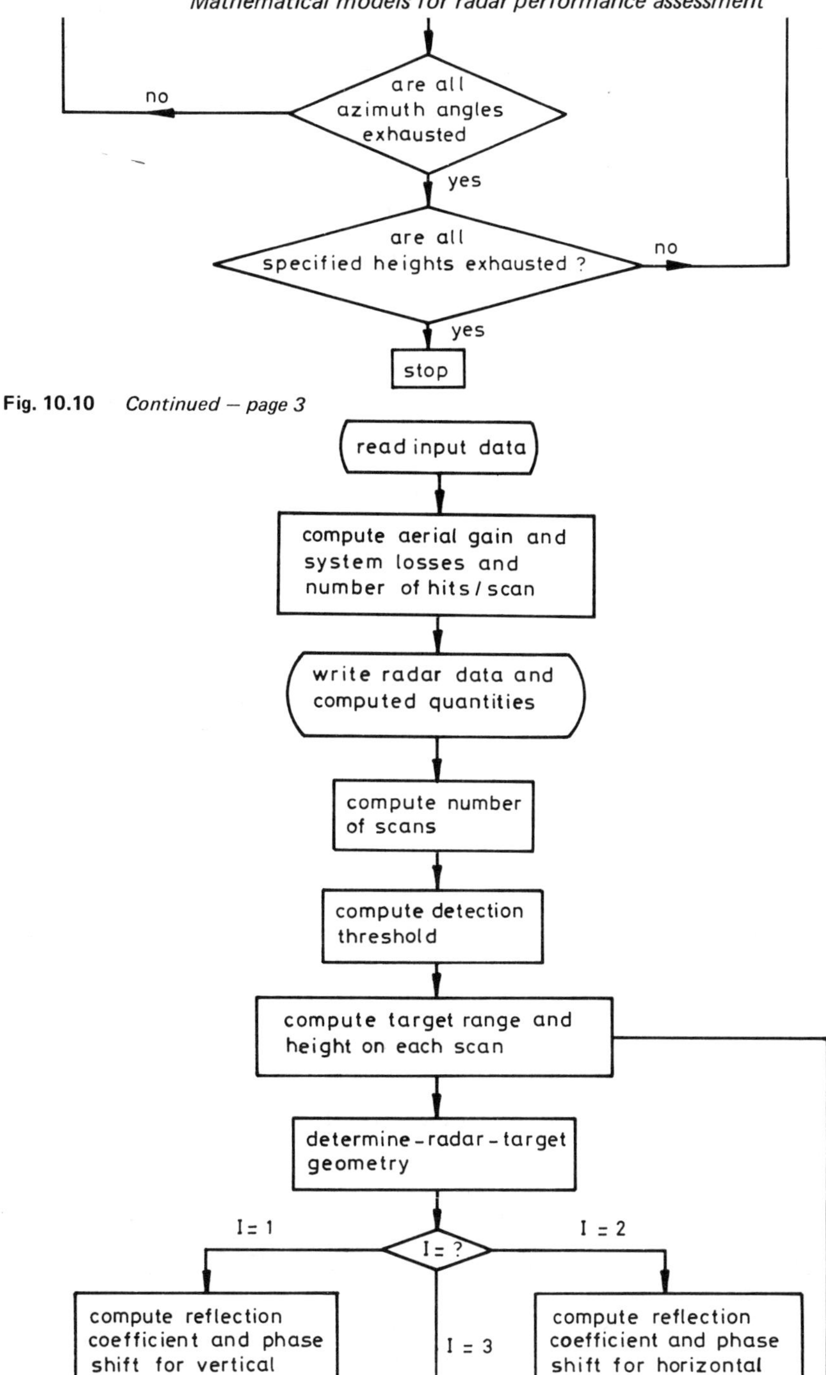

Fig. 10.10 *Continued – page 3*

Fig. 10.11 *Continued – page 2*

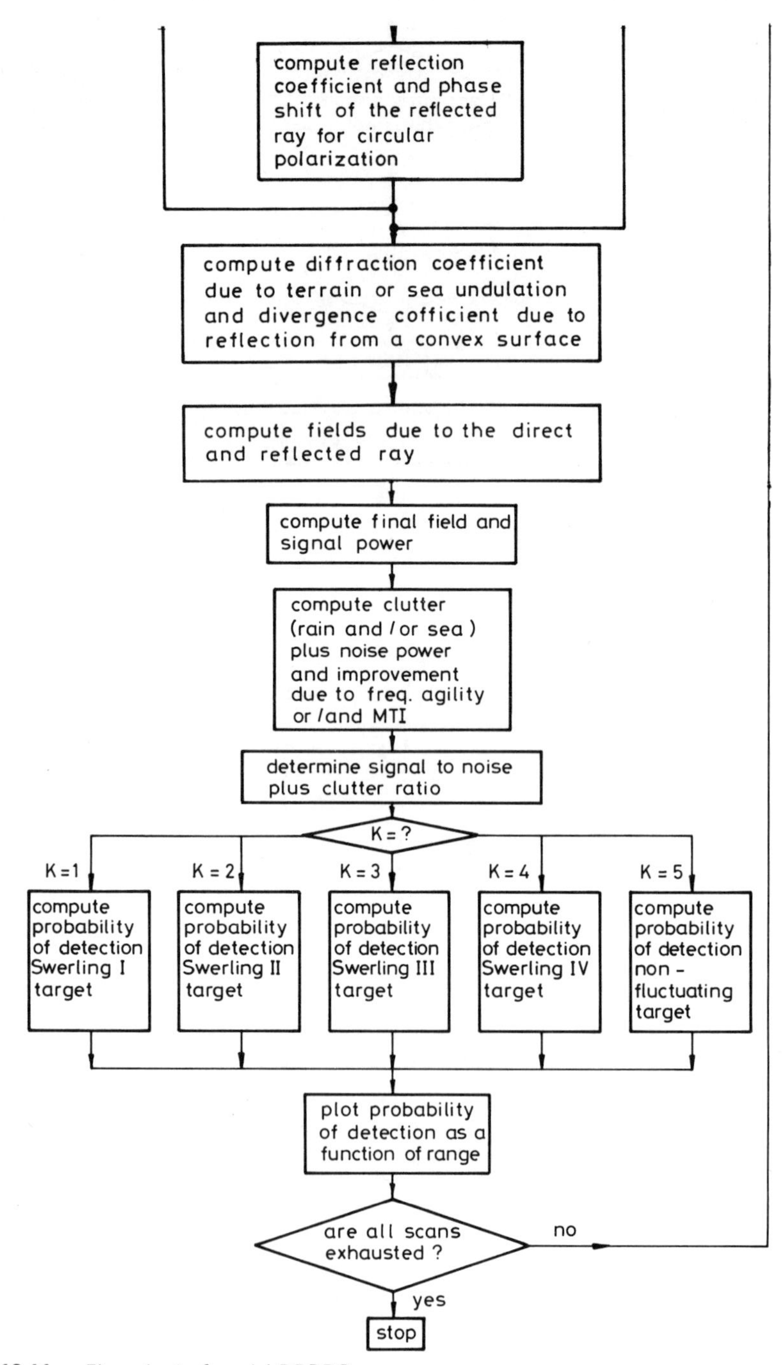

Fig. 10.11 *Flow-chart of model RDPRO*

IPROB	0 – scan-to-scan	probability of detection,
IPROB	1 – cumulative	
PIA	maximum pitch angle in degrees,	
PWVEL	velocity of the pitch causing wave in km,	
PWVL	wavelength of the above wave,	
VPWV	the angle between the ships trajectory and the above wave, degrees,	
SHV	the ships velocity in km/h.	

Looking at the listing one notices that after reading the input data, computing some auxiliary quantities and writing some of the input and computed data, the program uses subroutine ARRAY, listed in Fig. C.2, to determine the number of scans intercepting the target moving from the horizon towards the radar. It determines then the detection threshold in subroutine BIAS and DXCGAM, written by Tenne-Sens (1971) and listed in Figs. C.4 and C.8, and continues determining the radar–ground–target geometry on each scan by using the formerly given formulae. The necessary quantities for the determination of signal power are then computed in the main program; the clutter power and its reduction due to the used clutter combating circuitry is determined in subroutine CLUTTR, listed in Fig. C.3, and then returning to the main program the probability of detection of the specified Swerling or Marcum target is computed.

That part of the main program, i.e. lines 717 to 935, which determines the detection probability together with the subroutines shown by Figs. C.5 and C.6, were taken from Blake (1972) but written by White (1968) who transcribed Fehlner's formulae, given in Chapter 3, into expressions computable by a digital computer.

The signal power in the diffraction region may be computed by subroutine DIFREG, listed in Fig. C.9, which uses the function POLAGR, as listed in Fig. C.10, to obtain the values of Kerr's (1951) Fig. 2.20 by a Lagrange polynomial approximation.

The subroutines AERPOD and AERPLD used for the assessment of the Adelaide airport's radar are identical with those listed in Figs. A.4 and A.5. Subroutine SDLB, used for the computation of jammer interference through sidelobe reception, is identical with that listed in Fig. A.3 of Appendix A.

The formulae used in the CLUTTR subroutine, listed in Fig. C.3, are partly those given by Barton (1967) and partly by Nathanson (1969) and others as shown in the comments in the listing.

Figures 11.12 and 11.13 are examples of the model's output. The double lobing of the probability of detection plots, annotated by *PD*, is due to the two frequency operation of the assessed radar, as described in Chapter 11. Curve S shows the dependence of signal power scaled by the coordinate axis on the right, on range. The curve annotated by $C+N$ is the clutter plus noise level which is also read on the right vertical scale.

Statements limiting the extent of the range over which the ground may be considered to be smooth at the used frequency, similar to those of lines 607 to 609 of the COVER model, may be inserted between lines 593 and 594 of model RDPRO, if required.

Statements calling for writing and plotting the computed data round off the program.

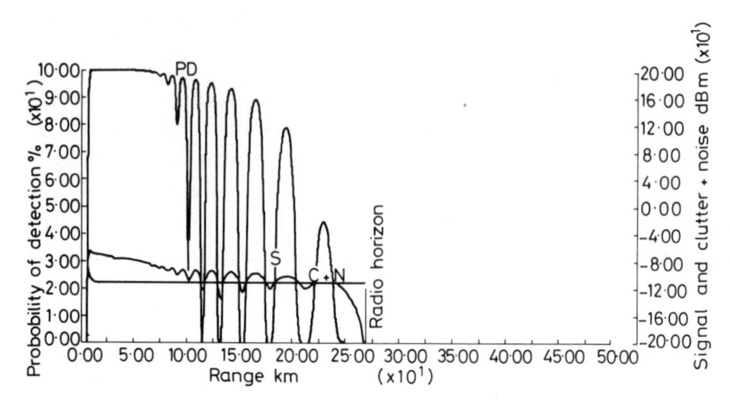

Fig. 10.12 *The performance of a shipborne radar in a sea state 5 environment*
Horizontal polarization, PFA $= 10^{-6}$, pulse width $= 2{\cdot}5\,\mu s$, aerial scan rate $=$ 5 r.p.m., radar height $= 19{\cdot}8$ m, REA $= 25\,m^2$, Swerling case I, target height $=$ 3658 m

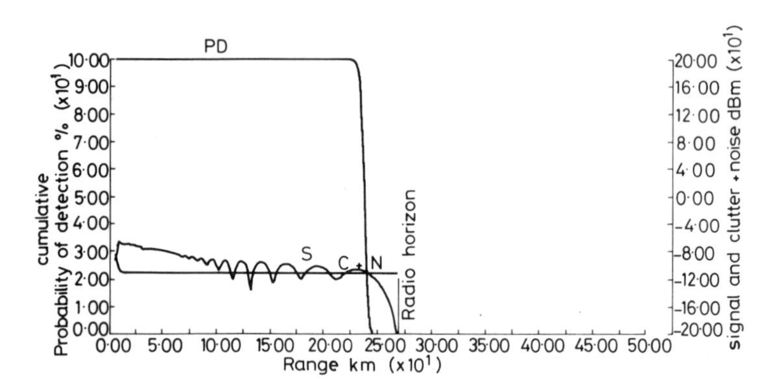

Fig. 10.13 *The cumulative probability of detection of a shipborne radar*
Horizontal polarization, PFA $= 10^{-6}$, pulse width $= 2{\cdot}5\,\mu s$, aerial scan rate $=$ 5 r.p.m., radar height $= 19{\cdot}8$ m, REA $= 25\,m^2$, Swerling case I, target height $=$ 3658 m, sea state 5.

Besides sea and ground clutter, the effect of rain, fog and hail clutter and attenuation may be studied by this model.

Figure 10.12 shows the effect of sea clutter in sea state 5 conditions for a shipborne radar having parameters identical with those of the later assessed radar, except for single frequency operation only, while Fig. 10.13 shows the cumulative probability of detection for the same conditions. It will be noticed that as the grazing angle of the ray reaching the sea's surface at the target's range increases, the originally straight line showing the clutter-plus-noise level, designated by $C+N$, starts rising.

The model is capable of dealing with other than a constant-height target

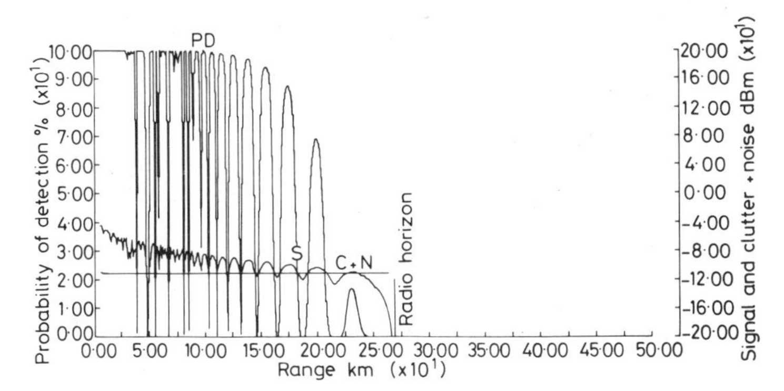

Fig. 10.14 *The performance of the shipborne radar in a sea state 3 environment against a target descending at a rate of 13·5 m/km*
Horizontal polarization, PFA $= 10^{-6}$, pulse width $= 2{\cdot}5\,\mu s$, aerial scan rate $=$ 5 r.p.m., radar height $= 19{\cdot}8$ m, REA $= 25\,m^2$, Swerling case I, target height $=$ 3658 m

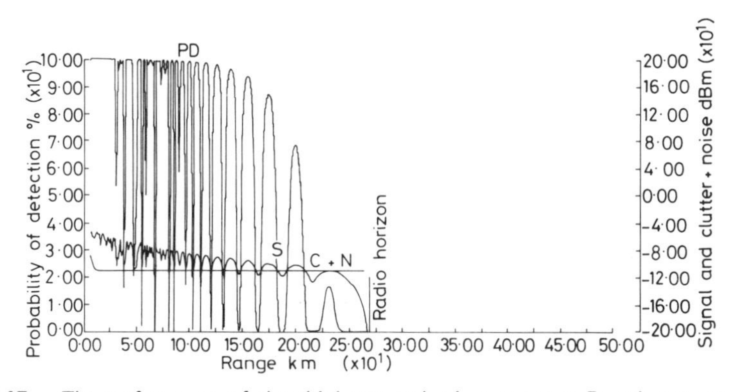

Fig. 10.15 *The performance of the shipborne radar in a sea state 5 environment against a target descending at a rate of 13·5 m/km*
Horizontal polarization, PFA $= 10^{-6}$, pulse width $= 2{\cdot}5\,\mu s$, aerial scan rate $=$ 5 r.p.m., radar height $= 19{\cdot}8$ m, REA $= 25\,m^2$, Swerling case I, target height $=$ 3658 m

trajectory. Figure 10.14 shows the radar, modelled in Fig. 10.12, in a sea state 3 environment when the target descends at a rate of 13·5 m/km as it approaches the radar. Figure 10.15 shows the same changing radar-target geometry in a sea state 5 environment in which the clutter plus noise level understandably starts rising at a greater range than it does in a sea state 3 environment. While the program can compute the dependence of the probability of detection on range of a target flying past the radar, it cannot plot it because the range diminishes while the target approaches the radar and increases while it recedes so that the second part of the plot would overlap the first one.

It should be noted that in the last case the number of hits per scan will be influenced by the relative velocity of the target and the beam of the scanning aerial, as it was described in Chapter 4. The number of hits per scan is given then by

$$k = \frac{\theta \text{ PRF}}{\omega + v \sin \zeta/R},$$

where θ is the aerial's horizontal beamwidth, ω its angular scanning velcoity, v the target's velocity, ζ the angle between the radius from the radar to the target at range R and the target's trajectory at the instant of interception. The statement in line 212 of the model listing should read then

```
451 ASR = SCA * RPM/60.+ VELOC * SIN(ZETA)/R.
```

Extended by a subroutine yielding the variation of the target's radar cross-section with aspect angle, as was proposed earlier, the model can account for the influence of the variation of this parameter on the detection probability.

A modification of the RDPRO model calling for the rejection of ranges at which the probability of detection falls below a specified limit may be used to determine the zones over which detection probability equals or exceeds this limit. An example of the plot produced by a so modified model is shown by Fig. 11.15.

10.4 Other known models

This section compared the described models with available models of other authors.

Two programmes, namely those written by Tenne-Sens (1971) and White (1968) have been mentioned earlier. Besides these, two other programmes due to Blake (1970, 1972) were available for comparison.

Blake's (1970) model computes and plots either the vertical coverage diagram or the signal strength against range of constant altitude targets for a computed or assumed free-space range. By using the approximate formulae obtained from Kerr (1951) it considers the radar–sea–target geometry for vertical and horizontal polarization. The influence of sea waves on dispersion is taken into account by Ament's formula (see Chapter 6, eqn. (6.16)), but scattering is assumed to be of negligible effect.

This model may be compared with the COVER model which, however, is more comprehensive since it is applicable to ground and shipborne radars at vertical, horizontal and circular polarization. COVER considers terrain features and slopes and their influence on the vertical coverage diagram, a feature not catered for by Blake's model.

Blake's second model computes the free-space range of a radar for given probabilities of detection and false alarm or for a given signal-to-noise ratio only. It considers a standard atmosphere in determining atmospheric, galactic and solar noise and atmospheric attenuation and treats post-detection integration only. Blake uses White's numerical solutions of Fehlner's formulae for the computation of detection probabilities.

All three described models consider all three used types of pulse integration and wave polarization and the COVER and MAP models also compute, besides the true slant and ground ranges for a given radar and its environment, the free-space ranges for the given probabilities and the signal-to-noise ratio appropriate to the signal level obtained from a specified target. They compute the approximate atmospheric attenuation by using formulae of eqns. (4.22) to (4.25).

The previously mentioned Tenne-Sens's (1971) model computes and plots the probability of detection of a non-fluctuating, Swerling I and Swerling II type target against the free-space range only. It considers only an approximately cosec^2 aerial and, as published, it cannot use any other aerial types. It is capable of accounting for rain attenuation; it does not compute the various system losses and omits the influence, amongst others, of the collapsing and operator loss unless these are included as inputs under the heading of miscellaneous system losses.

The most comprehensive of the four mentioned models of other authors is that due to White (1968). It simulates clutter and target motion throughout the engagement. The results of the computations are presented in the form of tables or plots of target motion during the engagement, of target cross-section variations, of the radar–target geometry, single scan and cumulative detection probabilities and signal and clutter levels.

A simplified aerial characteristic with a constant gain over the solid angle defined by the half-power beamwidth is used only.

The model could be likened to the RDPRO program. However, it falls short of this model in its approximation of the aerial coverage diagram and of the reflection coefficient and phase shift on reflection, its omission of considering ground clutter and circular polarization, its accounting for two sea states only, and its omission of taking into consideration the influence of terrain slope which may affect the radar's performance to a great extent.

While the here described model was written for search/surveillance radars, White's model is applicable to these and also to tracking radars.

The above comparisons show that the proposed models are more versatile and more generally applicable since they consider a wider range of parameters which affect a search or surveillance radar's performance; they are therefore likely to yield more realistic results.

Besides that, no model comparable to MAP has, to the author's knowledge, been published so far; it is therefore unique and singularly suited to the previously outlined applications.

Chapter 11

Search radar performance assessment

Having dealt with the radar parameters and radar wave propagation used for the development of the three mathematical models described in the preceding chapters, this chapter demonstrates the application of the models to the assessment of an operational radar at Adelaide's West Beach Airport. Besides this, results of some other assessments are presented.

11.1 Specifications and description of the studied radar

The radar at Adelaide's West Beach airport will be used for demonstrating radar performance assessment.

11.1.1 Parameters

The parameters of the radar used at the above airport are as follows:

Transmitter	
Peak power, MW	2·2
Pulse repetition frequency (PRF), pps	400
Pulse width, μs	2·5
Operating frequency, GHz	1·3 and 1·345
Receiver	
Noise figure, dB	4·5
Intermediate frequency (i.f.), MHz	30
Bandwidth, MHz	0·60
Aerial	
Gain, dB	31·7
Polarization	Linear or circular (switchable)
Horizontal beamwidth, degrees	1·3
Vertical pattern	cosec^2 (reinforced above 15°) to 45°
Scan rate, r.p.m.	5
Aerial height, m	19·8

Only the currently used parameters are listed although a wider range is available.

There are two identical transmitters operating alternately; one transmitting a pulse at a frequency of 1·3 GHz followed by a pulse transmitted by the other transmitter at a frequency of 1·345 GHz. Besides partly filling the gaps between lobes by frequency diversity (see Chapter 4), the two transmitter operation offers a measure of safety since the break-down of one transmitter does not curtail radar cover of the air space of interest and thus deprive the airport authority of information about approaching or departing aircrafts.

A secondary radar, sharing the primary radar's aerial mount, has been installed a few years ago. However, only the primary radar is of interest and therefore no data on the secondary one will be given.

11.1.2 Site

The airport at West Beach, Adelaide lies between the city of Adelaide and the sea. It is almost due west of the city and about 3 km from the sea shore which is running in a roughly NNW–SSE direction. While the airport site was originally bordered by unbuilt-up areas of land, e.g. swamps in the north, the inevitable suburban sprawl caused its being almost completely encircled by buildings, both industrial premises and dwellings. Even the swamps were reclaimed and built upon.

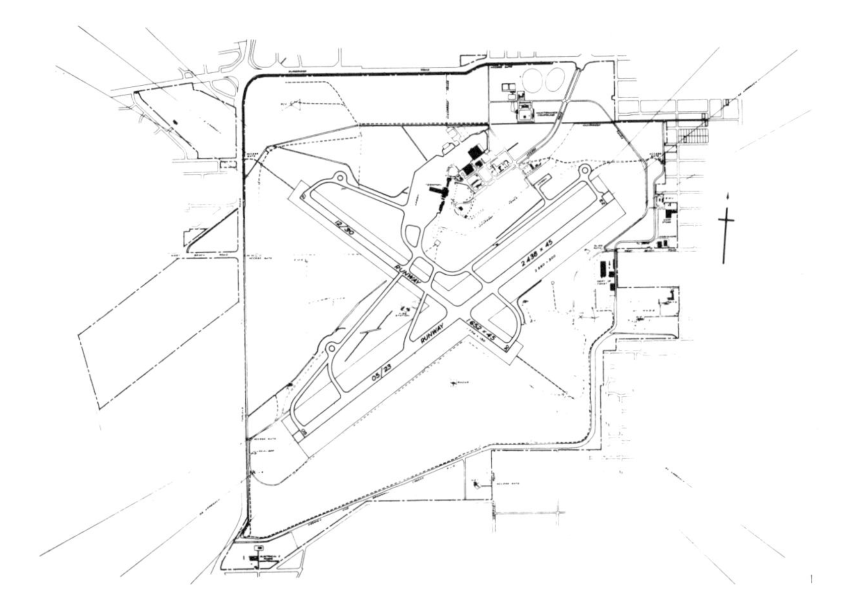

Fig. 11.1 *The Adelaide West Beach Airport*

The airport itself, at an altitude of about 2·5 m above sea level, occupies an irregular area shown by Fig. 11.1 (scale 1 : 50 000). There are two runways, the longer one in a SW–NE direction and a shorter one, intersecting the first one and, running in a NW–SE direction. The administration and passenger terminal buildings are in the northern wedge between the intersecting runways, while the radar is in the southern one. The great variation of the distance between the radar and the airport's boundary within which the ground may be considered to be smooth, is of importance, as it will transpire later.

Grass covers the soil except for the runways and the tarmac. The soil may be considered to be fairly dry except in the southern winter, the rainy season in Adelaide.

A mesh wire fence, high enough to prevent unauthorized entry to the area but not high enough to affect the radar's performance, runs around the whole boundary.

Radar visibility in the east and south-east is impaired by the Mt. Lofty Ranges, a chain of hills, with Mt. Lofty its highest peak at a height of about 726 m, at a distance of approximately 16 km from the airport.

11.1.3 Search radar performance assessment

A few performance figures which depend on the listed parameters only may be ascertained by short calculations; accuracy and resolution are amongst these. Mathematical models, which have been described earlier, are used for the assessment proper.

Using some data computed by one of the models one may show that, with 17 hits per scan, a free space range of 162 km and a signal-to-noise ratio of 5·85 dB for a probability of detection of 0·5 and a false alarm probability of 10^{-5} is obtained. These figures change to a free space range of 268 km and a 2·19 dB signal-to-noise ratio when a probability of detection of 0·2 at the same probability of false alarm is considered.

The range accuracy for these two cases, using eqn. (8.1), is 65 m for the higher probability of detection and 150 m for the lower one, while the angular accuracy, from eqn. (8.6) for the rectangular aperture aerial in the horizontal plane, is 3·7 and 8·6 milliradians respectively.

The resolution in range and azimuth with the safety factor $m = 1{\cdot}5$, half-way within its recommended range, when using eqns. (8.12) and (8.13), is 563 m and 0·034 radians respectively. The formula expressing the bandwidth dependence of the range resolution is not applicable for this radar since the bandwidth–pulse-width product $B\tau$ is much higher than unity. Two identical targets approaching from the same direction must be separated by 563 m in range and the same targets when equidistant from the radar must have an angular separation of 0·034 radians or 3·4 km at a range of 100 km in order to be resolved as separate targets.

Three mathematical models are used, with the previously listed parameters, for the assessment of the performance of the Adelaide airport's radar.

In order to achieve realistic results it is important to use for the assessment aerials with vertical coverage diagrams closely resembling those actually used.

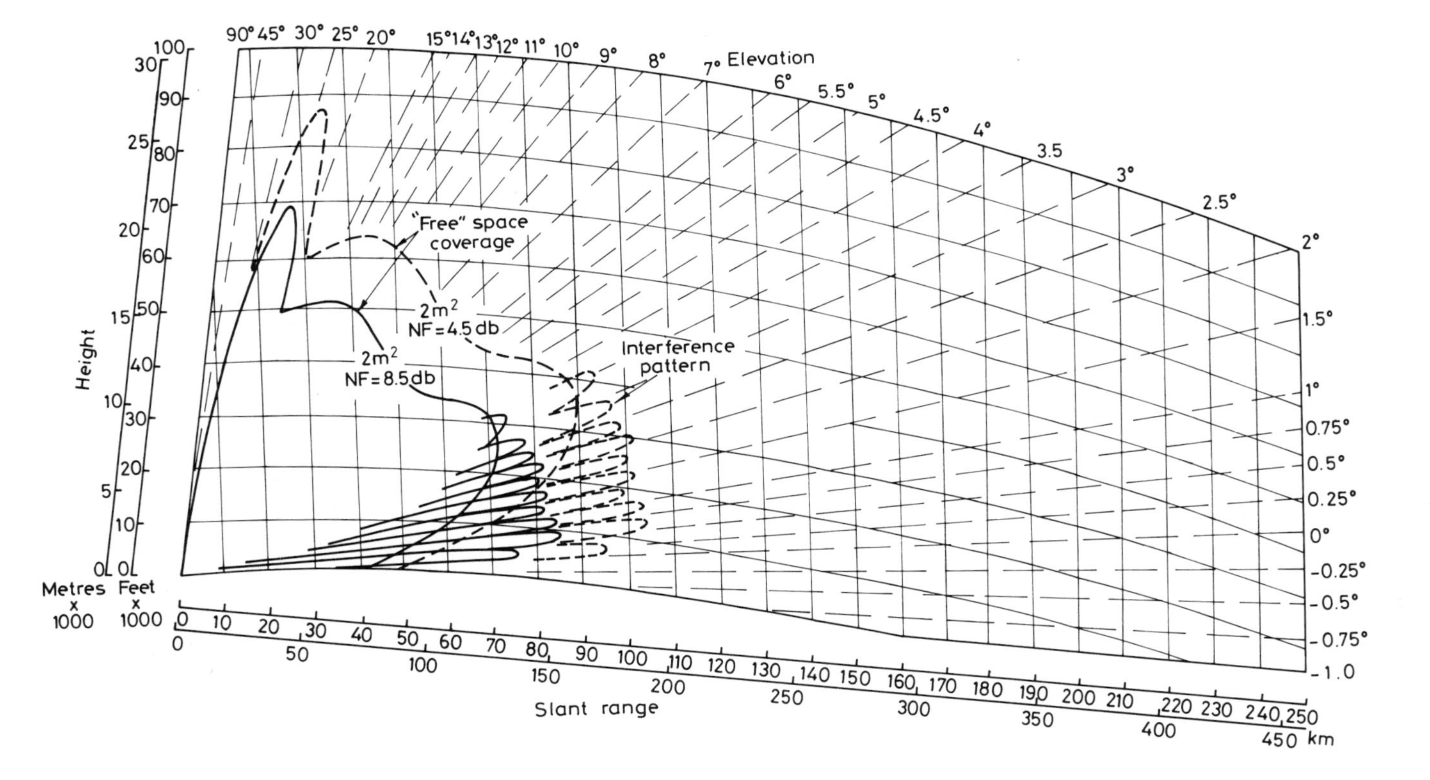

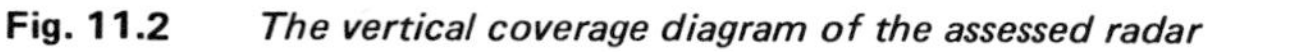

Fig. 11.2 *The vertical coverage diagram of the assessed radar*

Unfortunately manufacturers' specifications are seldom adequate for writing algorithms leading to the sought coverage diagrams unless the aerial is of a classical configuration such as a paraboloidal reflector with a circular or rectangular aperture and given aperture illumination. In the majority of cases the modeller has to accept either the often inadequately defined coverage diagram sketches in sale literature or equipment handbooks or try to fit a curve to a coverage diagram obtained by flight tests performed in connection with the radar installed on site. Since range is a function of the signal-to-noise ratio and this depends on the probability of detection and false alarm, the latter method of coverage diagram derivation, depending on the subjective assessment of paint brightness on the display by the operators, is not accurate.

A plot of the vertical coverage diagram of the radar of interest, shown in Fig. 11.2 obtained by courtesy of the Department of Aviation, was used for the following assessment. An algorithm for use in all three mathematical models was derived by a method developed for this purpose as described earlier. Figure 11.3 shows by a dashed line the given free space coverage diagram redrawn in orthogonal coordinates in a convenient scale from that given by the previous figure and by a full line the computer plot of the coverage diagram based on the derived algorithm.

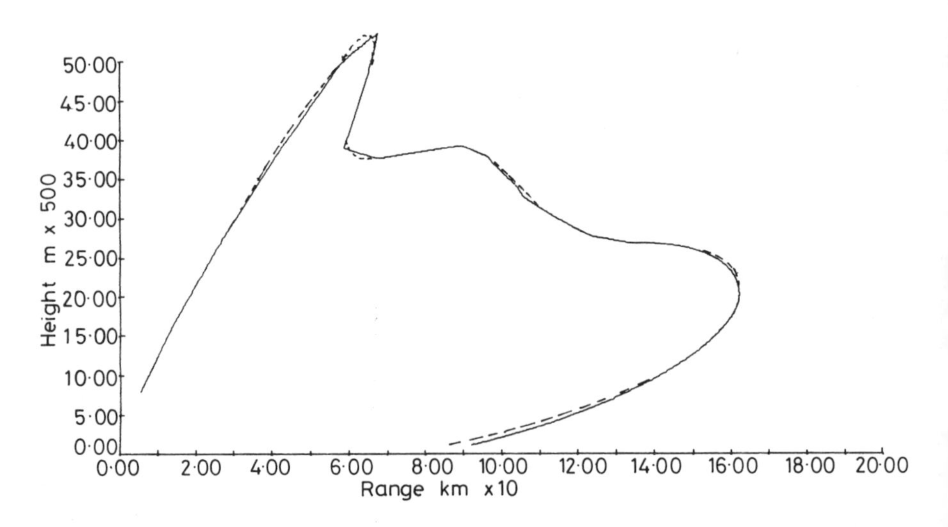

Fig. 11.3 *The given and the modelled vertical coverage diagram*

The deviation of the computed from the given curves is less than 10 per cent in range, an accuracy considered to be adequate for the task at hand.

Figure 11.4 shows the computer plot of the vertical coverage diagram of the radar for a probability of detection of 0·2, the choice of which probability of detection will be justified later, and a probability of false alarm of 10^{-5}. A Swerling type I target having an echoing area of 25 m^2 corresponding to that of a Fokker F27 aircraft was used in the model and the aerial was tilted 1·5° upwards to simulate the 'nominal' 2° tilt of the aerial at the airport.

Lobing extends to a ray elevation angle of approximately 10° only. Beyond that the aerial gain for higher depression (negative elevation) angles decreases and causes the reflected ray's field to diminish to such negligible levels that it does not support lobing.

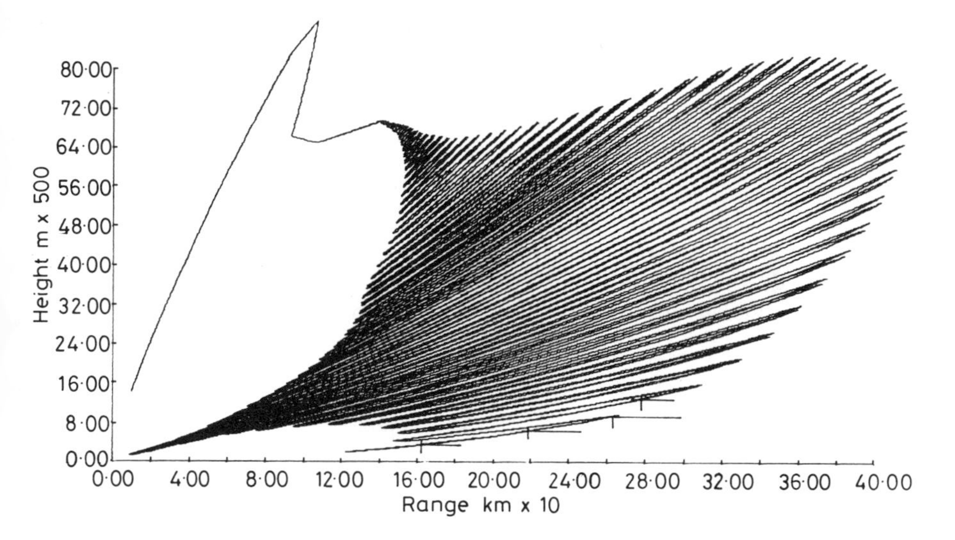

Fig. 11.4 *The lobing pattern of the assessed radar looking westward. Horizontal polarization, $P_d = 0{\cdot}2$, $P_{fa} = 10^{-5}$, $\delta = 25\,m^2$, aerial height = 19·8 m, aerial tilt = 15°, Swerling I target*

The effect of frequency agility is shown by the double lobing pattern. The difference of frequencies of 1·3 and 1·345 GHz at which the transmitters operate alternately is not adequate to affect the coverage diagram at low angles of ray elevation and no benefit from frequency diversity is obtained where it would be most desirable.

The total phase difference between the direct ray and the reflected one for producing lobe maxima, assuming a plane earth, is

$$2\pi n = \mu + \Delta R \frac{2\pi}{\lambda},$$

where n is the lobe number counting upwards from the bottom, as described in Chapter 5, μ is the phase shift on reflection and

$$\Delta R = 2h \sin \epsilon$$

with h the aerial height and ϵ the ray elevation angle.

From here

$$\epsilon = \sin^{-1}\left[\frac{\lambda}{2h}\left(n - \frac{\mu}{\lambda}\right)\right]$$

and since

$$\sin^{-1}(x) = x + \frac{x^3}{6} + \frac{3x^5}{40} + \ldots \quad \text{for} -1 < x < 1,$$

the difference of the elevation angles of the lobe maxima at the two used frequencies and low elevation angles becomes

$$\Delta\epsilon \doteq x_1 - x_2 + \frac{1}{6}(x_1^3 - x_2^3) \ldots$$

$$\doteq \frac{1}{2h}[n(\lambda_1 - \lambda_2) - (\mu_1 - \mu_2)]$$

with μ_1 and μ_2 the phase shifts on reflection at the wavelengths λ_1 resp. λ_2. However, since for horizontal and circular polarization the phase shift on reflection is

$$\mu_1 = \mu_2 \doteq \pi = \lambda/2,$$

the above expression becomes

$$\Delta\epsilon = \frac{n}{2h}(\lambda_1 - \lambda_2)$$

showing that for the used aerial height of 19·8 m the difference in the angles of the lobe maxima, when viewed from the radar's aerial, are for the first lobe $\Delta\epsilon = 1{\cdot}9 \times 10^{-4}$ rad and for the twentieth lobe $\Delta\epsilon = 3{\cdot}8 \times 10^{-3}$ rad. Doubling of lobes starts to be noticeable at the third lobe where $\Delta\epsilon = 5{\cdot}7 \times 10^{-4}$ rad, however, proper gap-filling seems to start only at about the tenth lobe.

At the specified probabilities detection of targets approaching the radar at heights of 1524 m (5000 ft), 3048 m (10 000 ft), 4572 m (15 000 ft) and 6096 m (20 000 ft) is predicted to occur at about 162, 220, 264 resp. 278 km as shown by Fig. 11.4.

Figure 11.5, also supplied by courtesy of the Department of Aviation, shows the detection ranges of an identical inbound target flying at the above altitudes to be 162, 220, 269 and 280 km respectively. The exact determination of the detection ranges from this figure is again dependent on subjective cathode ray tube paint intensities.

The coverage diagram was originally computed and plotted for a probability of detection of 0·5 at which the lower lobes do not extend far enough to intersect the target trajectories at the remote ranges shown by Fig. 11.5 and it was re-modelled therefore at the low probability of detection of 0·2.

The legend states that circular polarization was used during the operations yielding results on which Fig. 11.5 is based. With circular polarization, however,

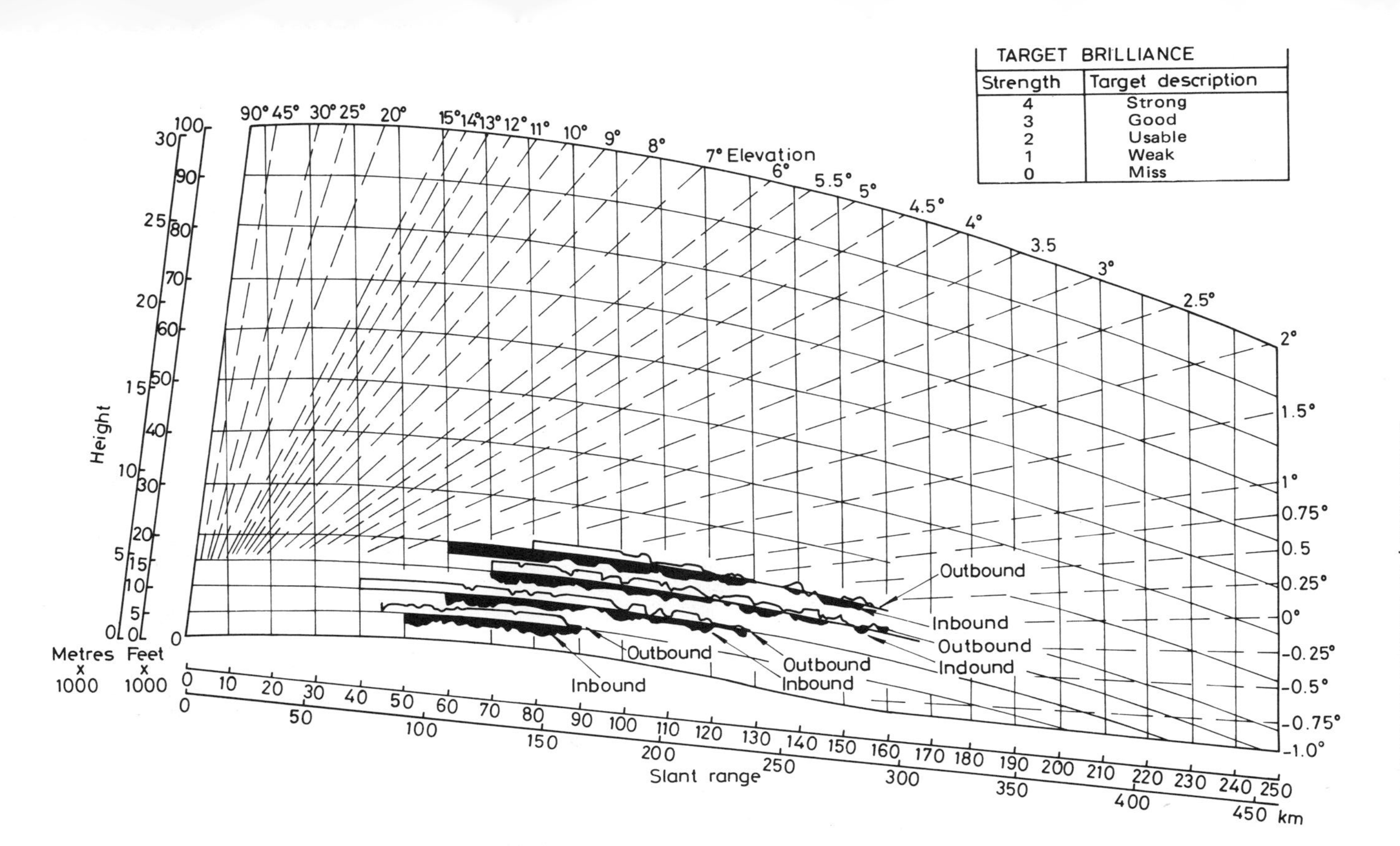

Fig. 11.5 *The detection ranges of the assessed radar looking westward*

the target cross-section is reduced by 6 to 8 dB, as stated in Chapter 4, so that the original 25 m^2 cross-section becomes about 5 m^2 only and causes the reduction of the detection ranges by about 1·75 dB, or a range of, e.g. 150 km obtained with horizontal polarization is reduced to about 100 km with circular one. It could be argued though that in these trials the operators knew when and from what direction targets will arrive so that operator loss, which for a probability of detection of 0·5 is about 7 dB, counter-balances the target cross-section reduction. With this modification ranges approaching those on Fig. 11.5 could be achieved as shown by Fig. 11.6 where ranges of about 155, 214, 260 resp. 273 are computed for the given target at the heights of interest and circular polarization.

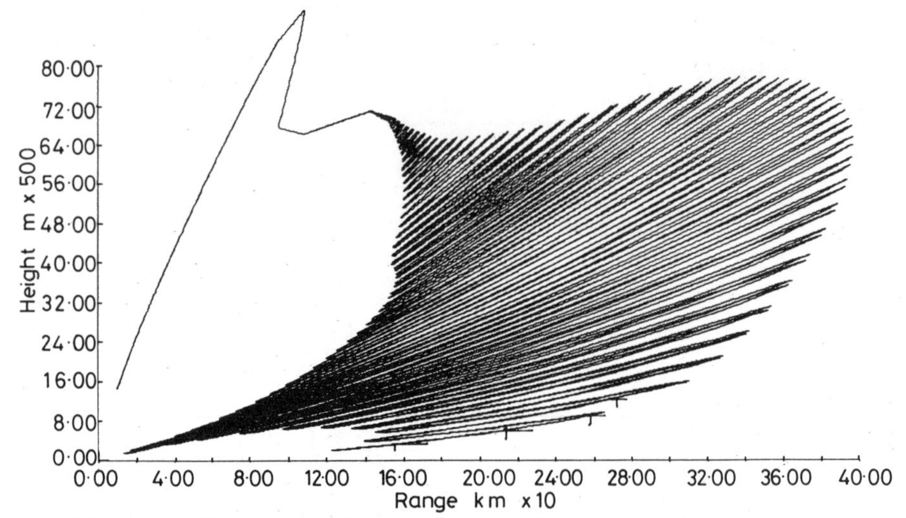

Fig. 11.6 *The lobing pattern of the assessed radar looking westward. (Circular polarization) For data see Fig. 11.4*

All the above ranges are at such an azimuth angle that the radar is looking westwards towards the sea. Looking eastwards, the bottom part of the coverage diagram is curtailed by the Mt. Lofty ranges, as shown by Fig. 11.7 computed and plotted for an azimuth angle of about 94° geographic. A plot of the detection ranges obtained by the Department of Aviation, and reproduced here as Fig. 11.8 by courtesy of the above authority, for targets at 6 and 12 kft altitudes confirm the results of modelling.

While reduction in detection ranges due to terrain features might be tolerable in some radar applications, there are others where it would be objectionable and where the necessity of a thorough survey of the intended site is intimated.

The second model, the MAP model, which was described earlier, simulates such a survey based on topographical data obtained from maps of the area of interest.

Figure 11.9 shows, for the previously used probability of false alarm and a probability of detection of 0·5, plots of the detection ranges of the 25 m^2 radar cross-section target at the earlier used four height of 5, 10, 15 and 20 kft when viewed by the assessed radar at all azimuth angles starting at 0° due north and progressing clockwise in 1° steps to 360°, while Fig. 11.10 shows the same for the target at 6 and 12 kft altitudes. Both figures are for horizontally polarized radiation.

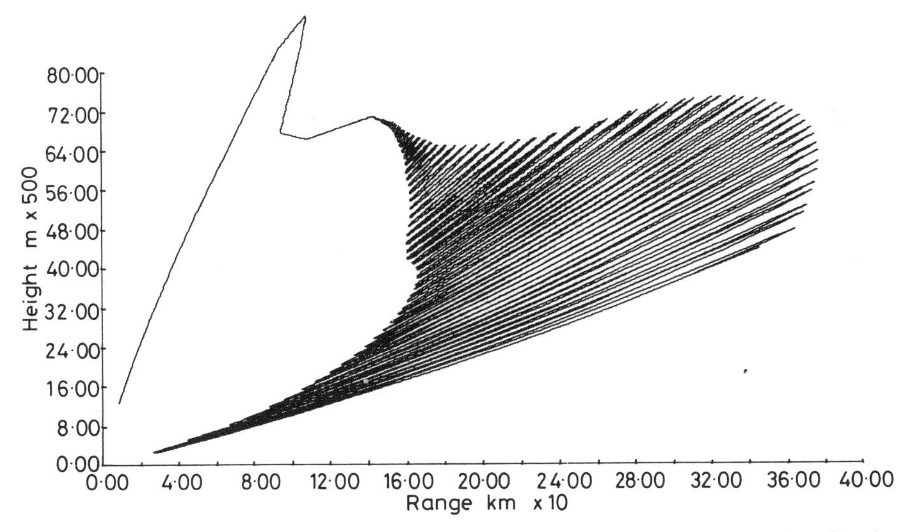

Fig. 11.7 *The lobing pattern of the assessed radar looking towards Mt. Lofty. Circular polarization. For data see Fig. 11.4*

These plots superposed on a map drawn in the appropriate scale permit the determination of the geographical locations above which the target flies at the likely instant of its detection with the specified probabilities.

The shown contours are rather irregular in shape. Their irregularities on the eastern side are due to the varying heights and distances of the previously mentioned hills. The step-wise irregularities in the south-westerly, westerly and north-westerly directions are due to the irregular shape of the airport's area within which reflection occurs from fairly level ground and the reflected ray is well developed. Since irregularities in the terrain rise rapidly beyond the airport's boundary, the reflected ray is scattered when reflection occurs there. The reflected ray is diminished in intensity then and the resultant field which is a vector sum of the direct and reflected contributions is diminished and the detection ranges are shortened. This is particularly noticeable on the 20 kft contour which yields shorter detection ranges than does the 15 kft one. The reason for this has been touched upon earlier when it was stated that the coverage diagram (Fig. 11.6) has to be modelled for a probability of detection of 0·2 in order to achieve the detection ranges shown by Fig. 11.5. The shorter ranges of the 20 kft contour of Fig. 11.9 are caused by the target's being intercepted at this altitude by a lobe of higher order. This is also borne out by ranges indicated for an inbound target at this altitude and a signal level slightly above half the maximum, possibly close to a

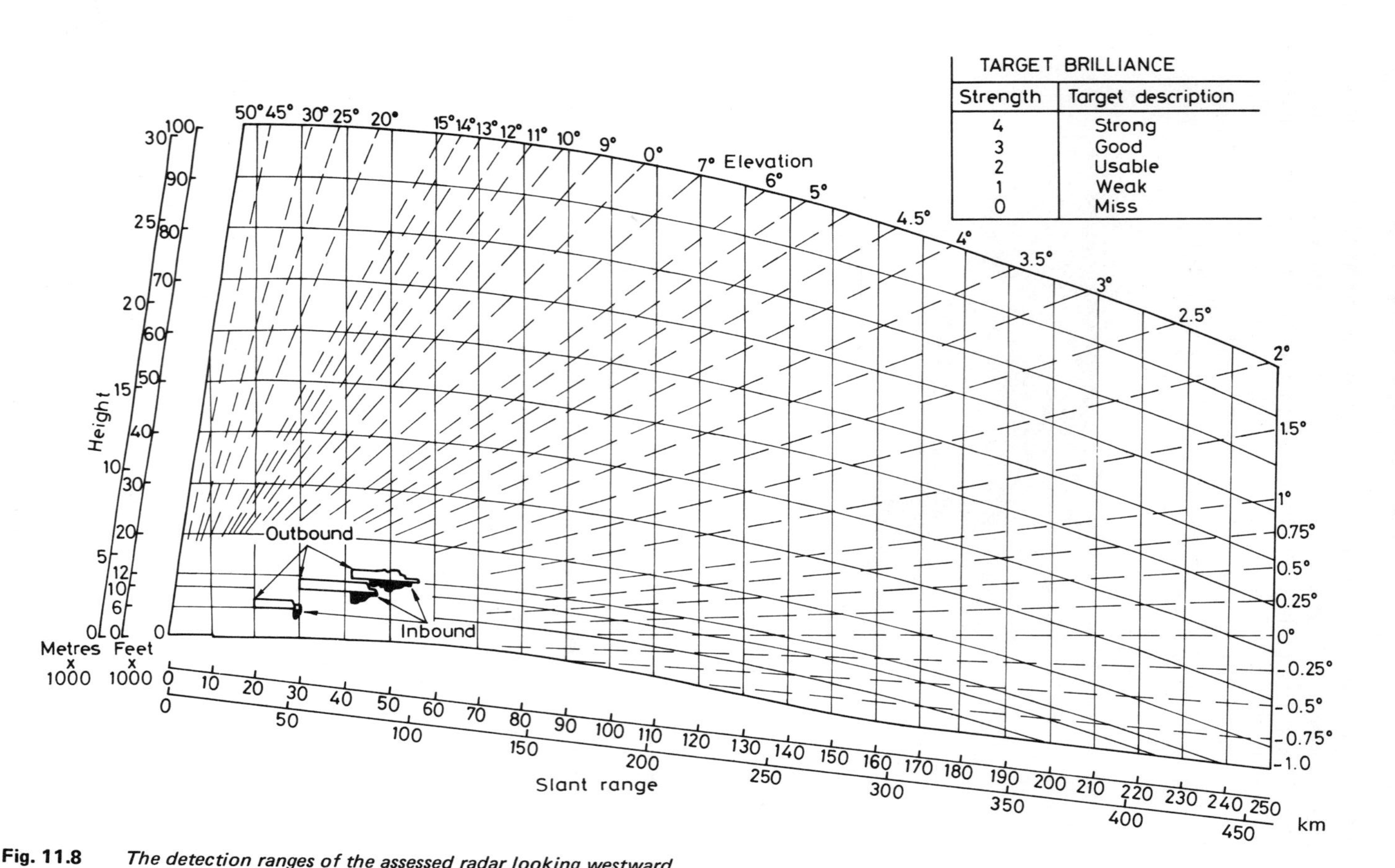

Fig. 11.8 *The detection ranges of the assessed radar looking westward*

probability of detection of 0·5, shown by Fig. 11.5 and used for the computation of Fig. 11.9.

The variation of the probability of detection of a specified target flying at a given altitude from a westerly direction towards the radar is shown by Fig. 11.11 produced by the third model, the RDPRO model. There are three curves on this plot. One, annotated by PD, demonstrates the dependence of the probability of detection at the specified probability of false alarm on range. The second curve, marked by S, shows the signal level in dBm as a function of range, while the third curve, marked $C + N$, depicts the noise level and the dependence of clutter, both in dBm, on range.

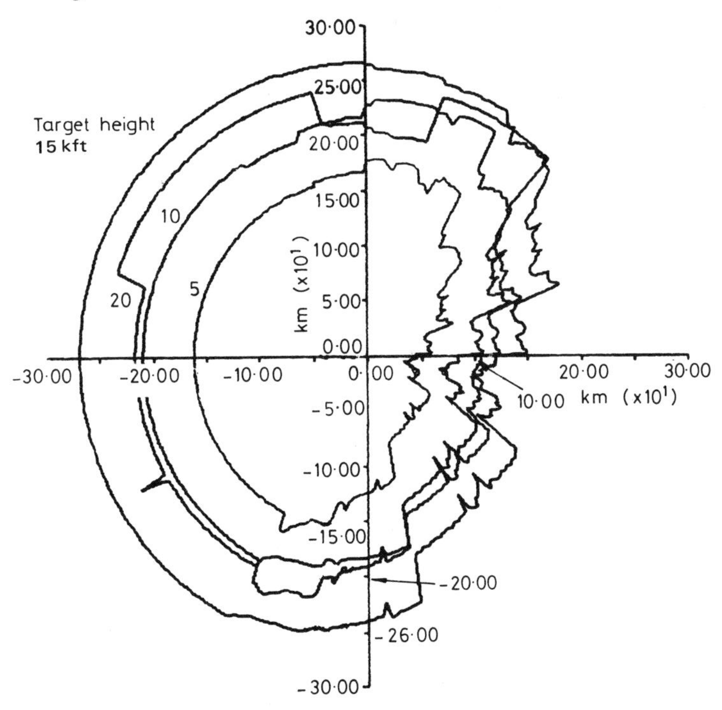

Fig. 11.9 *Detection map contours for the target of interest at the four considered altitudes Probability of detection 0.5*

When the radar looks westwards the signal level starts rising almost at the radio horizon, at 341 km for a target height of 20 kft, and as it approaches the noise level the probability of detection reaches, in the first lobe, the value of about 0·055 to drop again to zero and rise again to about 0·37 when the target crosses the second lobe. After that the probability of detection follows the lobing pattern, rising when the target crosses the lobes and falling when it is within the gaps between them. It is only within the third lobe that the probability of detection reaches a value of 0·5 proving the necessity of choosing the low value of 0·2 for modelling the coverage diagram shown earlier.

Figure 11.12 shows the above dependencies for the radar looking eastward at a target at an altitude of 12 kft. The target cannot be detected while it is shadowed by the previously mentioned hills, but the signal level and hence the probability of detection rise rapidly as soon as the target becomes visible as shown by the truncated coverage diagram of Fig. 11.7.

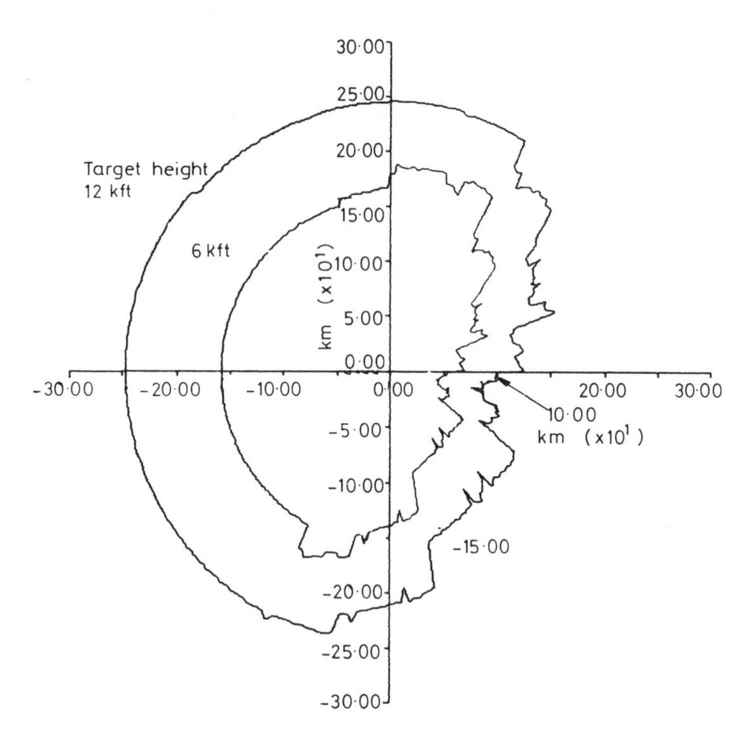

Fig. 11.10 *Detection map contours for the considered target at 6 and 12 kft altitudes Probability of detection 0.5.*

The choice of target heights used in the assessment was dictated by those used by the Department of Aviation (then Department of Civil Aviation) in Figs. 11.5 and 11.8 in order to permit the comparison of the measured and the computed ranges.

Having computed data necessary for the assessment of the performance, one may ask whether the radar fulfills the requirements placed on a search or airport surveillance radar.

While Ward, Fowler and Lipson (1974) have published typical airport surveillance radar requirements, examining the published data of some 25 search and surveillance radars shows that the performance figures vary to a great extent and the requirements of individual users will therefore depend on the intended application of the radar.

The average range resolution of the above range of radars is about 695 m when using the previously mentioned safety factor $m = 1{\cdot}5$, while the average azimuth resolution, for the same safety factor, is $2{\cdot}1°$. On this basis, the assessed radar with its range resp. azimuth resolution of 563 m and $1{\cdot}95°$, is well within the range of the above published resolutions of surveillance and search radars.

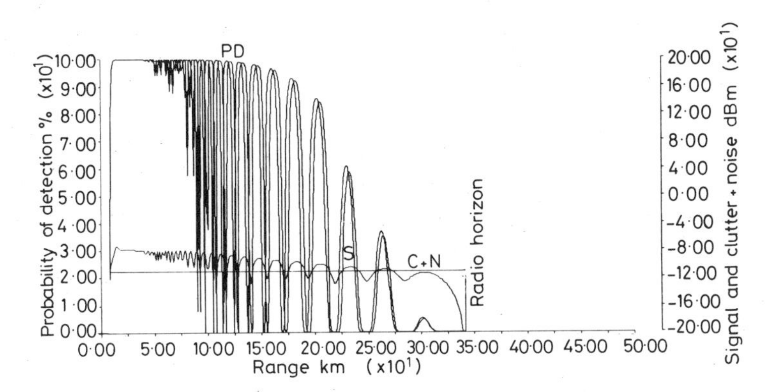

Fig. 11.11 *The probability of detection and signal level versus range for a target approaching from the west. For data see Fig. 11.12*

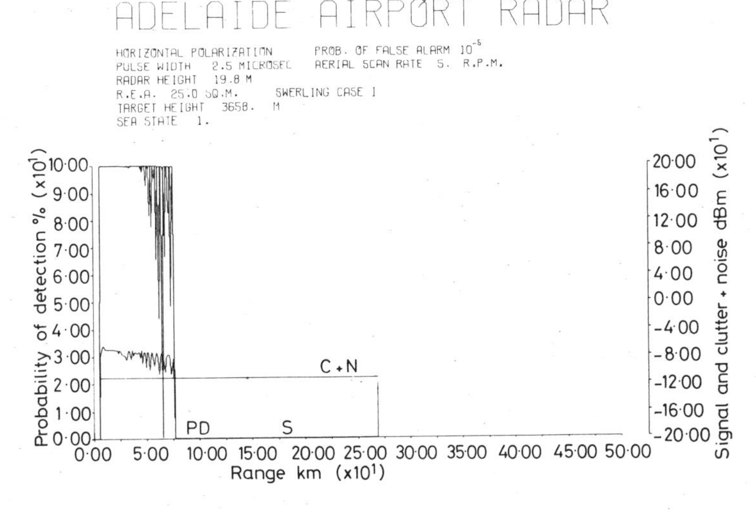

Fig. 11.12 *The probability of detection and signal level versus range for a target approaching from the east*

While the listed range performance of seven of the above 25 radars is for targets specified by their radar cross-section, that of the others is not, thus making it impossible to use data of all these radars for obtaining an average range performance specification of search and surveillance radars. The average detection range of the mentioned seven radars, considering a target having a 25 m^2 radar cross-section, is 574 km which is greater than that of the Adelaide airport's radar. However, the above figure is inadequately defined since no indication of the probabilities affecting detection is available. The detection ranges of the assessed Adelaide airport radar are such that with the current velocities of targets of interest adequate time for alerting ground staff is available and for this reason it may be assumed that the range performance is adequate for the operating authority's requirements.

Judging by the coverage diagram shown, Figs. 11.2 and 11.4, and considering that the ceiling height of targets of interest is below about 12 km, it becomes obvious that the assessed radar has adequate altitude coverage for requirements envisaged in this respect for a number of years.

Summing up it may be said that the assessed radar, as far as accuracy, resolution and coverage is concerned, fulfills all requirements placed on an airport surveillance radar used at a not very busy commercial airport. Higher resolution would be required from a military radar or a radar at a busier civilian airport in order to cope with denser air traffic.

11.1.4 Polarization

The assessed radar has provision for selecting either linear or circular polarization for its operation. Since the legends of Figs. 11.5 and 11.8 state that circular polarization was used during the measurements yielding the indicated ranges, this section investigates whether this mode of operation is justified considering Adelaide's weather pattern and circular polarization's effect on target cross-section.

11.2 Meteorology of the investigated radar's environment

Adelaide which lies at 138°35′ Eastern Longitude and 34°58′39″ Southern Latitude and is about 45 m above sea level enjoys a Mediterranean climate with a dry summer and a winter which, due to the influence of westerly winds containing cold air masses from the polar regions, is humid (Blüthgen, 1964).

Adelaide is the capital of South Australia, the driest state of the Commonwealth. 82.8% of South Australia's area of 973 000 km^2 receives less than 250 mm of rain yearly. Adelaide itself has a mean annual rainfall of about 535 mm distributed over 122 days with rain of at least 0·25 mm with maximum rainfall in the southern autumn and winter between May and October. The mean yearly rainfall in June, the wettest month, as based on 119 years' observation, is about 76 mm (Bureau of Meteorology, 1961), while it is about 66 mm in July and 63 mm in August.

Bridgewater, a small township in the hills at the east to south-east of Adelaide, situated at 138°48′ Eastern Longitude and 35°00′ Southern Latitude at a height of about 400 m above sea level and lying close to one of the examined flight paths has a mean yearly rainfall of about 1090 mm spread over 131 days with rain of at least 0·25 mm as observed over 80 years, with about 170 mm June, 157 July and 152 mm August mean rainfall.

Based on the published percentage frequency distribution of monthly rai over the 119 years of observation (Bureau of Meteorology, 1961) 48% of Adelai June precipitation was in the range of 50 to 100 mm, 27% within 25 to 50 mm and 18% within 100 to 150 mm, while the percentages for July were 53, 23 and 16 and for August they were 54, 33 and 7 respectively. Those for Bridgewater, over 80 years of observation, were 29% within the range of 200 to 300 mm, 25% in the range of 100 to 150 mm, 16% between 50 and 100 mm and 15% between 150 and 200 mm. The percentages for July, in the same order as those quoted, were 24, 32, 15, 25 and for August they were 19, 24, 20 and 33.

11.2.1 Meteorological influences on the detection of targets at various heights

As it was pointed out earlier meteorological phenomena affect radar performance by attenuating electromagnetic waves and by causing clutter. Attenuation by weather affects to a certain extent the detection of targets flying at all altitudes since weather manifests its effects mainly at low altitudes and the radar's radiation to a target must pass through a layer of precipitation, cloud, hail or fog and so must the echo signal propagating from the target to the radar. The path-length affected by these phenomena depends, besides the thickness of the prevailing meteorological phenomenon, on the slant range of the target from the radar. However, since the magnitude of the attenuation depends on the density and extent of the precipitation, cloud or fog, and this in turn is a function of altitude above ground, it is not possible to compute the exact attenuation without the knowledge of the exact relevant meteorological data so that weather attenuation computations have more the meaning of a correction of the computed ranges than their exact determination.

The radar at Adelaide's West Beach Airport operates at frequencies of 1·3 and 1·345 GHz or wavelengths of 23 resp. 22·3 cm for which attenuation is not given in any of the available relevant publications. The closest wavelength for which the attenuation by rain is given is 15 cm (Bean, Dutton and Warner, 1970) where, at a heavy precipitation rate of 15·2 mm/h, the attenuation is $1{\cdot}69 \times 10^{-3}$ dB/km. The attenuation at the used wavelength is well below this figure and therefore even if propagation to an arbitrary range of 300 km were all in rain of the above rate, the total attenuation at a wavelength of 15 cm would amount to about 1 dB and even less at a wavelength of 22 cm.

It is then weather clutter and not attenuation which is objectionable at the used frequencies.

Some radars, including the investigated one, use circularly polarized radiation for combating rain clutter. Microwaves alter their polarization when propagating through rain. Depolarization occurs when the incident energy is scattered by spherical or oblate spheroid shaped raindrops. The magnitude of depolarization depends on the rain rate, spatial extent and direction of the rain and the size and shape of the raindrops. In a first approximation leading to an easily understandable explanation of the mechanism of depolarization, one may assume that circularly polarized waves consist of two orthogonal components of equal magnitude. The radar cross-section of single raindrops is negligible at the commonly used radar frequencies, so it is the ensemble of raindrops which may be approximated by a partly transparent sheet of discontinuity in the propagation medium which affects the propagation of electromagnetic waves.

This sheet is vertical when there is no wind and both components, i.e. the horizontal one and the vertical one, reach it at a grazing angle of $\pi/2$ radians. On reflection, both components will remain equal in magnitude but the horizontal one will be delayed in phase by π radians, as shown by Figs. 6.1 and 6.2, or the phase difference between the two components after reflection will be again $\pi/2$ radians but of a sense opposite to that prevailing before reflection.

If $E_z = E \cos \omega t$ is the vertical component, while

$$E_y = E \cos (\omega t - \pi/2) = E \sqrt{1 - \left(\frac{E_z}{E}\right)^2}$$

the horizontal one, then before reflection the magnitude of the circularly polarized wave is

$$E = \sqrt{E_y^2 + E_z^2}$$

with the horizontal component lagging the vertical one by $\pi/2$ radians.

After reflection the vertical component remains $E_z = E \cos \omega t$, while the horizontal one becomes

$$E_y = E \cos (\omega t - 3\pi/2) = E \cos (\omega t + \pi/2)$$

which, when treated similarly to the components of the wave before reflection, yields a wave of an equal magnitude, however, with the horizontal component leading the vertical one by $\pi/2$ radians, i.e. a circularly polarized wave of a sense opposite to the one before reflection.

The radar aerial cannot receive opposite sense circularly polarized waves thus the effect of rain clutter is reduced.

When wind prevails, the reflecting imaginary sheet is no longer vertical and the grazing angle differs from $\pi/2$ radians. This causes not only a difference in the magnitude of the two components as shown by the above two figures but it also changes the phase shift on reflection from π radians to some other angle so that on reflection the incident circularly polarized wave becomes elliptically polarized.

Some surfaces of the target will also change the sense of polarization and are lost as far as radar detection is concerned, others will reflect one or the other of the linearly polarized components as linearly polarized waves of the incident polarization and these will contribute to the signal received by the radar receiver.

The above considerations apply to the direct ray only. The reflected ray becomes elliptically polarized on reflection from the ground since at a grazing angle which is much smaller than $\pi/2$ radians the initial $\pi/2$ radians phase difference of the two components will change due to the π radians shift of the horizontal one and some other phase shift μ of the vertical one. On reflection from the sheet-like discontinuity caused by the rain the horizontal component will be further shifted by π radians, the vertical component, however, will undergo a small phase shift α since the grazing angle of the reflected ray doesn't differ very much from $\pi/2$ radians at this discontinuity.

The phase difference between the two components will, at this instant, be $5\pi/2-(\mu+\alpha)$. Due to another ground reflection of the two components of the reflected ray on their passage back to the radar, the total phase difference between the horizontal and vertical components, when reaching the radar, will be $7\pi/2-(2\mu+\alpha)$.

Circular polarization increases the ratio of wanted target to unwanted rain echo by about 15 to 20 dB. However, it reduces the effective target echoing area by 6 to 8 dB (Skolnik, 1962) and accordingly the detection range in fair weather by 1·5 to 2 dB.

The question is then whether circular polarization should be used at the location and frequencies of the radar in question.

In order to be able to answer it one has to determine whether rain clutter at the rain rates experienced in Adelaide may reach objectionable proportions.

The interfering clutter displayed simultaneously with the target is caused by energy back-scattered by a volume of rain within the half-power beamwidths of the radar which, due to two-way propagation, extends from a distance equivalent to a quarter of a pulse duration ahead of the target to an equal distance behind it.

Assuming a conical beam bounded by a base shaped as a spherical segment of an area $\pi R^2\theta^2/4$ and having a height equal to $c\tau/2$, where θ is the halfpower beamwidth, R the range, c the velocity of propagation of electromagnetic waves and τ the pulse duration, the intercepted volume is the product of these two expressions or, in case of a rectangular beam of beamwidths θ and ϕ, it is approximately

$$\frac{\pi R^2\theta\phi}{4}\,\frac{c\tau}{2}.$$

The scattering cross-section of the rain within this volume is

$$\sigma \doteq \frac{\pi R^2\theta\phi}{4}\,\frac{c\tau}{2}\,\eta$$

with η the total back-scatter cross-section of the particles per unit volume which for rain and matched polarization is (Barton, 1967)

$$\eta = 6\times 10^{-14}\lambda^{-4}r^{1\cdot 6}\quad \mathrm{m}^2/\mathrm{m}^3,$$

while for opposite polarization it is only about one percent of this quantity. The symbol r designates rain rate in mm/h and λ the wavelength.

Figure 11.13 computed for the investigated radar using its parameters of 1·3° for the horizontal 3 dB beamwidth, 6° vertical beamwidth for the lower part of the cosec² vertical coverage diagram, a mean wavelength of 0·227 m and a pulse duration of 2·5 μs shows the rainfall rate required to produce scattering cross-sections or echoing areas of 1 to 1000 m² at distances of 30, 60, 100 and 300 km.

The figure also shows, by a dashed line, Adelaide's maximum heavy rainfall rate averaged over 118 entries spread over 20 years (Bureau of Meteorology, 1918) which, however, due to large temporal and locational changes in rainfall rates, is no indication of the true rate since, e.g. on the 13th March 1917, a rate of 122 mm/h prevailed for 2 minutes and similarly on the 11th May of the same year a heavy downpour of the same rate lasted for 3 minutes. The average maximum rainfall rate shown in the figure yields echoing areas of 3·5, 14, 37 resp. 330 m^2 at the four shown ranges.

No similar data on rainfall rates are available for Bridgewater.

The F-27 Fokker-Friendship aircraft used in the performance study has a radar cross-section of 25 m^2 and is the smallest aircraft on regular passenger flight services of the two major airways using Adelaide's West Beach airport. Besides these there are smaller airway companies providing charter flights and private users with aircrafts having echoing areas far below that given above.

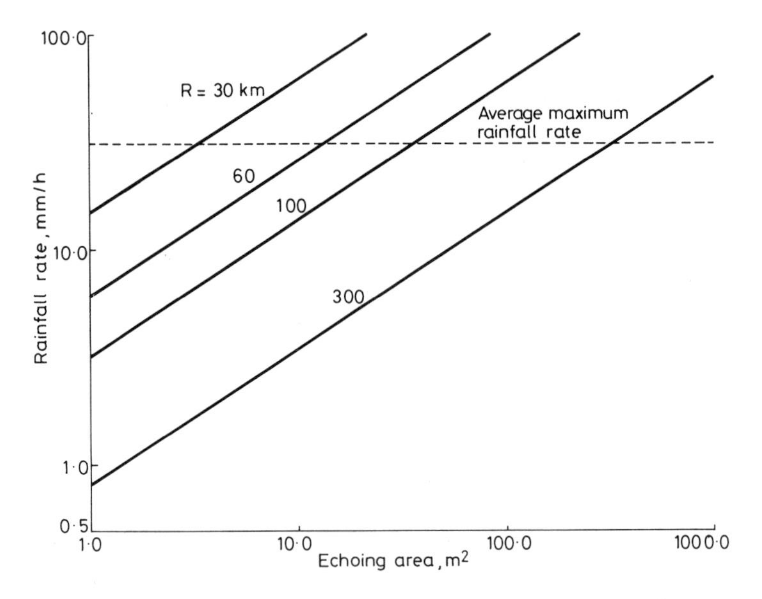

Fig. 11.13 *Required rainfall rate for clutter cross-section of 1 to 1000 m^2 for the assessed radar's parameters*

On the basis of these deliberations it is concluded that there is no justification for using circular polarization in fair weather and, since in the Adelailde area heavy rainfalls are rare and of short duration, care should be exercised with circular polarization due to its deleterious effect on the radar cross-section of targets though this is counteracted to a great extent by the improved target echo to rain clutter discrimination during inclement weather and the reduced operator loss due to the known arrival times and directions of targets of interest.

11.3 Other radar performance assessments

11.3.1 A coastal surveillance radar

A federal authority concerned with the safety of small surface vessels in the near coastal waters north of Adelaide called for tenders for a radar capable of detecting at a range of 20·5 km (11·0 n.m.) a 1 m^2 target at a height of 30 cm above sea with a probability of at least 0·5. Three manufacturers submitted their offers claiming that their radar, when elevated to about 30 m (100 ft), is able to meet the tender's requirements. The data of the accepted tender, designated as Radar A, are shown in Table 11.1.

Table 11.1 *Parameters of the tendered radar*

	Radar A
Peak power, kW	55
Pulse width, μs	0·5 0·2
Pulse repetition frequency, pps	1000 2000
Aerial gain, dB	41·5
Horizontal beamwidth, degrees	0·55
Vertical beamwidth, degrees	2·5
Aerial revolution rate, r.p.m.	40 20
Polarisation	Circular
Receiver noise figure, dB	10
Bandwidth, MHz	2 10

On assessing the radar it was found that it would not fulfill the specified performance requirements when installed as recommended by its manufacturer. The results of this assessment are presented in Table 11.2.

Table 11.2 *Performance of the tendered radar with a 30 m high aerial*

	Pulse					
Radar	Width (μs)	Repetition frequency (pps)	Aerial scan rate (r.p.m.)	Polarisation	Target Swerling	Range (km)
A	0·5	1000	20	Circular	1	10·33
					3	10·46
			40		1	9·31
					3	9·59
	0·2	2000	20		1	9·82
					3	10·00
			40		1	8·67
					3	8·85

The third model, model RDPRO, was then modified in order to determine the height from which the tendered radar might be able to cope with the set task. Assuming the base of the aerial to be about 3 m (10 ft) above sea level, the aerial height was raised in 50 ft steps and the ranges over which the specified target may be detected with at least the required probability of 0·5 when the probability of false alarm is 10^{-6} were computed. When elevated to a height of 500 ft (+ 10 ft), the radar was predicted to just meet the requirements set by the tender.

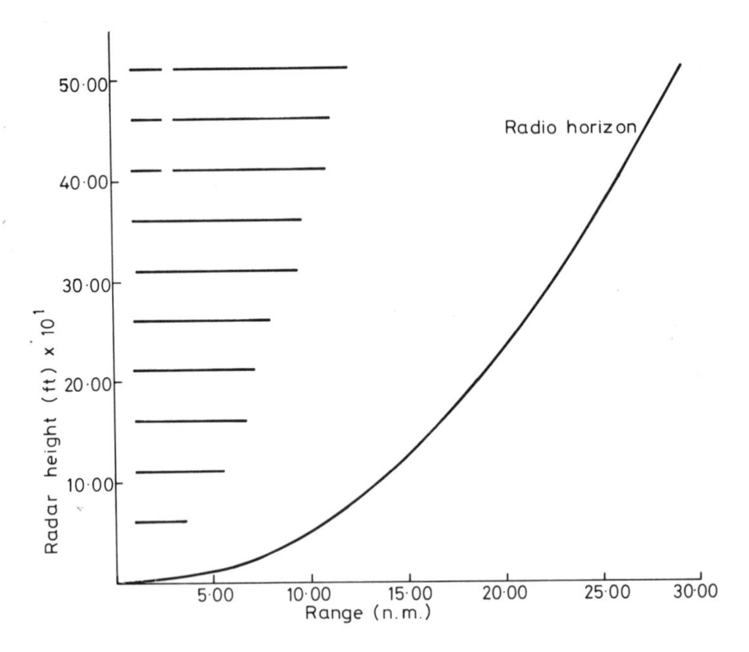

Fig. 11.14 *Detection zones for various heights of radar A. Probability of detection > 0.5, probability of false alarm 10^{-6}; sea state 2*

Figure 11.14 shows the plot produced by the modified model for radar A which was subsequently purchased and installed according to recommendation on a 500 ft (152·4 m) high tower.

On testing this radar in approximately the specified environment (see state 2) it was found to yield a unity blip-scan ratio* at 11·5 n.m. or 21·3 km, while only a 0·1 ratio at 13·85 n.m. (26·65 km). The target was lost in sea clutter at quite short range during a trial in what was estimated to be sea state 5.

* The blip-scan ratio represents the probability of detection on a single scan of the radar's aerial, independent of possible previous detections. It is the ratio of the number of PPI paints produced by a target at a given range to the total number of aerial scans which could have lead to PPI paints.

It is considered that the proving trials demonstrated radar performance showing excellent agreement with the predictions made by the described models, within the limitations imposed by available experimental observational methods.

11.4 Conclusions

Many factors influencing the detection range of radars have to be considered in its computation if an acceptable result is to be obtained. While there are explicitly stated parameters in the radar equation, there are other factors which are not obvious, yet they have a considerable influence on the accuracy of the calculations. Results obtained from the radar equation, and hence based on free-space propagation, have to be corrected to account for, e.g. reflection, unevenness of the surface from which reflection occurs, environmental effects, clutter, sky noise which adds to system noise and thus influences the signal-to-noise ratio, etc.

It is considered that the preceding chapters have adequately covered those topics which must be taken into consideration when predicting or assessing a surveillance radar's performance.

The examples of performance assessment shown above prove that by a careful choice of environmental parameters and attention to details, such as coverage diagram, terrain slopes, terrain features, etc., it is possible to predict and assess a radar's performance with an accuracy compatible with most practical requirements.

It must be borne in mind that environmental conditions which change not only from season to season but may change from hour to hour and are not exactly predictable will affect the accuracy of the radar performance predictions. Other factors unknown with adequate accuracy are the target cross-section and the coverage diagram, besides others, so that to expect a generally applicable prediction accuracy better than 10 per cent is unrealistic. The good agreement of the measured and computed ranges shown for the Adelaide airport's radar is due probably to experience and hence a judicious combination of calculated and assumed parameters as is probably the case in the other prediction.

As it has been shown above, careful attention to details makes acceptably accurate predictions of radar performance by the models described in Chapter 10 possible. Some modifications of the models may be necessary to suit some special radar features not accounted for by the models or some operational requirements of some computers, however, as it has been demonstrated, the models are directly applicable to the performance prediction and assessment of most surveillance radars.

Appendix A

Listing of model 'COVER'

```
 1          PROGRAM COVER(INPUT,OUTPUT,TAPE5=INPUT,TAPE6=OUTPUT,DEBUG=OUTPUT
           1)
        C   USE L EQUAL TO 1 FOR THE COMPUTATION AND PLOTTING OF THE FREE
        C        SPACE COVERAGE DIAGRAM.
 5      C   L EQUAL TO 2 IS USED FOR THE COMPUTATION AND PLOTTING OF THE LOB-
        C        ING PATTERN IN A UNIFORM ATMOSPHERE.THE EARTH RADIUS IS
        C        MODIFIED TO SUIT THE SPECIFIED REFRACTIVITY.
            DIMENSION  W(91),V(91)
            DIMENSION  RC(1981),HEIGHT(1981)
10          DIMENSION  XA(1983),YA(1983)
            DIMENSION TITLE(3)
            COMMON  PI,DEG,V,VW,W
            READ(5,1)   (W(I),I=1,91)
          1 FORMAT(16F4.2)
15          READ(5,2)    (V(I),I=1,91)
          2 FORMAT(10E7.1)
            READ(5,10) L
         10 FORMAT(2X,I1)
        C   GRAD IS THE N.B.S. CONSTANT CE
20          READ(5,20) GRAD
         20 FORMAT(2X,E15.8)
        C   REF IS THE N.B.S. CONSTANT NS * 10 TO MINUS 6
            READ(5,25) REF
         25 FORMAT(2X,E14.8)
25          READ(5,30) PERM
         30 FORMAT(2X,F4.1)
            READ(5,32) RHO
         32 FORMAT(2X,F6.3)
            READ(5,34) ATM
30       34 FORMAT(2X,F8.6)
            READ(5,40) UNDUL
         40 FORMAT(2X,F5.1)
            READ(5,50) SHORE
         50 FORMAT(2X,F8.2)
35          READ(5,52) AERSEA
         52 FORMAT(2X,F9.2)
            READ(5,70) SEA
         70 FORMAT(2X,F5.1)
        C   IPOL EQUALS 1 FOR VERTICAL POLARIZATION,2 FOR HORIZONTAL AND 3 FOR
40      C   CIRCULAR POLARIZATION.
            READ(5,80) IPOL
         80 FORMAT(2X,I1)
            READ(5,82) AERHIT
         82 FORMAT(2X,F6.1)
45          READ(5,84) HBMWTH
         84 FORMAT(2X,F5.2)
            READ(5,86) VBMWTH
         86 FORMAT(2X,F5.2)
            READ(5,110) TILT
50      110 FORMAT(2X,F5.2)
        C   IF A SECTOR SCAN IS USED AND NEITHER THE ANGULAR SCAN RATE ASR NOR
        C   THE APPROPRIATE RPM IS GIVEN,RPM IS TO BE DETERMINED FROM THE FULL
        C   SCAN RPM AS 360. * RPM/SCA AND USED AS INPUT IN THE PROGRAMME.
            READ(5,454) ASR
55      454 FORMAT(2X,F7.2)
        C   RPM IS THE AERIAL SCAN RATE WHICH IN CASE OF A SECTOR SCAN
        C       MAY DIFFER FROM THAT USED FOR A 360 DEGREE SCAN (SEE THE PRE -
        C       CEDING COMMENT).
            READ(5,120) RPM
60      120 FORMAT(2X,F7.2)
        C   SCA IS THE ANGLE OF A SCAN CYCLE,E.G. IT IS 360.00 FOR A CIRCULAR
        C        SCAN AND E.G. 80.00 FOR A PLUS-MINUS 40 DEGREE SCAN.
            READ(5,31) SCA
         31 FORMAT(2X,F6.2)
65          READ(5,130) PKW
        130 FORMAT(2X,F8.2)
            READ(5,132) FREQ
        132 FORMAT(2X,F10.3)
            READ(5,134) FDB
```

Fig. A.1 *Listing of model COVER*

```
70          134 FORMAT(2X,F5.2)
                READ(5,136) BMHZ
            136 FORMAT(2X,F6.2)
                READ(5,138) ELOSS
            138 FORMAT(2X,F5.2)
75              READ(5,131) FIF
            131 FORMAT(2X,F6.2)
                READ(5,140) PRF
            140 FORMAT(2X,F7.2)
                READ(5,100) AEGDB
80          100 FORMAT(2X,F5.2)
                READ(5,150) PFA
            150 FORMAT(2X,E7.1)
                READ(5,160) PRODET
            160 FORMAT(2X,F6.2)
85              READ(5,170) SIGMA
            170 FORMAT(2X,F7.2)
                READ(5,171) SLST,TERHTS
            171 FORMAT(2X,F6.2,2X,F9.2)
                READ(5,172) SLF,TEHF
90          172 FORMAT(2X,F6.2,2X,F9.2)
                READ(5,173) K
            173 FORMAT(2X,I1)
          C     NOPER IS ZERO FOR POSTDETECTION INTEGRATION,IT IS 1 FOR OPERATOR -
          C     CRT INTEGRATION AND 2 FOR PREDETECTION INTEGRATION.
95              READ(5,176) NOPER
            176 FORMAT(2X,I1)
                READ(5,453) SIT
            453 FORMAT(2X,F3.0)
                READ(5,179) PCR
100         179 FORMAT(2X,F7.2)
          C     ANGRD IS THE GRADIENT AND ANREF THE SURFACE VALUE OF REFRACTIVITY
          C     DURING ANOMALOUS ATMOSPHERIC CONDITIONS.
                READ(5,900) ANGRD
            900 FORMAT(2X,E15.8)
105             READ(5,910) ANREF
            910 FORMAT(2X,E14.8)
          C     DMS IS THE DISTANCE IN KM BETWEEN THE LOCATIONS OF THE HEIGHTS
          C     HD1 AND HD2 WHICH ARE SPECIFIED IN METRES.
                READ(5,905) HD1,HD2,DMS
110         905 FORMAT(2X,3F7.2)
          C     JAM IS ZERO FOR A TARGETBORNE JAMMER AND ONE FOR A STANDOFF ONE.
                READ(5,1009) JAM
           1009 FORMAT(2X,I1)
                READ(5,1100) PJ
115        1100 FORMAT(2X,F8.2)
                READ(5,1105) GJ
           1105 FORMAT(2X,F5.2)
                READ(5,1110) BJ
           1110 FORMAT(2X,F6.2)
120             READ(5,1115) DJ
           1115 FORMAT(2X,F5.2)
                READ(5,1120) RJ
           1120 FORMAT(2X,F6.2)
                READ(5,1125) ANTJAM
125        1125 FORMAT(2X,F8.2)
                READ(5,1130) DECLJ
           1130 FORMAT(2X,F6.2)
                READ(5,985) ICIR
            985 FORMAT(2X,I1)
130             READ(5,950) TITLE
            950 FORMAT(1X,3A10)
                WRITE(6,180) TITLE
            180 FORMAT(1X,3A10)
                IF(IPOL - 2)         200,220,246
135         200 WRITE(6,210)
            210 FORMAT(1X,21HVERTICAL POLARIZATION)
                GO TO 240
            220 WRITE(6,230)
            230 FORMAT(1X,23HHORIZONTAL POLARIZATION)
140             GO TO 240
            246 WRITE(6,248)
            248 FORMAT(1X,21HCIRCULAR POLARIZATION)
            240 CONTINUE
                IF(K - 1)        810,250,252
145         250 WRITE(6,251)
            251 FORMAT(1X,15HSWERLING CASE I)
                GO TO 265
            252 IF(K - 2)        810,253,255
            253 WRITE(6,254)
150         254 FORMAT(1X,16HSWERLING CASE II)
                GO TO 265
            255 IF(K - 3)        810,256,258
            256 WRITE(6,257)
            257 FORMAT(1X,17HSWERLING CASE III)
155             GO TO 265
            258 IF(K - 4)          810,259,262
            259 WRITE(6,261)
            261 FORMAT(1X,16HSWERLING CASE IV)
                GO TO 265
160         262 IF(K - 5)          810,263,265
            263 WRITE(6,264)
            264 FORMAT(1X,24HNON - FLUCTUATING TARGET)
            265 CONTINUE
                  WRITE(6,260) PKW
165         260 FORMAT(1X,7HPOWER =,F8.3,6H     KW)
                WRITE(6,270) FREQ
            270 FORMAT(1X, 11HFREQUENCY =,F10.3,7H     MHZ)
                WRITE(6,280) BMHZ
            280 FORMAT(1X, 11HBANDWIDTH =,F6.2,7H     MHZ)
```

Fig. A.1 *Continued*

```
170                  WRITE(6,290) FDB
                 290 FORMAT(1X, 14HNOISE FIGURE =,F6.2,7H      DB)
                     WRITE(6,300) PRF
                 300 FORMAT(1X, 28HPULSE REPETITION FREQUENCY =,F7.2,14H    PULSES/SEC.)
                     WRITE(6,310) AEGDB
175              310 FORMAT(1X, 13HAERIAL GAIN =,F8.2,7H      DB)
                     WRITE(6,320) HBMWTH
                 320 FORMAT(1X, 22HHORIZONTAL BEAMWIDTH =,F6.2,10H    DEGREES)
                     WRITE(6,330) VBMWTH
                 330 FORMAT(1X, 20HVERTICAL BEAMWIDTH =,F6.2,10H    DEGREES)
180                  WRITE(6,340) AERHIT
                 340 FORMAT(1X, 15HAERIAL HEIGHT =,F8.2,4H    M)
                     WRITE(6,350) RPM
                 350 FORMAT(1X, 18HAERIAL SCAN RATE =,F7.2,8H   R.P.M.)
                     WRITE(6,360) TILT
185              360 FORMAT(1X, 13HAERIAL TILT =,F5.2,10H    DEGREES)
                     PI = 3141592.653589E-06
                     DEG = 5729.577951E-02
                     AEGAIN = 10. ** (AEGDB/10.)
               C     COMPUTE THE NUMBER OF HITS/SCAN,THE BEAM SHAPE LOSS AND THE TOTAL
190            C     SYSTEM LOSS.
                     IF(ASR - 0.)        810,451,455
                 451 ASR = SCA * RPM/60.
                 455 CONTINUE
                     PN = HBMWTH * PRF/ASR
195                  IF(SIT - 0.)        810,459,457
                 457 PN = SIT * PN * RPM/60.
                 459 CONTINUE
                     DELPHI = HBMWTH/PN
                     SUMONE = 0.0
200                  NX = (PN - 1.0)/2.0
                     DO 370      KA = 1,NX
                     XK = KA
                 370 SUMONE = EXP(-5.55 * ((XK * DELPHI) ** 2)/(HBMWTH ** 2)) + SUMONE
                     BEAMLS = PN/(1. + 2. * SUMONE)
205                  IF(RPM - 0.)        380,380,390
                 380 BEAMLS = 1.0
                 390 CONTINUE
                     BEALOS = 10. * ALOG10(BEAMLS)
                     XLOSS = 10. ** (ELOSS/10.)
210                  PD = PRODET/100.
                     MM = PN
                     PNN = SQRT(PN)
                     TA = 0.0299 + 0.241 * EXP(0.1799 * ALOG10(PFA))
                     TB = 0.0505 + 0.64 * EXP(0.27 * ALOG10(PFA))
215                  TC = 0.4599 - TB * EXP(- 0.681 * PD)
                     TD = 0.241 + TA * EXP(- 1.905 * PD)
                     IF(NOPER - 1)        401,400,399
                 399 SPI = 1.
                     GO TO 405
220              400 SPI = PNN
                     GO TO 405
                 401 IF(MM - 1)        810,402,403
                 402 SPI = 1.
                     GO TO 405
225              403 CONTINUE
                     SIMP = 0.5 + TC * EXP(- TD * ALOG10(PN))
                     SPI = PN/PN ** SIMP
                 405 CONTINUE
                     XMN = 577350. * PFA * SQRT(3. * FIF * FIF + BMHZ * BMHZ/4.)
230                  ZMN = XMN + PN
                     IF(NOPER - 1)        503,502,503
                 502 SXY = SQRT(ZMN)
                     GO TO 504
                 503 SMN = 0.5 + TC * EXP(- TD * ALOG10(ZMN))
235                  SXY = ZMN/ZMN ** SMN
                 504 CONTINUE
                     COLOS = SXY/SPI
                     COLAP = 10. * ALOG10(COLOS)
                     SUMDB = 10. * ALOG10(SPI)
240                  OPLOSS = 1./(0.7 * PD * PD)
                     OPLDB = 10. * ALOG10(OPLOSS)
                     TOTLOS = XLOSS * BEAMLS * COLOS * SPI * OPLOSS
                     BLOSS = BEALOS + COLAP + ELOSS + SUMDB + OPLDB
                     WRITE(6,406) MM
245              406 FORMAT(1X, 15HHITS PER SCAN =,I4)
                     WRITE(6,410) SUMDB
                 410 FORMAT(1X,18HINTEGRATION LOSS =,F5.2,5H    DB)
                     WRITE(6,412) OPLDB
                 412 FORMAT(1X,15HOPERATOR LOSS =,F6.2,5H    DB)
250                  WRITE(6,420) BLOSS
                 420 FORMAT(1X,17HTOTAL OF LOSSES =,F5.2,7H      DB)
                     WRITE(6,430) PFA
                 430 FORMAT(1X, 28HPROBABILITY OF FALSE ALARM =,E7.1)
                     WRITE(6,440) PRODET
255              440 FORMAT(1X, 26HPROBABILITY OF DETECTION =,F6.2,10H    PERCENT)
                     WRITE(6,450) SIGMA
                 450 FORMAT(1X, 21HTARGET ECHOING AREA =,F7.2,16H    SQUARE METRES)
                     WRITE(6,460) GRAD
                 460 FORMAT(1X, 34HGRADIENT OF THE REFRACTIVE INDEX =,E15.8,7H      /KM)
260                  WRITE(6,465) REF
                 465 FORMAT(1X, 22HSURFACE REFRACTIVITY =,F14.8)
                     FGHZ = FREQ/1000.
                     FLG = ALOG10(FGHZ)
                     IF(FGHZ - 1.)        651,655,655
265              651 EOX = 0.64866 * FLG * FLG - 0.67359 * FLG + 2.37675
                     AOX = 10. ** (- EOX)
                     EWV = 0.33667 * FLG * FLG + 2.66667 * FLG - 5.49485
                     AWV = 10. ** EWV
                     GO TO 665
```

Fig. A.1 *Continued*

```
270          655 IF(FGHZ - 10.)        660,660,671
             660 EOX = 0.11967 * FLG * FLG - 0.69782 * FLG + 2.37675
                 AOX = 10. ** (- EOX)
                 EWV = 0.61905 * FLG * FLG + 2.49095 * FLG - 5.49485
                 AWV = 10. ** EWV
275          665 ATM = AOX + AWV
             671 CONTINUE
                 WRITE(6,470) ATM
             470 FORMAT(1X, 25HATMOSPHERIC ATTENUATION =,F8.6,8H   DB/KM)
                 WRITE(6,480) PERM
280          480 FORMAT(1X, 37HRELATIVE PERMITTIVITY OF THE GROUND =,F4.1)
                 WRITE(6,490) RHO
             490 FORMAT(1X, 28HCONDUCTIVITY OF THE GROUND =,F6.3,12H   MHO/METRE)
                 WRITE(6,500) UNDUL
             500 FORMAT(1X, 20HTERRAIN UNDULATION =,F4.1,3H  M)
285              WRITE(6,510) SHORE
             510 FORMAT(1X, 21HDISTANCE TO THE SEA =,F6.2,5H   KM)
                  WDLCO = 80.
                  WCOND = 4.3
                 IF(FREQ - 1500.)       304,304,301
290          301 IF(FREQ - 3000.)       302,302,303
             302 WDLCO = 80. - 0.00733 * (FREQ - 1500.)
                 WCOND = 4.3 + 0.001467 * (FREQ - 1500.)
                 GO TO 304
             303 WDLCO = 69. - 0.000628 * (FREQ - 3000.)
295              WCOND = 6.5 + 0.001492 * (FREQ - 3000.)
             304 CONTINUE
                 WRITE(6,520) WDLCO
             520 FORMAT(1X, 36HRELATIVE PERMITTIVITY OF SEA WATER =,F4.1)
                 WRITE(6,530) WCOND
300          530 FORMAT(1X, 27HCONDUCTIVITY OF SEA WATER =,F6.3,12H   MHO/METRE)
                 WRITE(6,540) SEA
             540 FORMAT(1X, 17HSEA WAVE HEIGHT =,F5.1,4H   M)
                 WRITE(6,272) PJ
             272 FORMAT(1X,14HJAMMER POWER =,F8.2,6H    KW)
305              WRITE(6,274) GJ
             274 FORMAT(8X,13HAERIAL GAIN =,F5.2,7H     DB)
                 WRITE(6,276) BJ
             276 FORMAT(8X,11HBANDWIDTH =,F6.2,7H    MHZ)

                 WRITE(6,278) DJ
310          278 FORMAT(8X,8HLOSSES =,F5.2,7H     DB)
                 WRITE(6,282) PJ
             282 FORMAT(8X,10HDISTANCE =,F6.2,5H   KM)
                 WRITE(6,284) ANTJAM
             284 FORMAT(8X,15HAERIAL HEIGHT =,F8.2,4H   M)
315              WRITE(6,286) DECLJ
             286 FORMAT(8X,11HDIRECTION =,F6.2,10H   DEGREES)
                 WRITE(6,550)
             550 FORMAT(//)
                 CALL PLOT25
320              CALL PAUPLOT(29HBLANK PAPER,BLACK INK,0.3 NIB,29)
                 CALL PLOT(0.,0.,-3)
                 EARTH = 6371.228
                 COER = 1./(1. - GRAD * REF * EARTH)
                 EARAD = EARTH * COER * 1000.0
325              ANEAR = EARTH * 1000.0/(1. - ANGRD * ANREF * EARTH)
                 STEAR = EARAD
                 SREF = REF
                 SGRD = GRAD
                 SLOST = SLST * 1000.
330              SLOF = SLF * 1000.
                 AMBDA = 299.793/FREQ
                 DELTA = 0.
                 PWR = PKW * PCR * 1.0E3
                 VW = VBMWTH/DEG
335              HW = HBMWTH/DEG
                 TST = (TERHTS - TEHF)/(SLOF - SLOST)
                 IF(SLOST.EQ.SLOF)        TST = 0.
                 AX = AERHIT + AERSEA
                 A = EARAD + AX
340              TLT = TILT/DEG
                 IF(IPOL - 2)        586,586,584
             584 SIGMA = SIGMA/5.
             586 CONTINUE
           C     COMPUTE THE REQUIRED SIGNAL TO NOISE RATIO FOR THE SPECIFIED PRO-
345        C     BABILITY OF DETECTION AND FALSE ALARM.
                 XM = 3.11573 - 3.48772*PD + 0.26522*(PD**2) + 0.11477/PD
                 DO 600      KZ = 1,91
             600 CALL PHIPFA(KZ,PFA,PHI)
                 WRITE(6,1001) PHI
350         1001 FORMAT(1X,5HPHI =,F4.2)
                 IF(PD - 0.01)        613,616,614
             613 YM = 3.73
                 GO TO 612
             614 IF(PD - 0.99)        616,616,615
355          615 YM = - 3.73
                 GO TO 612
             616 CONTINUE
                 IF(PD - 0.5)     588,588,587
             588 YM = 0.084786E-01/PD - 3.1881 * PD + 1.534021
360              GO TO 612
             587 DP = 1. - PD
                 YM = 3.1881 * DP - 0.084786E-01/DP - 1.534021
             612 CONTINUE
                 IF(K - 1)        810,701,702
365          701 SN = 2. * PHI/((ALOG(1./PD)) * PNN)
                 IF(PN - 100.)          830,835,835
             830 CSW1 = 0.06245 * (ALOG(PN)) ** 2 - 0.572 * ALOG(PN) + 2.435
                 SN = SN * CSW1
             835 CONTINUE
```

Fig. A.1 *Continued*

```
370                  GO TO 709
              702    IF(K - 2)          810,825,703
              825    SN = 2. * (PHI - YM)/PNN
                     IF(PN - 100.)          840,845,845
              840    CSW2 = 0.03681 * (ALOG(PN)) ** 2 - 0.40362 * ALOG(PN) + 2.25712
375                  SN = SN * CSW2
              845    CONTINUE
                     GO TO 709
              703    IF(K - 3)          810,704,705
              704    SN = (4. * PHI)/(XM * PNN)
380                  IF(PN - 100.)          850,855,855
              850    CSW3 = 0.00052 * (ALOG(PN)) ** 2 - 0.18429 * ALOG(PN) + 1.93131
                     SN = SN * CSW3
              855    CONTINUE
                     GO TO 709
385           705    CONTINUE
                     IF(K - 4)          810,706,707
              706    SN = 2. * (PHI - YM)/PNN
                     IF(PN - 100.)          860,865,865
              860    CSW4 = 0.05981 * (ALOG(PN)) ** 2 - 0.63253 * ALOG(PN) + 2.80934
390                  SN = SN * CSW4
              865    CONTINUE
                     GO TO 709
              707    IF(K - 5)          810,708,810
              708    SN = 2. * (PHI - YM)/PNN
395                  IF(PN - 100.)          820,709,709
              820    CNF = 0.05551 * (ALOG(PN)) ** 2 - 0.56485 * ALOG(PN) + 2.62388
                     SN = SN * CNF
              709    CONTINUE
                     SNDB = 10. * ALOG10(SN)
400                  WRITE(6,312) SNDB
              312    FORMAT(1X,14HSIGNAL/NOISE =,F7.2,5H    DB)
                     FA = FDB/10.
                     FB = 10.**FA
                     TRH = 0.
405                  XN = 4.0E-15 * BMHZ * FB
                     IF(PJ.EQ.0.)      GO TO 725
                     PJAM = PJ * 1.0E3
                     AERJAM = 10. ** (GJ/10.)
                     XLJAM = 10. ** (DJ/10.)
410                  BJAM = BJ * 1.0E6
                     DECL = DECLJ/DEG
                     RJAM = RJ * 1000.
                     ZJAM = EXP(-1.15E-4 * ATM * RJAM)
                     CJAM = PJAM * AERJAM * AEGAIN * ZJAM/(XLJAM* BJAM)
415                  HJAM = EARAD + ANTJAM
              725    CONTINUE
                     RADCON = (PWR*SIGMA*((AEGAIN*AMBDA)**2))/(((4.*PI)**3)*XN*TOTLOS)
                     CONST = SQRT(RADCON)
                     RACO = SQRT(CONST)
420                  SIGSE = 1./SQRT(SN)
                     SNR = SQRT(SIGSE)
                     R = RACO * SNR
                     RFS = R/1000.
                     WRITE(6,1000) RFS
425          1000    FORMAT(1X,18HFREE SPACE RANGE =,F7.2,5H    KM)
                     IF(L - 2)          305,307,810
              305    WRITE(6,306)
              306    FORMAT(1X,33HL = 1,FREE SPACE COVERAGE DIAGRAM)
                     GO TO 309
430           307    WRITE(6,308)
              308    FORMAT(1X,26HL = 2,INTERFERENCE PATTERN)
              309    CONTINUE
                     WRITE(6,632)
              632    FORMAT(//)
435                  WRITE(6,560)
              560    FORMAT(3X,51HELEVATION    RADAR RANGE    GROUND RANGE          HEIGHT
                    1)
                     WRITE(6,570)
              570    FORMAT(1X,52HANGLE DEGREES       KM                KM                METRE)
440                  WRITE(6,580)
              580    FORMAT(///)
                     IT1 = 1000
                     IT2 = 2700
                     ETAD = 0.01
445                  DO 3000       II = 1,2
                     IF(II - 1)          3100,3100,3200
             3100    DO 800        IXX = 1,2
                     DO 790        ITA = IT1,IT2
                     N = ITA - 999
450                  IF(IXX - 2)          1020,1010,790
             1010    N = ITA + 1431
             1020    CONTINUE
                     ETA = ITA
                     ETA = ETA * ETAD - 10.
455                  E = ETA/DEG
                     BEAMLS = 1.
                     IF(RPM - 0.)          385,385,395
              385    BEAMLS = EXP(5.55 * (E/VW) ** 2)
              395    CONTINUE
460                  R = R/(BEAMLS ** 0.25)
                     IF(L - 2)          750,605,790
              605    CONTINUE
                     EARAD = ANEAR
                     REF = ANREF
465                  GRAD = ANGRD
            C        DETERMINE THE GEOMETRY OF THE DIRECT AND REFLECTED RAYS FOR THE
            C        GIVEN TERRAIN PROFILE.
                     A = EARAD + AX
                     B = SQRT(R * R + A * A + 2. * A * R * SIN(E))
```

Fig. A.1 *Continued*

```
470           XAB = A - B
              XI = SQRT((R * R - XAB * XAB)/(A * B))
              BX = B - EARAD
              IF(HD2 - HD1)         917,915,912
          912 SDH = HD1 + HD2 * EARAD * XI/(DMS * 1000.)
475           GO TO 918
          915 SDH = HD1
              GO TO 918
          917 SDH = HD2 + HD1 * EARAD * XI/(DMS * 1000.)
          918 CONTINUE
480           IF(BX - SDH)        940,920,920
          920 IF(REF - ANREF)       930,940,930
          930 EARAD = STEAR
              REF = SREF
              GRAD = SGRD
485           A = EARAD + AX
              B = SQRT(R * R + A * A + 2. * A * R * SIN(E))
              XAB = A - B
              XI = SQRT((R * R - XAB * XAB)/(A * B))
              BX = B - EARAD
490       940 CONTINUE
              RXC = XI * EARAD
              PAX = 2. * SQRT(EARAD * (AX + BX) + RXC * RXC/4.)/SQRT(3.)
              IF(BX - AX)         90,92,92
           90 CPH = 2. * EARAD * (AX - BX) * RXC/(PAX ** 3)
495           GO TO 94
           92 CPH = 2. * EARAD * (BX - AX) * RXC/(PAX ** 3)
           94 CONTINUE
              SPH = SQRT(1. - CPH * CPH)
              PHA = ATAN(SPH/CPH)
500           RAX = RXC/2. + PAX * COS((PHA + PI)/3.)
              IF(BX - AX)         112,114,114
          112 ALPHA = RAX/EARAD
              BETA = XI - ALPHA
              GO TO 116
505       114 BETA = RAX/EARAD
              ALPHA = XI - BETA
          116 CONTINUE
              X = SQRT(AX * AX + EARAD * A * ALPHA * ALPHA)
              Y = SQRT(BX * BX + EARAD * B * BETA * BETA)
510           CODEL = EARAD * ALPHA/X
              SIDEL = SQRT(1. - CODEL * CODEL)
              DELTA = ATAN(SIDEL/CODEL)
              DIST = SHORE * 1000.
              GDR = EARAD * ALPHA * AERHIT/AX
515           GRT = ALPHA * (EARAD + TRH)
              GAM = PI/2. - ABS(DELTA - ALPHA)
              TSL = ATAN(TST)
              IF(TERHTS.EQ.TEHF)      TSL = 0.
              IF(GRT.GT.SLOF)      TSL = 0.
520           IF(GRT - SLOST)        102,102,641
          641 IF(TSL - ABS(DELTA - ALPHA))        105,105,103
          103 IF(GDR - SLOST)       102,104,104
          102 TRH = AERSEA
          104 TSL = 0.
525           GO TO 644
          105 TRH = (SLOF - GRT) * TAN(TSL)
          644 CONTINUE
              GAM = GAM + TSL
              DIELCO = PERM
530           COND = RHC
              IF(GRT - DIST)        680,670,670
          670 DIELCO = WDLCO
              COND = WCOND
              TRH = AERSEA
535       680 CONTINUE
              QC = 60. * AMBDA * COND
              IF(IPOL - 2)        690,700,690
          690 CONTINUE
        C     VERTICAL POLARIZATION
540           AO = DIELCO * COS(GAM)
              OZ = SQRT((DIELCO - (SIN(GAM)) ** 2) **2 + QC * QC)
              OY = SQRT(0.5 + 0.5 * (DIELCO - (SIN(GAM)) **2)/OZ)
              OX = SQRT(0.5 - 0.5 * (DIELCO - (SIN(GAM)) ** 2)/OZ)
              O = AO - OY * SQRT(OZ)
545           P = QC * COS(GAM) - OX * SQRT(OZ)
              Q = AO + OY * SQRT(OZ)
              PA = QC * COS(GAM) + OX * SQRT(OZ)
              OA = O * Q + P * PA
              OB = P * Q - O * PA
550           OC = Q * Q + PA * PA
              RHOVER = SQRT((OA * OA + OB * OB)/(OC * OC))
              AC = OB/OA
              IF(ABS(AC) - 1.E-07)       721,710,710
          721 AMU = PI
555           IF(IPOL - 2)          715,715,717
          710 IF(AC)        713,713,712
          712 AMU = ATAN(AC)
              IF(IPOL - 2)          715,715,717
          713 AMU = PI - ABS(ATAN(AC))
560       717 AMUVER = AMU
              IF(IPOL - 2)        715,715,700
          700 CONTINUE
        C     HORIZONTAL POLARIZATION
              QB = DIELCO - (SIN(GAM)) ** 2
565           HOZ = SQRT(QB * QB + QC * QC)
              HOY = SQRT(0.5 + 0.5 * QB/HOZ)
              HOX = SQRT(0.5 - 0.5 * QB/HOZ)
              HO = COS(GAM) - HOY * SQRT(HOZ)
              HP = HOX * SQRT(HOZ)
```

Fig. A.1 *Continued*

```
570                   HQ = COS(GAM) + HOY * SQRT(HOZ)
                      HA = HC * HC - HP * HP
                      HB = HP * (HC + HQ)
                      HC = HQ * HC + HP * HP
                      RHOHOR = SQRT((HA * HA + HB * HB)/(HC * HC))
575                   HAC = HB/HA
                      AC = HAC
                      AMU = PI + ABS(ATAN(AC))
                      AMUHOR = AMU
                      IF(IPOL - 2)        715,715,722
580               722 CONTINUE
               C      CIRCULAR POLARIZATION
                      RVT = RHOVER * RHOVER
                      RHT = RHOHOR * RHOHOR
                      RVH = 2. * RHOVER * RHOHOR * COS(AMUHOR - AMUVER)
585                   SICA = RHOVER * SIN(AMUVER)
                      SICB = RHOHOR * SIN(AMUHOR)
                      COCA = RHOVER * COS(AMUVER)
                      COCB = RHOHOR * COS(AMUHOR)
                      IF(ICIR - 1)        714,716,715
590               714 RHOCIR = 0.5 * SQRT(RVT + RHT - RVH)
                      AMUCIR = PI + ATAN((SICA - SICB)/(COCA - COCB))
                      AMU = AMUCIR
                      GO TO 715
                  716 RHOCIR = 0.5 * SQRT(RVT + RHT + RVH)
595                   AMUCIR = PI + ATAN((SICA + SICB)/(COCA + COCB))
                      AMU = AMUCIR
                  715 CONTINUE
                      ZY = X + Y
                      S1 = (2. * X * Y)/(EARAD * ZY)
600                   S2 = (1. + S1)/COS(GAM)
                      DA = SQRT(1. + S1 * (1. + S2))
                      DX = 1./DA
                      WHT = SEA
                      IF(GRT - DIST)        720,730,730
605               720 WHT = UNDUL
                  730 CONTINUE
                      IF(GRT - 2000.)        620,620,625
                  620 WHT = 0.
                  625 CONTINUE
610                   RUFS = EXP(- 8. * (PI * WHT * COS(GAM)/(4. * AMBDA)) ** 2)
               C      DETERMINE THE PATTERN PROPAGATION FACTOR FOR THE GIVEN AERIAL CO-
               C      VERAGE DIAGRAM AND TERRAIN.
                      PADI = X * (1. - COS(E + DELTA))
                      CALL AERPLD(E,DELTA,TLT,FORM1)
615                   IF(IPOL - 2)        735,740,745
                  735 EX = FORM1 * RHOVER * DX * RUFS

                      GO TO 750
                  740 EX = FORM1 * RHOHOR * DX * RUFS
                      GO TO 750
620               745 EX = FORM1 * RHOCIR * DX * RUFS
                  750 CONTINUE
                      CALL AERPOD(E,DELTA,TLT,FORM)
                      IF(L - 2)        342,344,790
                  342 FIELD = ABS(FORM)
625                   GO TO 346
                  344 CONTINUE
                      E1 = FORM
                      FIELD = SQRT(E1 ** 2 + EX ** 2 + 2. * E1 * EX * COS(AMU + 2.*PI*
                     1PADI/AMBDA))
630               346 CONTINUE
               C      COMPUTE THE SLANT RANGE,GROUND RANGE AND TARGET HEIGHT.
                      ZR = EXP(-1.15E-4 * ATM * 2. * R * FIELD)
                      RC(N) = R * FIELD * (ZR ** 0.25)
                      IF(PJ.EQ.0)        GO TO 1175
635                   IF(JAM - 0)        790,1160,1165
                 1160 RJAM = R
                      FIJ = FIELD
                      GO TO 1150
                 1165 CONTINUE
640                   CXIJ = (A * A + HJAM * HJAM - RJAM * RJAM)/(2. * A * HJAM)
                      SXIJ = SQRT(1. - CXIJ * CXIJ)
                      XIJ = ATAN(SXIJ/CXIJ)
                      CEJ = HJAM * SIN(XIJ)/RJAM
                      SEJ = SQRT(1. - CEJ * CEJ)
645                   EJ = ATAN(SEJ/CEJ)
                      RXCJ = XIJ * EARAD
                      AJAM = ANTJAM
                      PAXJ = 2.*SQRT(EARAD*(AX + AJAM)+RXCJ*RXCJ/4.)/SQRT(3.)
                      CPJ = 2. * EARAD * (AX - AJAM) * RXCJ/(PAXJ ** 3)
650                   SPJ = SQRT(1. - CPJ * CPJ)
                      PHAJ = ATAN(SPJ/CPJ)
                      RAXJ = RXCJ/2. + PAXJ * COS((PHAJ + PI)/3.)
                      BETJ = RAXJ/EARAD
                      ALJ = XIJ - BETJ
655                   DELJ = ATAN((A - EARAD * COS(ALJ))/(EARAD * SIN(ALJ)))
                      CALL SDLB(TLT,HW,VW,DECL,EJ,DELJ,HORZ,HORZ1)
                      XJ = SQRT((AX * AX) + EARAD * A * (ALJ * ALJ))
                      YJ = SQRT((AJAM ** 2) + EARAD * HJAM * (BETJ ** 2))
                      PDJ = (XJ + YJ - RJAM)/AMBDA
660                   PHJ = COS(PI * (1. + 2. * PDJ))
                      FIJ = SQRT(HORZ * HORZ + HORZ1 * HORZ1 + 2. * HORZ * HORZ1 * PHJ)
                 1150 CONTINUE
                      CONJAM = CJAM * (AMBDA/(4. * PI * RJAM)) ** 2
                      RXN = XN + CONJAM * FIJ * FIJ
665                   RC(N) = RC(N) * ((XN/RXN) ** 0.25)
                 1175 CONTINUE
                      BC = SQRT(RC(N) * RC(N) + A * A + 2. * A * RC(N) * SIN(E))
                      HEIGHT(N) = BC - EARAD
                      IF(L - 2)        770,755,790
```

Fig. A.1 *Continued*

```
670              755 CONTINUE
                     EST = PI/2. + E
                     BF = EARAD + TEHF
                     XIF = SLOF/EARAD
                     RSF = SQRT(A * A + BF * BF - 2. * A * BF * COS(XIF))
675                  CTF = (A * A + RSF * RSF - BF * BF)/(2. * A * RSF)
                     STF = SQRT(1. - CTF * CTF)
                     FAG = ATAN(STF/CTF)
                     IF(CTF - 0.)          324,326,326
                 324 FAG = PI + FAG
680              326 CONTINUE
                     IF(EST - FAG)         315,322,322
                 315 RC(N) = RSF
                 322 CONTINUE
                     BC = SQRT(RC(N) * RC(N) + A * A + 2. * A * RC(N) * SIN(E))
685                  HEIGHT(N) = BC - EARAD
                     ZB = SCRT(AERHIT * (2. * EARAD + AERHIT))
                     ZC = SQRT(HEIGHT(N) * (2. * EARAD + HEIGHT(N)))
                     ZD = ZB + ZC
                     IF(RC(N) - ZD)        770,770,760
690              760 RC(N) = ZC
                     BC = SCRT(RC(N) * RC(N) + A * A + 2. * A * RC(N) * SIN(E))
                     HEIGHT(N) = BC - EARAD
                 770 CONTINUE
                     H = A - BC
695                  GR = EARAD * SQRT((RC(N) * RC(N) - H * H)/(A * BC))
                     RC(N) = RC(N)/1000.
                     GR = GR/100C.
                     XA(N) = RC(N)/10.
                     YA(N) = HEIGHT(N) * 0.002
700                  WRITE(6,780) ETA,RC(N),GR,HEIGHT(N)
                 780 FORMAT(5X,F5.2,7X,F7.2,8X,F7.2,7X,F10.2)
                 790 CONTINUE
                     IT1 = 271
                     IT2 = 550
705                  ETAD = 0.1
                 800 CONTINUE
                3200 CONTINUE
                     XO = 0.
                     XS = 2.0
710                  YO = 0.
                     YS = 8.0
                     CALL AXIS(0.,0.,13HRANGE KM X 10,-13,20.,0.,XO,XS,0)
                     CALL AXIS(0.,0.,14HHEIGHT M X 500,14,10.,90.,YO,YS,-1)
                     CALL SYMBOL(2.,14.,1.,TITLE,0.,30)
715                  CALL SYMBOL(2.,12.,0.5,46HPD =             ,PFA = 10    ,RCS =
                   1 SQ.M.,0.,46)
                     CALL SYMBCL(12.,12.5,0.3,2H-5,0.,2)
                     CALL NUMBER(4.,12.,0.5,PRODET,0.,4HF6.2)
                     CALL SYMBOL(7.2,12.25,0.5,21,0.,-1)
720                  CALL NUMBER(16.,12.,0.5,SIGMA,0.,4HF7.2)
                     CALL SYMBOL(2.,11.,0.5,24HAERIAL HEIGHT =          M,0.,24)
                     CALL NUMBER(9.,11.,0.5,AERHIT,0.,4HF6.1)
                     CALL SYMBOL(12.5,11.,0.5,21H,TILT =         DEGREES,0.,21)
                     CALL NUMBER(15.5,11.,0.5,TILT,0.,4HF5.2)
725                  IF(L - 2)         3700,3800,3800
                3700 CALL SYMBCL(2.,13.,0.5,19HFREE SPACE COVERAGE,0.,19)
                     GO TO 3600
                3800 CONTINUE
                     IF(IPOL - 2)          3300,3400,3500
730             3300 CALL SYMBOL(2.,13.,0.5,21HVERTICAL POLARIZATICN,0.,21)
                     GO TO 3600
                3400 CALL SYMBCL(2.,13.,0.5,23HHORIZONTAL POLARIZATICN,0.,23)
                     GO TC 3600
                3500 CALL SYMBOL(2.,13.,0.5,21HCIRCULAR POLARIZATICN,0.,21)
735             3600 CONTINUE
                     IF(PJ - 0.)           1185,1185,1180
                1180 IF(JAM - 0)           1185,1182,1184
                1182 CALL SYMBCL(12.5,10.,0.5,19H,TARGETBORNE JAMMER,0.,19)
                     GO TO 1185
740             1184 CALL SYMBCL(12.5,10.,0.5,17H,STAND-OFF JAMMER,0.,17)
                1185 CONTINUE
                     IF(REF - ANREF)       945,970,960
                 945 CALL SYMBOL(12.,13.,0.5,16HSUPER-REFRACTION,0.,16)
                     GC TC 970
745              960 CALL SYMBOL(12.,13.,0.5,14HSUB-REFRACTION,0.,14)
                 970 CONTINUE
                     IF(K - 2)      4000,4100,4200
                4000 CALL SYMBOL(2.,10.,0.5,17HSWERLING 1 TARGET,0.,17)
                     GO TC 4600
750             4100 CALL SYMBCL(2.,10.,0.5,17HSWERLING 2 TARGET,0.,17)
                     GO TO 4600
                4200 IF(K - 4)             4300,4400,4500
                4300 CALL SYMBCL(2.,10.,0.5,17HSWERLING 3 TARGET,0.,17)
                     GO TO 4600
755             4400 CALL SYMBCL(2.,10.,0.5,17HSWERLING 4 TARGET,0.,17)
                     GC TC 4600
                4500 CALL SYMBOL(2.,10.,0.5,22HNON-FLUCTUATING TARGET,0.,22)
                4600 CONTINUE
                     CALL PLCT(XA,YA,3)
760                  IF(0. - ETA)          783,783,782
                 782 CALL PLCT(XA,YA,3)
                     GO TO 784
                 783 CALL PLOT(XA,YA,2)
                     XA(1982) = XO
765                  XA(1983) = XS
                     YA(1982) = YO
                     YA(1983) = YS
                 784 CALL LINE(XA,YA,1981,1,0,0)
                3000 CONTINUE
770                  CALL PLCT(30.,0.,-3)
```

Fig. A.1 *Continued*

```
  810 CONTINUE
      STOP
      END
```

Fig. A.1 *Continued*

```
 1                      SUBROUTINE PHIPFA(KZ,PFA,PHI)
                        DIMENSION  W(91),V(91)
                        COMMON  PI,DEG,V,VW,W
                C       THIS SUBROUTINE DETERMINES THE INVERSE OF THE FALSE ALARM PROBA-
 5              C       BILITY INTEGRAL AND HENCE THE DETECTION THRESHOLD.
                C       VALID FOR PFA BETWEEN 10 TO -12 AND 10 TO -2.
                        AE = V(KZ) - PFA
                        IF(ABS(AE) - 1.E-04 * V(KZ))      10,10,20
                     10 PHI = W(KZ)
10                   20 CONTINUE
                        RETURN
                        END
```

Fig. A.2 *Listing of subroutine PHIPFA*

```
 1                      SUBROUTINE SDLB(TLT,HW,VW,DECL,EJ,DELJ,HORZ,HORZ1)
                C       THIS SUBROUTINE COMPUTES THE RADAR TO JAMMER PATTERN PROPAGATION
                C       FACTORS.
                        HE = DECL
 5                      EE = ABS(EJ - TLT)
                        DE = DELJ + TLT
                        YY = SIN(VW/2.)
                        YW = SIN(HW/2.)
                        YV = 1391.5789E-03 * SIN(HE)/YW
10                      DIM = SIN(YV)/YV
                        IF(YV)      20,10,20
                     10 DIM = 1.
                     20 CONTINUE
                        YU = 1391.5789E-03 * SIN(EE)/YY
15                      YT = 1391.5789E-03 * SIN(DE)/YY
                        HORZ = DIM * SIN(YU)/YU
                        HORZ1 = DIM * SIN(YT)/YT
                        RETURN
                        END
```

Fig. A.3 *Listing of subroutine SDLB*

```
 1          SUBROUTINE AERPOD(E,DELTA,TLT,FORMZ)
            DIMENSION  W(91),V(91)
            COMMON  PI,DEG,V,VW,W
      C     THIS SUBROUTINE COMPUTES THE PATTERN PROPAGATION FACTOR FOR THE
 5    C     DIRECT RAY.
            EE = ABS(TLT - E)
            BW = VW/2.
            IF(E - (25./DEG + TLT))          26,26,25
         25 XA = 0.006715689
10          YA = 4.586212663
            GO TO 8
         26 IF(E - (23./DEG + TLT))          24,24,23
         23 XA = 0.063967167
            YA = 1.969344441
15          GO TO 8
         24 IF(E - (22./DEG + TLT))          22,22,21
         21 XA = 0.708840527
            YA = - 0.590163378
            GO TO 8
20       22 IF(E -(19./DEG + TLT))        20,20,19
         19 XA = 0.913521106
            YA = - 0.848521013
            GO TO 8
         20 IF(E - (16./DEG + TLT))          18,18,17
25       17 XA = 0.144772578
            YA = 0.793048724
            GO TO 8
         18 IF(E - (12.4/DEG + TLT))          16,16,15
         15 XA = 0.089684039
30          YA = 1.164650112
            GO TO 8
         16 IF(E - (11./DEG + TLT))          14,14,13
         13 XA = 0.198685790
            YA = 0.647581623
35          GO TO 8
         14 IF(E - (9./DEG + TLT))        12,12,11
         11 XA = 0.293027512
            YA = 0.413023216
            GO TO 8
40       12 IF(E - (8.5/DEG + TLT))          9,9,10
         10 XA = 0.372448083
            YA = 0.283742417
            GO TO 8
          9 IF(E - (6./DEG + TLT))        7,7,6
45        6 XA = 0.263730845
            YA = 0.464285828
            GO TO 8
          7 IF(E - (5.3/DEG + TLT))          5,5,3
          3 XA = 0.172597515
50          YA = 0.652023264
          8 FORMZ = XA/((SIN(EE)) ** YA)
            GO TO 30
          5 CONTINUE
            IF(E - (BW + TLT))        1,1,2
55        1 XK = 214.8264143
            GO TO 29
          2 XK = ALOG(2.)/(2. * (BW ** 2))
         29 CONTINUE
            FORMZ = EXP(- XK * (E - (BW + TLT)) ** 2)
60       30 RETURN
            END
```

Fig. A.4 *Listing of subroutine AERPOD*

```
 1          SUBROUTINE AERPLD(E,DELTA,TLT,FORMZ)
            DIMENSION  W(91),V(91)
            COMMON  PI,DEG,V,VW,W
      C     THIS SUBROUTINE COMPUTES THE PATTERN PROPAGATION FACTOR FOR THE
 5    C     REFLECTED RAY
            DE = ABS(DELTA + TLT)
            BW = VW/2.
            XK = 214.8264143
            FORMZ = EXP(- XK * (DE - BW) ** 2)
10          RETURN
            END
```

Fig. A.5 *Listing of subroutine AERPLD*

```
1.0D-122.0D-123.0D-124.0D-125.0D-126.0D-127.0D-128.0D-129.0D-121.0D-11
2.0D-113.0D-114.0D-115.0D-116.0D-117.0D-118.0D-119.0D-111.0D-102.0D-10
3.0D-104.0D-105.0D-106.0D-107.0D-108.0D-109.0D-101.0D-092.0D-093.0D-09
4.0D-095.0D-096.0D-097.0D-098.0D-099.0D-091.0D-082.0D-083.0D-084.0D-08
5.0D-086.0D-087.0D-088.0D-089.0D-081.0D-072.0D-073.0D-074.0D-075.0D-07
6.0D-077.0D-078.0D-079.0D-071.0D-062.0D-063.0D-064.0D-065.0D-066.0D-06
7.0D-068.0D-069.0D-061.0D-052.0D-053.0D-054.0D-055.0D-056.0D-057.0D-05
8.0D-059.0D-051.0D-042.0D-043.0D-044.0D-045.0D-046.0D-047.0D-048.0D-04
9.0D-041.0D-032.0D-033.0D-034.0D-035.0D-036.0D-037.0D-038.0D-039.0D-03
1.0D-02
```

Fig. A.6 (*a*) *Listing of the variable* $V = P_{fa}$

```
7.036.906.846.806.766.746.726.706.696.686.576.506.436.406.386.36
6.326.316.306.206.136.096.076.046.016.005.965.935.835.765.725.70
5.665.645.625.605.595.505.405.365.315.305.255.205.185.165.064.97
4.904.854.824.794.754.724.714.594.504.434.384.354.304.294.264.23
4.104.003.903.853.813.773.743.703.693.513.403.323.263.213.163.13
3.113.082.852.722.642.592.512.462.412.382.33
```

Fig. A.6 (*b*) *Listing of the variable* $W = \phi(P_{fa})$

```
      BESSEL FUNCTION OF THE FIRST KIND AND ORDER N (FROM 0 TO 3)
C     BJ IS THE BESSEL J-FUNCTION OF THE N-TH ORDER
C     FOR OTHER N,THE STATEMENT FOR AX MUST BE CHECKED AND MODIFIED
      N = 1
      PI = 3.141592653589
      DO 207    IX = 1,1550
      X = IX
      X = X * 0.01
      XN = N
      AX = 7.0
      IF(N - 2)     216,218,218
  218 AX = 8.0
  216 CONTINUE
      IF(X - AX)     220,220,219
  219 CONTINUE
C     COMPUTE BJ FOR ARGUMENTS LARGER THAN AX (W.R.E.TECH.MEMO 807(AP)
C     EXPRESSION 4)
      AI = (4. *(XN**2) - 1.)*(4. *(XN**2) - 9.)
      AJ = AI * (4.*(XN**2) -25.) * (4.*(XN**2) - 49.)
      AK = AI/(2. * ((8. * X)**2))
      AL = AJ/(24. * ((8.*X)**4))
      AM = (4. *(XN**2) - 1.)/(8. * X)
      AN = AI * (4. *(XN**2) - 25.)/(6. * ((8.*X)**3))
      AO = SQRTF(2./(PI * X))
      AP = X - (PI/2.) * (XN + 0.5)
      AQ = COSF(AP)
      AR = SINF(AP)
      BJ = AO*AQ*(1. - AK + AL) - AO * AR * (AM - AN)
      GO TO 221
  220 CONTINUE
C     COMPUTE BJ FOR ARGUMENTS SMALLER THAN AX (W.R.E.TECH.MEMO 807(AP)
C     EXPRESSION 3)
C     IPROD IS THE NUMBER OF TERMS OF THE SERIES EXPANSION
      IPROD = 50
      BZ = 0.0
      KMAX = IPROD - 1
      XIP = IPROD
      DO 200    K=1,KMAX
C     DETERMINE THE SIGN OF THE TERM BY RAISING -1 TO THE POWER OF K
      I = (-1) ** K
      XK = K
      ZI = (X/2.)**(2. * XK)
      ZZ = I
      ZY = ZZ * ZI
C     COMPUTE THE GAMMA FUNCTION OF THE ARGUMENT (N + K + 1)
      LMAX = N + K - 1
      IF(N - 1)     222,223,223
  222 IF(K - 1)     223,224,223
  224 LMAX = 1.
      XIM = 1.
      GO TO 202
  223 CONTINUE
      XIM = N + K
      DO 202    L=1,LMAX
      XIN = N + K - L
      XIM = XIM * XIN
  202 CONTINUE
      ZX = ZY/XIM
C     COMPUTE K FACTORIAL
      JMAX = K - 1
      XKA = K
      DO 205    J= 1,JMAX
      XL = K - J
      XKA = XKA * XL
      IF(K -.1)    204,204,205
  204 XKA = 1.
```

Fig. A.7 *Programme listing for the computation of the Bessel function of the first kind and orders 0, 1, 2 and 3*

```
  205 CONTINUE
      BX = ZX/XKA
      BZ = BZ + BX
  200 CONTINUE
C     COMPUTE N-FACTORIAL FOR THE FIRST TERM OF THE SERIES EXPANSION
C     (WHEN K=0)
      IF(N - 1)     210,210,211
  210 ZM = 1.
      GO TO 212
  211 CONTINUE
      MAX = N - 1
      ZM = N
      DO 212     M = 1,MAX
      ZN = N - M
      ZM = ZM * ZN
  212 CONTINUE
      CX = (X/2.) ** N
C     BJ IS THE BESSEL FUNCTION OF THE FIRST KIND AND ORDER N
      BJ = CX * ((1./ZM) + BZ)
  221 CONTINUE
      WRT  3,206,X,BJ
  206 FORMAT(2X,2HX=,F5.2,2X,3HBJ=,F10.7)
  209 CONTINUE
      CALL EXIT
      END
```

Fig. A.7 *Continued*

Appendix B

Listing of model 'MAP'

```
 1            PROGRAM MAP(INPUT,OUTPUT,TAPE5=INPUT,TAPE6=OUTPUT,DEBUG=OUTPUT)
              DIMENSION  W(91),V(91)
              DIMENSION  SHC(361),SLO(361),TER(361),SF(361),TEF(361)
              DIMENSION  RC(1444),GR(1444),HEIGHT(1444)
 5            DIMENSION  XA(4,363),YA(4,363)
              DIMENSION  XB(363),YB(363)
              DIMENSION TITLE(3)
              COMMON  PI,DEG,V,VW,W
              READ(5,1)   (W(I),I=1,91)
10          1 FORMAT(16F4.2)
              READ(5,2)    (V(I),I=1,91)
            2 FORMAT(10E7.1)
              READ(5,5) (SHC(N),N=1,361)
            5 FORMAT(14F5.1)
15            READ(5,6) (SLO(N),N=1,361)
            6 FORMAT(14F5.1)
              READ(5,7) (TER(N),N=1,361)
            7 FORMAT(14F5.1)
              READ(5,8) (SF(N),N=1,361)
20          8 FORMAT(14F5.1)
              READ(5,9) (TEF(N),N=1,361)
            9 FORMAT(14F5.1)
        C     GRAD IS THE N.B.S. CONSTANT CE
              READ(5,20) GRAD
25         20 FORMAT(2X,E15.8)
        C     REF IS THE N.B.S. CONSTANT NS * 10 TO MINUS 6
              READ(5,25) REF
           25 FORMAT(2X,E14.8)
              READ(5,30) PERM
30         30 FORMAT(2X,F4.1)
              READ(5,32) RHO
           32 FORMAT(2X,F6.3)
              READ(5,34) ATM
           34 FORMAT(2X,F8.6)
35            READ(5,40) UNDUL
           40 FORMAT(2X,F5.1)
              READ(5,52) AERSEA
           52 FORMAT(2X,F9.2)
              READ(5,70) SEA
40         70 FORMAT(2X,F5.1)
        C     IPOL EQUALS 1 FOR VERTICAL POLARIZATION,2 FOR HORIZONTAL AND 3 FOR
        C     CIRCULAR POLARIZATION.
              READ(5,80) IPOL
           80 FORMAT(2X,I1)
45            READ(5,82) AERHIT
           82 FORMAT(2X,F6.1)
              READ(5,84) HBMWTH
           84 FORMAT(2X,F5.2)
              READ(5,86) VBMWTH
50         86 FORMAT(2X,F5.2)
              READ(5,110) TILT
          110 FORMAT(2X,F5.2)
        C     IF A SECTOR SCAN IS USED AND NEITHER THE ANGULAR SCAN RATE ASR NOR
        C     THE APPROPRIATE RPM IS GIVEN,RPM IS TO BE DETERMINED FROM THE FULL
55      C     SCAN RPM AS 360. * RPM/SCA AND USED AS INPUT IN THE PROGRAMME.
              READ(5,454) ASR
          454 FORMAT(2X,F7.2)
        C     RPM IS THE AERIAL SCAN RATE WHICH IN CASE OF A SECTOR SCAN
        C          MAY DIFFER FROM THAT USED FOR A 360 DEGREE SCAN (SEE THE PRE -
60      C          CEDING COMMENT).
              READ(5,120) RPM
          120 FORMAT(2X,F7.2)
        C     SCA IS THE ANGLE OF A SCAN CYCLE,E.G. IT IS 360.00 FOR A CIRCULAR
        C          SCAN AND E.G. 80.00 FOR A PLUS-MINUS 40 DEGREE SCAN.
65            READ(5,31) SCA
           31 FORMAT(2X,F6.2)
              READ(5,130) PKW
          130 FORMAT(2X,F8.2)
              READ(5,132) FREQ
```

Fig. B.1 *Listing of model MAP*

```
 70        132 FORMAT(2X,F10.3)
               READ(5,134) FDB
           134 FORMAT(2X,F5.2)
               READ(5,136) BMHZ
           136 FORMAT(2X,F6.2)
 75            READ(5,138) ELOSS
           138 FORMAT(2X,F5.2)
               READ(5,131) FIF
           131 FORMAT(2X,F6.2)
               READ(5,140) PRF
 80        140 FORMAT(2X,F7.2)
               READ(5,100) AEGDB
           100 FORMAT(2X,F5.2)
               READ(5,150) PFA
           150 FORMAT(2X,E7.1)
 85            READ(5,170) SIGMA
           170 FORMAT(2X,F7.2)
               READ(5,173) K
           173 FORMAT(2X,I1)
         C     NOPER IS ZERO FOR POSTDETECTION INTEGRATION,IT IS 1 FOR OPERATOR -
 90      C     CRT INTEGRATION AND 2 FOR PREDETECTION INTEGRATION.
               READ(5,176) NOPER
           176 FORMAT(2X,I1)
               READ(5,453) SIT
           453 FORMAT(2X,F3.0)
 95            READ(5,179) PCR
           179 FORMAT(2X,F7.2)
         C     ANGRD IS THE GRADIENT AND ANREF THE SURFACE VALUE OF REFRACTIVITY
         C     DURING ANOMALOUS ATMOSPHERIC CONDITIONS.
               READ(5,900) ANGRD
100        900 FORMAT(2X,E15.8)
               READ(5,910) ANREF
           910 FORMAT(2X,E14.8)
         C     DMS IS THE DISTANCE IN KM BETWEEN THE LOCATIONS OF THE HEIGHTS
         C     HD1 AND HD2 WHICH ARE SPECIFIED IN METRES.
105            READ(5,905) HD1,HD2,DMS
           905 FORMAT(2X,3F7.2)
         C     JAM IS ZERO FOR A TARGETBORNE JAMMER AND ONE FOR A STANDOFF ONE.
               READ(5,1009) JAM
          1009 FORMAT(2X,I1)
110            READ(5,1100) PJ
          1100 FORMAT(2X,F8.2)
               READ(5,1105) GJ
          1105 FORMAT(2X,F5.2)
               READ(5,1110) BJ
115       1110 FORMAT(2X,F6.2)
               READ(5,1115) DJ
          1115 FORMAT(2X,F5.2)
               READ(5,1120) RJ
          1120 FORMAT(2X,F6.2)
120            READ(5,1125) ANTJAM
          1125 FORMAT(2X,F8.2)
               READ(5,1130) DECLJ
          1130 FORMAT(2X,F6.2)
               READ(5,95) TARGHT
125         95 FORMAT(2X,F8.2)
               READ(5,96) PRCDET
            96 FORMAT(2X,F6.2)
               READ(5,985) ICIR
           985 FORMAT(2X,I1)
130            READ(5,950) TITLE
           950 FORMAT(1X,3A10)
               WRITE(6,180) TITLE
           180 FORMAT(1X,3A10)
               IF(IPOL - 2)       200,220,246
135        200 WRITE(6,210)
           210 FORMAT(1X,21HVERTICAL POLARIZATION)
               GO TO 240
           220 WRITE(6,230)
           230 FORMAT(1X,23HHORIZONTAL POLARIZATION)
140            GO TO 240
           246 WRITE(6,248)
           248 FORMAT(1X,21HCIRCULAR POLARIZATION)
           240 CONTINUE
               IF(K - 1)      810,250,252
145        250 WRITE(6,251)
           251 FORMAT(1X,15HSWERLING CASE I)
               GO TO 265
           252 IF(K - 2)      810,253,255
           253 WRITE(6,254)
150        254 FORMAT(1X,16HSWERLING CASE II)
               GO TO 265
           255 IF(K - 3)      810,256,258
           256 WRITE(6,257)
           257 FORMAT(1X,17HSWERLING CASE III)
155            GO TO 265
           258 IF(K - 4)      810,259,262
           259 WRITE(6,261)
           261 FORMAT(1X,16HSWERLING CASE IV)
               GO TO 265
160        262 IF(K - 5)      810,263,265
           263 WRITE(6,264)
           264 FORMAT(1X,24HNON - FLUCTUATING TARGET)
           265 CONTINUE
                WRITE(6,260) PKW
165        260 FORMAT(1X,7HPOWER =,F8.3,6H     KW)
               WRITE(6,270) FREQ
           270 FORMAT(1X, 11HFREQUENCY =,F10.3,7H     MHZ)
               WRITE(6,280) BMHZ
           280 FORMAT(1X, 11HBANDWIDTH =,F6.2,7H     MHZ)
```

Fig. B.1 *Continued*

```
170          WRITE(6,290) FDB
         290 FORMAT(1X, 14HNOISE FIGURE =,F6.2,7H      DB)
             WRITE(6,300) PRF
         300 FORMAT(1X, 28HPULSE REPETITION FREQUENCY =,F7.2,14H   PULSES/SEC.)
             WRITE(6,310) AEGDB
175      310 FORMAT(1X, 13HAERIAL GAIN =,F6.2,7H      DB)
             WRITE(6,320) HBMWTH
         320 FORMAT(1X, 22HHORIZONTAL BEAMWIDTH =,F6.2,10H   DEGREES)
             WRITE(6,330) VBMWTH
         330 FORMAT(1X, 20HVERTICAL BEAMWIDTH =,F6.2,10H   DEGREES)
180          WRITE(6,340) AERHIT
         340 FORMAT(1X, 15HAERIAL HEIGHT =,F8.2,4H   M)
             WRITE(6,350) RPM
         350 FORMAT(1X, 18HAERIAL SCAN RATE =,F7.2,8H  R.P.M.)
             WRITE(6,360) TILT
185      360 FORMAT(1X, 13HAERIAL TILT =,F5.2,10H   DEGREES)
             PI = 3141592.653589E-06
             DEG = 5729.577951E-02
             AEGAIN = 10. ** (AEGDB/10.)
      C      COMPUTE THE NUMBER OF HITS/SCAN,THE BEAM SHAPE LOSS AND THE TOTAL
190   C      SYSTEM LOSS.
             IF(ASR - 0.)        810,451,455
         451 ASR = SCA * RPM/60.
         455 CONTINUE
             PN = HBMWTH * PRF/ASR
195          IF(SIT - 0.)        810,459,457
         457 PN = SIT * PN * RPM/60.
         459 CONTINUE
             DELPHI = HBMWTH/PN
             SUMONE = 0.0
200          NX = (PN - 1.0)/2.0
             DO 370      KA = 1,NX
             XK = KA
         370 SUMONE = EXP(-5.55 * ((XK * DELPHI) ** 2)/(HBMWTH ** 2)) + SUMONE
             BEAMLS = PN/(1. + 2. * SUMONE)
205          IF(RPM - 0.)        380,380,390
         380 BEAMLS = 1.0
         390 CONTINUE
             BEALOS = 10. * ALOG10(BEAMLS)
             XLOSS = 10. ** (ELOSS/10.)
210          PD = PRODET/100.
             PNN = SQRT(PN)
             MM = PN
             TA = 0.0299 + 0.241 * EXP(0.1799 * ALOG10(PFA))
             TB = 0.0505 + 0.64 * EXP(0.27 * ALOG10(PFA))
215          TC = 0.4599 - TB * EXP(- 0.881 * PD)
             TD = 0.241 + TA * EXP(- 1.905 * PD)
             IF(NCPER - 1)       401,400,399
         399 SPI = 1.
             GO TO 405
220      400 SPI = PNN
             GO TO 405
         401 IF(MM - 1)        810,402,403
         402 SPI = 1.
             GO TO 405
225      403 CONTINUE
             SIMP = 0.5 + TC * EXP(- TD * ALOG10(PN))
             SPI = PN/PN ** SIMP
         405 CONTINUE
             XMN = 577350. * PFA * SQRT(3. * FIF * FIF + BMHZ * BMHZ/4.)
230          ZMN = XMN + PN
             IF(NCPER - 1)       503,502,503
         502 SXY = SQRT(ZMN)
             GO TO 504
         503 SMN = 0.5 + TC * EXP(- TD * ALOG10(ZMN))
235          SXY = ZMN/ZMN ** SMN
         504 CONTINUE
             COLOS = SXY/SPI
             COLAP = 10. * ALOG10(COLOS)
             SUMDB = 10. * ALOG10(SPI)
240          OPLOSS = 1./(0.7 * PD * PD)
             OPLDB = 10. * ALOG10(OPLOSS)
             TOTLOS = XLOSS * BEAMLS * COLOS * SPI * OPLOSS
             BLOSS = BEALOS + COLAP + ELOSS + SUMDB + OPLDB
             WRITE(6,406) MM
245      406 FORMAT(1X, 15HHITS PER SCAN =,I4)
             WRITE(6,412) OPLDB
         412 FORMAT(1X,15HOPERATOR LOSS =,F6.2,5H    DB)
             WRITE(6,420) BLOSS
         420 FORMAT(1X,17HTOTAL OF LOSSES =,F5.2,7H      DB)
250          WRITE(6,430) PFA
         430 FORMAT(1X, 28HPROBABILITY OF FALSE ALARM =,E7.1)
             WRITE(6,450) SIGMA
         450 FORMAT(1X, 21HTARGET ECHOING AREA =,F7.2,16H   SQUARE METRES)
             WRITE(6,460) GRAD
255      460 FORMAT(1X, 24HGRADIENT OF THE REFRACTIVE INDEX =,E15.8,7H     /KM)
             WRITE(6,465) REF
         465 FORMAT(1X, 22HSURFACE REFRACTIVITY =,F14.8)
             FGHZ = FREQ/1000.
             FLG = ALOG10(FGHZ)
260          IF(FGHZ - 1.)       651,655,655
         651 EOX = 0.64866 * FLG * FLG - 0.67359 * FLG + 2.37675
             AOX = 10. ** (- EOX)
             EWV = 0.33667 * FLG * FLG + 2.66667 * FLG - 5.49485
             AWV = 10. ** EWV
265          GO TO 665
         655 IF(FGHZ - 10.)      660,660,675
         660 EOX = 0.11967 * FLG * FLG - 0.69782 * FLG + 2.37675
             AOX = 10. ** (- EOX)
             EWV = 0.61905 * FLG * FLG + 2.49095 * FLG - 5.49485
```

Fig. B.1 *Continued*

```
270          AWV = 10. ** EWV
         665 ATM = ACX + AWV
         675 CONTINUE
             WRITE(6,470) ATM
         470 FORMAT(1X, 25HATMOSPHERIC ATTENUATION =,F8.6,8H   DB/KM)
275          WRITE(6,480) PERM
         480 FORMAT(1X, 37HRELATIVE PERMITTIVITY OF THE GROUND =,F4.1)
             WRITE(6,490) RHO
         490 FORMAT(1X, 28HCONDUCTIVITY OF THE GROUND =,F6.3,12H   MHO/METRE)
             WRITE(6,500) UNDUL
280      500 FORMAT(1X, 20HTERRAIN UNDULATION =,F4.1,3H  M)
              WDLCO = 80.
              WCOND = 4.3
             IF(FREQ - 1500.)       304,304,301
         301 IF(FREQ - 3000.)       302,302,303
285      302 WDLCO = 80. - 0.00733 * (FREQ - 1500.)
             WCOND = 4.3 + 0.001467 * (FREQ - 1500.)
             GO TO 304
         303 WDLCO = 69. - 0.000628 * (FREQ - 3000.)
             WCOND = 6.5 + 0.001492 * (FREQ - 3000.)
290      304 CONTINUE
             WRITE(6,520) WDLCO
         520 FORMAT(1X, 36HRELATIVE PERMITTIVITY OF SEA WATER =,F4.1)
             WRITE(6,530) WCOND
         530 FORMAT(1X, 27HCONDUCTIVITY OF SEA WATER =,F6.3,12H   MHO/METRE)
295          WRITE(6,540) SEA
         540 FORMAT(1X, 17HSEA WAVE HEIGHT =,F5.1,4H   M)
             WRITE(6,272) PJ
         272 FORMAT(1X,14HJAMMER POWER =,F8.2,6H    KW)
             WRITE(6,274) GJ
300      274 FORMAT(8X,13HAERIAL GAIN =,F5.2,7H     DB)
             WRITE(6,276) BJ
         276 FORMAT(8X,11HBANDWIDTH =,F6.2,7H    MHZ)
             WRITE(6,278) DJ
         278 FORMAT(8X,8HLOSSES =,F5.2,7H     DB)
305          WRITE(6,282) RJ
         282 FORMAT(8X,10HDISTANCE =,F6.2,5H   KM)
             WRITE(6,284) ANTJAM
         284 FORMAT(8X,15HAERIAL HEIGHT =,F8.2,4H   M)
             WRITE(6,286) DECLJ
310      286 FORMAT(8X,11HDIRECTION =,F6.2,10H   DEGREES)
             WRITE(6,550)
         550 FORMAT(//)
             WRITE(6,560)
         560 FORMAT(3X,85HAZIMUTH         SLANT RANGE         GROUND RANGE         TA
315         1RGET HEIGHT         TERRAIN HEIGHT)
             WRITE(6,570)
         570 FORMAT(3X,78HDEGREES                  KM                   KM
            1    M                     M)
             WRITE(6,572)
320      572 FORMAT(///)
             CALL PLOT25
             CALL PAUPLOT(29HBLANK PAPER,BLACK INK,0.3 NIB,29)
             CALL PLOT(0.,0.,-3)
             EARTH = 6371.228
325          COER = 1./(1. - GRAD * REF * EARTH)
             EARAD = EARTH * COER * 1000.0
             ANEAR = EARTH * 1000.0/(1. - ANGRD * ANREF * EARTH)
             STEAR = EARAD
             SREF = REF
330          SGRD = GRAD
             AMBDA = 299.793/FREQ
             DELTA = 0.
             PWR = PKW * PCR * 1.0E3
             VW = VBMWTH/DEG
335          HW = HBMWTH/DEG
             AX = AERHIT + AERSEA
             A = EARAD + AX
             TLT = TILT/DEG
             IF(IPOL - 2)       586,586,584
340      584 SIGMA = SIGMA/5.
         586 CONTINUE
             FA = FDB/10.
             FB = 10.**FA
             XN = 4.0E-15 * BMHZ * FB
345          IF(PJ.EQ.0.)       GO TO 725
             PJAM = PJ * 1.0E3
             AERJAM = 10. ** (GJ/10.)
             XLJAM = 10. ** (DJ/10.)
             BJAM = BJ * 1.0E6
350          DECL = DECLJ/DEG
             RJAM = RJ * 1000.
             ZJAM = EXP(-1.15E-4 * ATM * RJAM)
             CJAM = PJAM * AERJAM * AEGAIN * ZJAM/(XLJAM* BJAM)
             HJAM = EARAD + ANTJAM
355      725 CONTINUE
             WRITE(6,650)
         650 FORMAT(//)
             XM = 3.11573 - 3.48772*PD + 0.26522*(PD**2) + 0.11477/PD
             DO 600       KZ = 1,91
360      600 CALL PHIPFA(KZ,PFA,PHI)
             IF(PD - 0.01)       613,616,614
         613 YM = 3.73
             GO TO 612
         614 IF(PD - 0.99)       616,616,615
365      615 YM = - 3.73
             GO TO 612
         616 CONTINUE
             IF(PD - 0.5)     588,588,587
         588 YM = 0.084786E-01/PD - 3.1881 * PD + 1.534021
```

Fig. B.1 *Continued*

```
370          GO TO 612
         587 DP = 1. - PD
             YM = 3.1881 * DP - 0.084786E-01/DP - 1.534021
         612 CONTINUE
             IF(K - 1)        810,701,702
375      701 SN = 2. * PHI/((ALOG(1./PD)) * PNN)
             IF(PN - 100.)          830,835,835
         830 CSW1 = 0.06245 * (ALOG(PN)) ** 2 - 0.572 * ALOG(PN) + 2.435
             SN = SN * CSW1
         835 CONTINUE
380          GO TO 709
         702 IF(K - 2)        810,825,703
         825 SN = 2. * (PHI - YM)/PNN
             IF(PN - 100.)          840,845,845
         840 CSW2 = 0.03681 * (ALOG(PN)) ** 2 - 0.40362 * ALOG(PN) + 2.25712
385          SN = SN * CSW2
         845 CONTINUE
             GO TO 709
         703 IF(K - 3)       810,704,705
         704 SN = (4. * PHI)/(XM * PNN)
390          IF(PN - 100.)          850,855,855
         850 CSW3 = 0.00052 * (ALOG(PN)) ** 2 - 0.18429 * ALOG(PN) + 1.93131
             SN = SN * CSW3
         855 CONTINUE
             GO TO 709
395      705 CONTINUE
             IF(K - 4)        810,706,707
         706 SN = 2. * (PHI - YM)/PNN
             IF(PN - 100.)          860,865,865
         860 CSW4 = 0.05981 * (ALOG(PN)) ** 2 - 0.63253 * ALOG(PN) + 2.80934
400          SN = SN * CSW4
         865 CONTINUE
             GO TO 709
         707 IF(K - 5)        810,708,810
         708 SN = 2. * (PHI - YM)/PNN
405          IF(PN - 100.)        820,709,709
         820 CNF = 0.05551 * (ALOG(PN)) ** 2 - 0.56485 * ALOG(PN) + 2.62388
             SN = SN * CNF
         709 CONTINUE
             RADCON = (PWR*SIGMA*((AFGAIN*AMBDA)**2))/(((4.*PI)**3)*XN*TOTLOS)
410          CONST = SQRT(RADCON)
             RACO = SQRT(CONST)
             SIGSE = 1./SQRT(SN)
             SNR = SQRT(SIGSE)
             R = RACO * SNR
415          DO 795        IPD = 1,4
             TARG = TARGHT * IPD
             B = EARAD + TARG
             IF(IPD - 2)          293,293,292
         292 IF(IPD - 3)          295,293,294
420      293 SINE = AMBDA/(4. * AX)
             GO TO 295
         294 SINE = 3. * AMBDA/(4. * AX)
         295 CONTINUE
             COSE = SQRT(1. - SINE * SINE)
425          E = ATAN(SINE/COSE)
             BEAMLS = 1.
             IF(RPM - 0.)         385,385,395
         385 BEAMLS = EXP(5.55 * (E/VW) ** 2)
         395 CONTINUE
430          R = R/(BEAMLS ** 0.25)
             RFS = R/1000.
             ET = E * DEG
             WRITE(6,591) RFS,ET
         591 FORMAT(25X,18HFREE SPACE RANGE =,F8.3,6H    KM,10X,7HELEV. =,F7.3,
435         110H   DEGREES)
             WRITE(6,592)
         592 FORMAT(//)
             DO 790        N = 1,361
             IF(IPD - 2)          104,106,108
440      104 J = N
             GO TO 126
         106 J = 361 + N
             GO TO 126
         108 IF(IPD - 4)        122,124,126
445      122 J = 722 + N
             GO TO 126
         124 J = 1083 + N
         126 CONTINUE
             EARAD = ANEAR
450          REF = ANREF
             GRAD = ANGRD
       C     DETERMINE THE GEOMETRY OF THE DIRECT AND REFLECTED RAYS FOR THE
       C     GIVEN TERRAIN PROFILE.
             A = EARAD + AX
455          B = EARAD + TARG
             XAB = A - B
             XI = SQRT((R * R - XAB * XAB)/(A * B))
             BX = B - EARAD
             IF(HD2 - HD1)          917,915,912
460      912 SDH = HD1 + HD2 * EARAD * XI/(DMS * 1000.)
             GO TO 918
         915 SDH = HD1
             GO TO 918
         917 SDH = HD2 + HD1 * EARAD * XI/(DMS * 1000.)
465      918 CONTINUE
             IF(BX - SDH)         940,920,920
         920 IF(REF - ANREF)        930,940,930
         930 EARAD = STEAR
             REF = SREF
```

Fig. B.1 *Continued*

```
470          GRAD = SGRD
             XAB = A - B
             XI = SQRT((R * R - XAB * XAB)/(A * B))
             BX = B - EARAD
         940 CONTINUE
475          RXC = XI * EARAD
             PAX = 2. * SQRT(EARAD * (AX + BX) + RXC * RXC/4.)/SQRT(3.)
             IF(BX - AX)          90,92,92
          90 CPH = 2. * EARAD * (AX - BX) * RXC/(PAX ** 3)
             GO TO 94
480       92 CPH = 2. * EARAD * (BX - AX) * RXC/(PAX ** 3)
          94 CONTINUE
             SPH = SQRT(1. - CPH * CPH)
             PHA = ATAN(SPH/CPH)
             RAX = RXC/2. + PAX * COS((PHA + PI)/3.)
485          IF(BX - AX)          112,114,114
         112 ALPHA = RAX/EARAD
             BETA = XI - ALPHA
             GO TO 116
         114 BETA = RAX/EARAD
490          ALPHA = XI - BETA
         116 CONTINUE
             X = SQRT(AX * AX + EARAD * A * ALPHA * ALPHA)
             Y = SQRT(BX * BX + EARAD * B * BETA * BETA)
             CODEL = EARAD * ALPHA/X
495          SIDEL = SQRT(1. - CODEL * CODEL)
             DELTA = ATAN(SIDEL/CODEL)
             GAM = PI/2. - ABS(DELTA - ALPHA)
             DIST = SHO(N) * 1000.
             SLOF = SF(N) * 1000.
500          SLOST = SLO(N) * 1000.
             TERHTS = TER(N)
             TEHF = TEF(N)
             TST = (TERHTS - TEHF)/(SLOF - SLOST)
             TSL = ATAN(TST)
505          IF(TERHTS.EQ.TEHF)      TSL = 0.
             GRT = ALPHA * EARAD
             IF(GRT.GT.SLOF)      TSL = 0.
             IF(GRT - SLOST)         102,102,641
         641 IF(TSL - ABS(DELTA - ALPHA))          644,644,102
510      102 TSL = 0.
         644 CONTINUE
             GAM = GAM + TSL
             DIELCO = PERM
             COND = RHO
515          IF(GRT - DIST)       680,670,670
         670 DIELCO = WDLCO
             COND = WCOND
         680 CONTINUE
             QC = 60. * AMBDA * COND
520          IF(IPOL - 2)        690,700,690
         690 CONTINUE
       C     VERTICAL POLARIZATION
             AC = DIELCO * COS(GAM)
             OZ = SQRT((DIELCO - (SIN(GAM)) ** 2) **2 + QC * QC)
525          OY = SQRT(0.5 + 0.5 * (DIELCO - (SIN(GAM)) **2)/OZ)
             OX = SQRT(0.5 - 0.5 * (DIELCO - (SIN(GAM)) ** 2)/OZ)
             O = AC - OY * SQRT(OZ)
             P = QC * COS(GAM) - OX * SQRT(OZ)
             Q = AC + OY * SQRT(OZ)
530          PA = QC * COS(GAM) + OX * SQRT(OZ)
             OA = O * Q + P * PA
             OB = P * Q - O * PA
             OC = Q * Q + PA * PA
             RHOVER = SQRT((OA * OA + OB * OB)/(OC * OC))
535          AC = OB/OA
             IF(ABS(AC) - 1.E-07)        721,710,710
         721 AMU = PI
             IF(IPOL - 2)          715,715,717
         710 IF(AC)          713,713,712
540      712 AMU = ATAN(AC)
             IF(IPOL - 2)          715,715,717
         713 AMU = PI - ABS(ATAN(AC))
         717 AMUVER = AMU
             IF(IPOL - 2)        715,715,700
545      700 CONTINUE
       C     HORIZONTAL POLARIZATION
             QB = DIELCO - (SIN(GAM)) ** 2
             HOZ = SQRT(QB * QB + QC * QC)
             HOY = SQRT(0.5 + 0.5 * QB/HOZ)
550          HOX = SQRT(0.5 - 0.5 * QB/HOZ)
             HO = COS(GAM) - HOY * SQRT(HOZ)
             HP = HOX * SQRT(HOZ)
             HQ = COS(GAM) + HOY * SQRT(HOZ)
             HA = HO * HQ - HP * HP
555          HB = HP * (HO + HQ)
             HC = HQ * HQ + HP * HP
             RHOHOR = SQRT((HA * HA + HB * HB)/(HC * HC))
             HAC = HB/HA
             AC = HAC
560          AMU = PI + ABS(ATAN(AC))
             AMUHOR = AMU
             IF(IPOL - 2)          715,715,722
         722 CONTINUE
       C     CIRCULAR POLARIZATION
565          RVT = RHOVER * RHOVER
             RHT = RHOHOR * RHOHOR
             RVH = 2. * RHOVER * RHOHOR * COS(AMUHOR - AMUVER)
             SICA = RHOVER * SIN(AMUVER)
             SICB = RHOHOR * SIN(AMUHOR)
```

Fig. B.1 *Continued*

```
570            COCA = RHOVER * COS(AMUVER)
               COCB = RHOHOR * COS(AMUHOR)
               IF(ICIR - 1)        714,716,715
           714 RHOCIR = 0.5 * SQRT(RVT + RHT - RVH)
               AMUCIR = PI + ATAN((SICA - SICB)/(COCA - COCB))
575            AMU = AMUCIR
               GO TO 715
           716 RHOCIR = 0.5 * SQRT(RVT + RHT + RVH)
               AMUCIR = PI + ATAN((SICA + SICB)/(COCA + COCB))
               AMU = AMUCIR
580        715 CONTINUE
               ZY = X + Y
               S1 = (2. * X * Y)/(EARAD * ZY)
               S2 = (1. + S1)/COS(GAM)
               DA = SQRT(1. + S1 * (1. + S2))
585            DX = 1./DA
               WHT = SEA
               IF(GRT - DIST)        720,730,730
           720 WHT = UNDUL
           730 CONTINUE
590            IF(N - 17)        952,952,954
           952 RA = 1250.
               SA = 7.
               GO TO 980
           954 IF(N - 35)        956,956,958
595        956 RA = 1950.
               SA = 354.
               GO TO 980
           958 IF(N - 45)        960,960,962
           960 RA = 1130.
600            SA = 86.
               GO TO 980
           962 IF(N - 108)        964,964,966
           964 RA = 880.
               SA = 86.
605            GO TO 980
           966 IF(N - 240)        968,968,971
           968 RA = 350.
               SA = 173.
               GO TO 980
610        971 IF(N - 320)        972,972,974
           972 RA = 1350.
               SA = 265.
               GO TO 980
           974 IF(N - 349)        976,976,978
615        976 RA = 1950.
               SA = 354.
               GO TO 980
           978 IF(N - 361)        979,979,980
           979 RA = 1250.
620            SA = 7.
           980 CONTINUE
               AN = (SA - FLOAT(N))/DEG
               CR = RA/(COS(AN))
               IF(GRT - CR)        620,620,625
625        620 WHT = 0.
           625 CONTINUE
               RUFS = EXP(- 8. * (PI * WHT * COS(GAM)/(4. * AMBDA)) ** 2)
               PADI = X * (1. - COS(E + DELTA))
               CALL AERPLD(E,DELTA,TLT,FORM1)
630            IF(IPOL - 2)        735,740,745
           735 EX = FORM1 * RHOVER * DX * RUFS
               GO TO 750
           740 EX = FORM1 * RHOHOR * DX * RUFS
               GO TO 750
635        745 EX = FORM1 * RHOCIR * DX * RUFS
           750 CONTINUE
               CALL AERPCD(E,DELTA,TLT,FORM)
               E1 = FORM
               FIELD = SQRT(E1 ** 2 + EX ** 2 + 2. * E1 * EX * COS(AMU + 2.*PI*
640           1PADI/AMBDA))
           346 CONTINUE
               ZR = EXP(-1.15E-4 * ATM * 2. * R * FIELD)
               RC(J) = R * FIELD * (ZR ** 0.25)
               IF(PJ.EQ.0)        GO TO 1175
645            IF(JAM - 0)        790,1160,1165
          1160 RJAM = R
               FIJ = FIELD
               GO TO 1150
          1165 CONTINUE
650            CXIJ = (A * A + HJAM * HJAM - RJAM * RJAM)/(2. * A * HJAM)
               SXIJ = SQRT(1. - CXIJ * CXIJ)
               XIJ = ATAN(SXIJ/CXIJ)
               CEJ = HJAM * SIN(XIJ)/RJAM
               SEJ = SQRT(1. - CEJ * CEJ)
655            EJ = ATAN(SEJ/CEJ)
               RXCJ = XIJ * EARAD
               AJAM = ANTJAM
               PAXJ = 2.*SQRT(EARAD*(AX + AJAM)+RXCJ*RXCJ/4.)/SQRT(3.)
               CPJ = 2. * EARAD * (AX - AJAM) * RXCJ/(PAXJ ** 3)
660            SPJ = SQRT(1. - CPJ * CPJ)
               PHAJ = ATAN(SPJ/CPJ)
               RAXJ = RXCJ/2. + PAXJ * COS((PHAJ + PI)/3.)
               BETJ = RAXJ/EARAD
               ALJ = XIJ - BETJ
665            DELJ = ATAN((A - EARAD * COS(ALJ))/(EARAD * SIN(ALJ)))
               CALL SDLB(TLT,HW,VW,DECL,EJ,DELJ,HORZ,HORZ1)
               XJ = SQRT((AX * AX) + EARAD * A * (ALJ * ALJ))
               YJ = SQRT((AJAM ** 2) + EARAD * HJAM * (BETJ ** 2))
               PDJ = (XJ + YJ - RJAM)/AMBDA
```

Fig. B.1 *Continued*

```
670              PHJ = COS(PI * (1. + 2. * PDJ))
                 FIJ = SQRT(HORZ * HORZ + HORZ1 * HORZ1 + 2. * HORZ * HORZ1 * PHJ)
            1150 CONTINUE
                 CONJAM = CJAM * (AMBDA/(4. * PI * FJAM)) ** 2
                 RXN = XN + CONJAM * FIJ * FIJ
675              RC(J) = RC(J) * ((XN/RXN) ** 0.25)
            1175 CONTINUE
                 CEST = (A * A + RC(J) * RC(J) - B * B)/(2. * A * RC(J))
                 SEST = SQRT(1. - CEST * CEST)
                 EST = ATAN(SEST/CEST)
680              IF(CEST)       352,354,354
             352 EST = PI + EST
             354 CONTINUE
                 BF = EARAD + TEHF
                 XIF = SLOF/EARAD
685              RSF = SQRT(A * A + BF * BF - 2. * A * BF * COS(XIF))
                 CTF = (A * A + RSF * RSF - BF * BF)/(2. * A * RSF)
                 STF = SQRT(1. - CTF * CTF)
                 FAG = ATAN(STF/CTF)
                 IF(CTF - 0.)         324,326,326
690          324 FAG = PI + FAG
             326 CONTINUE
                 IF(EST - FAG)         315,322,322
             315 SCM = A * SIN(FAG)/B
                 COM = SQRT(1. - SCM * SOM)
695              TCM = ATAN(SCM/COM)
                 FXI = PI - FAG - TOM
                 RC(J) = SQRT(A * A + B * B - 2. * A * B * COS(FXI))
             322 CONTINUE
                 BC = B
700              HEIGHT(J) = BC - EARAD
                 ZB = SQRT(AERHIT * (2. * EARAD + AERHIT))
                 ZC = SQRT(HEIGHT(J) * (2. * EARAD + HEIGHT(J)))
                 ZD = ZB + ZC
                 IF(RC(J) - ZD)         770,770,760
705          760 RC(J) = ZD
                 BC = B
                 HEIGHT(J) = BC - EARAD
             770 CONTINUE
                 CGX = (A * A + B * B - RC(J) * RC(J))/(2. * A * B)
710              SGX = SQRT(1. - CGX * CGX)
                 GX = ATAN(SGX/CGX)
                 GR(J) = GX * EARAD
                 RC(J) = RC(J)/1000.
                 GR(J) = GR(J)/1000.
715              L = N - 1
                 AZ = L/DEG
                 XA(J) = GR(J) * SIN(AZ)
                 YA(J) = GR(J) * COS(AZ)
                 WRITE(6,780) L,RC(J),GR(J),HEIGHT(J),TEHF
720          780 FORMAT(5X,I3,11X,F7.2,11X,F7.2,12X,F8.2,12X,F8.2)
             790 CONTINUE
             795 CONTINUE
                 XO = - 300.
                 XS = 50.
725              YO = - 300.
                 YS = 50.
                 CALL AXIS(0.,6.,25H                         KM,-25,12.,0.,XO,XS,0)
                 CALL AXIS(6.,0.,25H                         KM,25,12.,90.,YO,YS,-1)
                 CALL SYMBOL(1.3,17.,0.4,28HTHE ADELAIDE AIRPORT RADAR-S,0.,28)
730              CALL SYMBOL(2.3,16.,0.4,22HDETECTION MAP CONTOURS,0.,22)
                 CALL SYMBOL(1.1,15.,0.3,39HPROBABILITY OF DETECTION          PERCENT
               1,0.,39)
                 CALL NUMBER(7.6,15.,0.3,PRODET,0.,4HF6.2)
                 XB(362) = XO
735              XB(363) = XS
                 YB(362) = YO
                 YB(363) = YS
                 DO 6000        I = 1,4
                 M = 361 * I - 361
740              DO 5500        KY = 1,361
                 XB(KY) = XA(M + KY)
            5500 YB(KY) = YA(M + KY)
                 CALL LINE(XB,YB,361,1,0,0)
            6000 CALL PLOT(0.,0.,3)
745              CALL PLOT(50.,0.,-3)
             810 CONTINUE
                 STOP
                 END
```

Fig. B.1 *Continued*

Appendix C

Listing of model 'RDPRO'

```
 1               PROGRAM RDPRO(INPUT,OUTPUT,TAPE5=INPUT,TAPE6=OUTPUT,DEBUG=OUTPUT)
                 DIMENSION  RANGE(500),SIGDBM(500),CNDBM(500),DEFRO(500)
                 DIMENSION  XPRO(500),YPRO(500)
                 DIMENSION  XSIG(500),YSIG(500)
 5               DIMENSION  XFAR(500),YFAR(500)
                 DIMENSION  TITLE(3)
                 COMMON  AGEX,AMBDA,BKB,BMHZ,DEG,DIST,FASTEP,GAM,GRT,HW,PCF,RFA,PI,
                1REC,RRT,SCV,SEA,TAU,TLT,VW,XAB,XN,ZB,ZR,ZX,
                2ICIR,IPOL,M,MM,MTI
10         C     IPOL EQUALS 1 FOR VERTICAL POLARIZATION,2 FOR HORIZONTAL AND 3 FOR
           C     CIRCULAR POLARIZATION.
                 READ(5,1) IPOL
               1 FORMAT(2X,I1)
           C     GRAD IS THE N.B.S. CONSTANT CE
15               READ(5,2) GRAD
               2 FORMAT(2X,E15.8)
           C     REF IS THE N.B.S. CONSTANT NS * 10 TO MINUS 6
                 READ(5,9) REF
               9 FORMAT(2X,E14.8)
20               READ(5,4) PERM,RHO
               4 FORMAT(2X,F4.1,2X,F6.3)
                 READ(5,5) ATM
               5 FORMAT(2X,F8.6)
                 READ(5,8) RAIN
25             8 FORMAT(2X,F8.5)
                 READ(5,34) RR
              34 FORMAT(2X,F7.3)
                 READ(5,32) RCC
              32 FORMAT(2X,F8.5)
30               READ(5,36) RF
              36 FORMAT(2X,F7.3)
                 READ(5,6) UNDUL
               6 FORMAT(2X,F5.1)
                 READ(5,7) SEA,BKB,STATE
35             7 FORMAT(2X,F5.1,2X,F3.0,2X,F2.0)
                 READ(5,10) RRT
              10 FORMAT(2X,F5.2)
                 READ(5,11) SHORE,AERSEA
              11 FORMAT(2X,F6.2,2X,F9.2)
40               READ(5,12) RKW
              12 FORMAT(2X,F8.2)
                 READ(5,13) FREQ
              13 FORMAT(2X,F10.3)
                 READ(5,14) RDB
45            14 FORMAT(2X,F5.2)
                 READ(5,15) BMHZ
              15 FORMAT(2X,F6.2)
                 READ(5,18) ELOSS
              18 FORMAT(2X,F5.2)
50               READ(5,16) PRF
              16 FORMAT(2X,F7.2)
                 READ(5,17) RFA
              17 FORMAT(2X,E7.1)
                 READ(5,40) AGEX
55            40 FORMAT(2X,F8.3)
                 READ(5,19) SCV
              19 FORMAT(2X,F4.1)
                 READ(5,20) FASTEP
              20 FORMAT(2X,F3.0)
60               READ(5,21) TAU
              21 FORMAT(2X,F5.2)
                 READ(5,22) ANTHIT
              22 FORMAT(2X,F8.2)
                 READ(5,25) AEGDB
65            25 FORMAT(2X,F5.2)
                 READ(5,23) HBMWTH
              23 FORMAT(2X,F4.2)
                 READ(5,24) VBMWTH
              24 FORMAT(2X,F4.1)
```

Fig. C.1 *Listing of model RDPRO*

```
 70          C      IF A SECTOR SCAN IS USED AND NEITHER THE ANGULAR SCAN RATE ASR NOR
             C      THE APPROPRIATE RPM IS GIVEN,RPM IS TO BE DETERMINED FROM THE FULL
             C      SCAN RPM AS 360. * RPM/SCA AND USED AS INPUT IN THE PROGRAMME.
                    READ(5,454) ASR
                454 FORMAT(2X,F7.2)
 75          C      RPM IS THE AERIAL SCAN RATE WHICH IN THE CASE OF A SECTOR SCAN
             C          MAY DIFFER FROM THAT USED FOR A 360 DEGREE SCAN (SEE THE PRE-
             C          CEDING COMMENT).
                    READ(5,26) RPM
                 26 FORMAT(2X,F7.2)
 80          C      SCA IS THE ANGLE OF A SCAN CYCLE,E.G. IT IS 360.00 FOR A CIRCULAR
             C          SCAN AND E.G. 80.00 FOR A PLUS-MINUS 40 DEGREE SCAN.
                    READ(5,31) SCA
                 31 FORMAT(2X,F6.2)
                    READ(5,27) TILT
 85              27 FORMAT(2X,F5.2)
                    READ(5,28) XAD
                 28 FORMAT(2X,F6.2)
                    READ(5,29) SIGMA
                 29 FORMAT(2X,F7.2)
 90          C      K ONE TO FOUR CORRESPONDS TO SWERLING CASES ONE TO FOUR,K FIVE IS
             C      FOR NON-FLUCTUATING TARGETS.
                    READ(5,30) K
                 30 FORMAT(2X,I1)
                    READ(5,33) VELOC
 95              33 FORMAT(2X,F7.2)
                    READ(5,37) DESRT
                 37 FORMAT(2X,F9.2)
                    READ(5,38) TARGHT
                 38 FORMAT(2X,F9.2)
100                 READ(5,39) SLCST,THST
                 39 FORMAT(2X,F6.2,2X,F9.2)
                    READ(5,45) SLCF,THF
                 45 FORMAT(2X,F6.2,2X,F9.2)
             C      NOPER IS ZERO FOR POSTDETECTION INTEGRATION,IT IS 1 FOR OPERATOR -
105          C      CRT INTEGRATION AND 2 FOR PREDETECTION INTEGRATION.
                    READ(5,46) NOPER
                 46 FORMAT(2X,I1)
                    READ(5,47) FIF
                 47 FORMAT(2X,F7.2)
110                 READ(5,48) FOR
                 48 FORMAT(2X,F7.2)
                    READ(5,453) SIT
                453 FORMAT(2X,F3.0)
             C      CHOOSE MTI = 0 WHEN NO MTI USED,= 1 FOR SINGLE COHERENT CANCELLER,
115          C      = 2 FOR DUAL COHERENT CANCELLER,= 3 FOR SINGLE NONCOHERENT CANCEL-
             C      LER AND = 4 FOR DUAL NONCOHERENT CANCELLER.
                    READ(5,49) MTI
                 49 FORMAT(2X,I1)
             C      IPROB IS ZERO FOR THE SCAN TO SCAN PROBABILITY OF DETECTION AND
120          C      UNITY FOR THE CUMULATIVE ONE.
                    READ(5,6000) IPROB
               6000 FORMAT(2X,I1)
             C      JAM IS ZERO FOR A TARGETBORNE JAMMER AND ONE FOR A STANDOFF ONE.
                    READ(5,1009) JAM
125            1009 FORMAT(2X,I1)
                    READ(5,1100) PJ
               1100 FORMAT(2X,F8.2)
                    READ(5,1105) GJ
               1105 FORMAT(2X,F5.2)
130                 READ(5,1110) BJ
               1110 FORMAT(2X,F6.2)
                    READ(5,1115) DJ
               1115 FORMAT(2X,F5.2)
                    READ(5,1120) RJ
135            1120 FORMAT(2X,F6.2)
                    READ(5,1125) ANTJAM
               1125 FORMAT(2X,F8.2)
                    READ(5,1130) DECLJ
               1130 FORMAT(2X,F6.2)
140                 READ(5,315) PIA,PWVEL,PWVL,VPWV
                315 FORMAT(2X,F4.1,2X,F4.0,2X,F5.0,2X,F5.1)
                    READ(5,319) SHV
                319 FORMAT(2X,F5.2)
             C      ICIR IS ZERO FOR OPPOSITE AND ONE FOR SAME CIRCULAR POLARIZATION.
145                 READ(5,985) ICIR
                985 FORMAT(2X,I1)
                    READ(5,950) TITLE
                950 FORMAT(1X,3A10)
             C      COMPUTE THE EARTH RADIUS CORRECTION FACTOR.
150                 EARTH = 6371.228
                    COER = 1./(1. - GRAD * REF * EARTH)
                    EARAD = EARTH * COER * 1000.0
                    V = VELOC * 1000./3600.
                    AAS = ANTHIT + AERSEA
155                 TARG = TARGHT
                    DHT = ANTHIT + AERSEA - TARGHT
                    ZB = SQRT(AAS * (2. * EARAD + AAS))
                    ZC = SQRT(TARG * (2. * EARAD + TARG))
                    ZD = ZB + ZC
160                 LMAX = 0
                    NA = LMAX + 2
                    CALL ARRAY(NA,ZD,RPM,V,RANGE,SIGDBM,CNDBM,DEPRC,XPRO,YPRO,
                   1DHT,TITLE,LMAX)
                    PI = 3141592.653589E-06
165                 DEG = 5729.57791E-02
                    AEGAIN = 10. ** (AEGDB/10.)
                    WRITE(6,50) TITLE
                 50 FORMAT(1X,3A10)
                    IF(IPOL - 2)        52,53,620
```

Fig. C.1 *Continued*

```
170           52 WRITE(6,54)
              54 FORMAT(1X,21HVERTICAL POLARIZATION)
                 GO TO 56
              53 WRITE(6,55)
              55 FORMAT(1X,23HHORIZONTAL POLARIZATION)
175              GO TO 56
             620 WRITE(6,630)
             630 FORMAT(1X,21HCIRCULAR POLARIZATION)
              56 CONTINUE
                 IF(K.EQ.1)       GO TO 57
180              IF(K.EQ.2)       GO TO 59
                 IF(K.EQ.3)       GO TO 61
                 IF(K.EQ.4)       GO TO 63
                 IF(K.EQ.5)       GO TO 65
              57 WRITE(6,58)
185           58 FORMAT(1X,15HSWERLING CASE I)
                 GO TO 350
              59 WRITE(6,60)
              60 FORMAT(1X,16HSWERLING CASE II)
                 GO TO 350
190           61 WRITE(6,62)
              62 FORMAT(1X,17HSWERLING CASE III)
                 GO TO 350
              63 WRITE(6,64)
              64 FORMAT(1X,16HSWERLING CASE IV)
195              GO TO 350
              65 WRITE(6,66)
              66 FORMAT(1X,22HNON-FLUCTUATING TARGET)
             350 CONTINUE
                 WRITE(6,67) PKW
200           67 FORMAT(1X,7HPOWER =,F10.3,6H     KW)
                 WRITE(6,68) FREQ
              68 FORMAT(1X,11HFREQUENCY =,F10.3,7H     MHZ)
                 WRITE(6,69) FDB
              69 FORMAT(1X,14HNOISE FIGURE =,F6.2,7H      DB)
205              WRITE(6,99) BMHZ
              99 FORMAT(1X,11HBANDWIDTH =,F6.2,7H     MHZ)
C                COMPUTE THE NUMBER OF HITS/SCAN,THE BEAM SHAPE LOSS AND THE TOTAL
C                SYSTEM LOSS.
                 PD = 0.5
210              SUMONE = 0.0
                 IF(ASR - 0.)       196,451,455
             451 ASR = SCA * RPM/60.
             455 CONTINUE
                 PN = HBMWTH * PRF/ASR
215              IF(SIT - 0.)       196,459,457
             457 PN = SIT * PN * RPM/60.
             459 CONTINUE
                 DELPHI = HBMWTH/PN
                 NX = (PN - 1.0)/2.0
220              MM = PN
                 TA = 0.0299 + 0.241 * EXP(0.1799 * ALOG10(PFA))
                 TB = 0.0505 + 0.640 * EXP(0.270 * ALOG10(PFA))
                 TC = 0.4599 - TB * EXP(- 0.881 * PD)
                 TD = 0.241 + TA * EXP(- 1.905 * PD)
225              IF(NOPER - 1)       201,200,199
             199 SPI = 1.
                 M = PN
                 GO TO 205
             200 SPI = SQRT(PN)
230              M = IFIX(SPI)
                 GO TO 205
             201 IF(MM - 1)       196,202,203
             202 SPI = 1.
                 M = IFIX(SPI)
235              GO TO 205
             203 CONTINUE
                 SIMP = 0.5 + TC * EXP(- TD * ALOG10(PN))
                 SPI = PN/PN ** SIMP
                 M = IFIX(PN ** SIMP)
240          205 CONTINUE
                 DO 70 KA = 1,NX
                 XK = KA
              70 SUMONE = EXP(-5.55 * ((XK * DELPHI) ** 2)/(HBMWTH ** 2)) + SUMONE
                 BEAMLS = PN/(1. + 2. * SUMONE)
245              IF(RPM - 0.)       71,71,72
              71 BEAMLS = 1.0
              72 CONTINUE
                 BEALOS = 10. * ALOG10(BEAMLS)
                 XLOSS = 10. ** (ELOSS/10.)
250              XMN = 577350. * PFA * SQRT(3. * FIF * FIF + BMHZ * BMHZ/4.)
                 ZMN = XMN + PN
                 IF(NOPER - 1)       632,634,632
             634 SXY = SQRT(ZMN)
                 GO TO 636
255          632 SMN = 0.5 + TC * EXP(- TD * ALOG10(ZMN))
                 SXY = ZMN/ZMN ** SMN
             636 CONTINUE
                 COLOS = SXY/SPI
                 OPLOSS = 1./(0.7 * PD * PD)
260              OPLDB = 10. * ALOG10(OPLOSS)
                 SUMDB = 10. * ALOG10(SPI)
                 COLAP = 10. * ALOG10(COLOS)
                 TOTLOS = XLOSS * BEAMLS * COLOS * SPI * OPLOSS
                 BLOSS = BEALOS + COLAP + ELOSS + SUMDB + OPLDB
265              WRITE(6,73) BLOSS
              73 FORMAT(1X,37HPLUMBING AND OTHER EQUIPMENT LOSSES =,F6.2,7H      DB)
                 IF(TAU - 0.)       96,96,101
              96 TAU = 1.2 * PCR/BMHZ
             101 CONTINUE
```

Fig. C.1 *Continued*

```
270               WRITE(6,84) TAU
               84 FORMAT(1X,13HPULSE WIDTH =,F5.2,15H       MICROSECS)
                  WRITE(6,74) PRF
               74 FORMAT(1X,28HPULSE REPETITION FREQUENCY =,F7.2,14H   PULSES/SEC.)
                  WRITE(6,75) PFA
275            75 FORMAT(1X,28HPROBABILITY OF FALSE ALARM =,E7.1)
                  WRITE(6,76) AGEX
               76 FORMAT(1X,19HFREQUENCY AGILITY =,F8.3,7H    MHZ)
                  WRITE(6,77) SCV
               77 FORMAT(1X,23HSUBCLUTTER VISIBILITY =,F4.1,7H      DB)
280               WRITE(6,78) ANTHIT
               78 FORMAT(1X,15HAERIAL HEIGHT =,F8.2,4H   M)
                  WRITE(6,79) HBMWTH
               79 FORMAT(1X,22HHORIZONTAL BEAMWIDTH =,F6.2,10H   DEGREES)
                  WRITE(6,81) AEGDB
285            81 FORMAT(1X,13HAERIAL GAIN =,F8.2,7H      DB)
                  WRITE(6,82) RPM
               82 FORMAT(1X,18HAERIAL SCAN RATE =,F7.2,8H  R.P.M.)
                  WRITE(6,83) TILT
               83 FORMAT(1X,13HAERIAL TILT =,F5.2,10H   DEGREES)
290               WRITE(6,85) SIGMA
               85 FORMAT(1X,21HTARGET ECHOING AREA =,F7.2,16H   SQUARE METRES)
                  WRITE(6,86) MM
               86 FORMAT(1X,15HHITS PER SCAN =,I4,16H  (RADIAL TRACK))
                  WRITE(6,87) GRAD
295            87 FORMAT(1X,34HGRADIENT OF THE REFRACTIVE INDEX =,E15.8,7H    /KM)
                  WRITE(6,298) REF
              298 FORMAT(1X,22HSURFACE REFRACTIVITY =,E14.8)
                  WDLCO = 80.
                  WCOND = 4.3
300               IF(FREQ - 1500.)        304,304,301
              301 IF(FREQ - 3000.)        302,302,303
              302 WDLCO = WDLCO - 0.00733 * (FREQ - 1500.)
                  WCOND = WCOND + 0.001467 * (FREQ - 1500.)
                  GO TO 304
305           303 WDLCO = 69. - 0.000628 * (FREQ - 3000.)
                  WCOND = 6.5 + 0.001492 * (FREQ - 3000.)
              304 CONTINUE
                  WRITE(6,88) WDLCO
               88 FORMAT(1X,36HRELATIVE PERMITTIVITY OF SEA WATER =,F4.1)
310               WRITE(6,89) WCOND
               89 FORMAT(1X,27HCONDUCTIVITY OF SEA WATER =,F6.3,12H   MHO/METRE)
                  WRITE(6,90) PERM
               90 FORMAT(1X,37HRELATIVE PERMEABILITY OF THE GROUND =,F4.1)
                  WRITE(6,91) RHO
315            91 FORMAT(1X,28HCONDUCTIVITY OF THE GROUND =,F6.3,12H   MHO/METRE)
                  FGHZ = FREQ/1000.
                  FLG = ALOG10(FGHZ)
                  IF(FGHZ - 1.)         650,655,655
              650 EOX = 0.64866 * FLG * FLG - 0.67359 * FLG + 2.37675
320               AOX = 10. ** (- EOX)
                  EWV = 0.33667 * FLG * FLG + 2.66667 * FLG - 5.49485
                  AWV = 10. ** EWV
                  GO TO 665
              655 IF(FGHZ - 10.)        660,660,670
325           660 EOX = 0.11967 * FLG * FLG - 0.69782 * FLG + 2.37675
                  AOX = 10. ** (- EOX)
                  EWV = 0.61905 * FLG * FLG + 2.49095 * FLG - 5.49485
                  AWV = 10. ** EWV
              665 ATM = AOX + AWV
330           670 CONTINUE
                  WRITE(6,92) ATM
               92 FORMAT(1X,25HATMOSPHERIC ATTENUATION =,F8.6,8H   DB/KM)
                  WRITE(6,93) UNDUL
               93 FORMAT(1X,20HTERRAIN UNDULATION =,F5.1,4H   M)
335               WRITE(6,94) SEA
               94 FORMAT(1X,17HSEA WAVE HEIGHT =,F5.1,4H   M)
                  WRITE(6,95) BKB
               95 FORMAT(1X,25HBEAUFORT WIND SCALE NO. =,F3.0)
                  WRITE(6,97) RRT
340            97 FORMAT(1X,15HRAINFALL RATE =,F5.2,9H     MM/HR)
                  WRITE(6,98) SHORE
               98 FORMAT(1X,21HDISTANCE TO THE SEA =,F6.2,5H    KM)
                  WRITE(6,103) VELOC
              103 FORMAT(1X,17HTARGET VELOCITY =,F9.2,9H     KM/HR)
345               WRITE(6,100) DESRT
              100 FORMAT(1X,14HDESCENT RATE =,F9.2,8H     M/KM)
                  WRITE(6,111)
              111 FORMAT(//)
                  IF(IPROB - 1)        6015,6005,196
350          6005 WRITE(6,6010)
             6010 FORMAT(93X,10HCUMULATIVE)
             6015 CONTINUE
                  WRITE(6,112)
              112 FORMAT(1X,109HRANGE    RADAR HEIGHT    TARGET HEIGHT    SIGNAL
355             1 CLUTTER    SIGNAL TO CLUTTER    PROBABILITY OF DETECTION)
                  WRITE(6,113)
              113 FORMAT(1X,101HKM             M                M          LEVEL DBM
                 1LEVEL DBM  PLUS NOISE RATIO DB              PERCENT)
                  WRITE(6,114)
360           114 FORMAT(///)
                  CALL PLOT25
                  CALL PAUPLOT(29HBLANK PAPER,BLACK INK,0.3 NIB,29)
                  CALL PLOT(0.,0.,-3)
                  ANT = ANTHIT
365               IF(DHT)        500,500,510
              500 AERHIT = AAS
                  XIN = TARG
                  GO TO 515
              510 AERHIT = TARG
```

Fig. C.1 *Continued*

```
370              XIN = AAS
             515 CONTINUE
                 RZD = ZD + 1.0E04
                 XUNIT = 1./(1000. * XAD)
                 XX = ZD * XUNIT
375              YY = 0.
                 PWR = PKW * PCR * 1000.
                 TLT = TILT/DEG
                 AMBDA = 299.793/FREQ
                 VEP = PWVEL * 1000./3600.
380              VPW = VPWV/DEG
                 SHV = SHV * 1000./3600.
                 PVEL = ABS(SHV - VEP * COS(VPW))
                 IF(VPW - PI/2.)          470,465,460
             460 IF(VPW - 3. * PI/2.)          470,465,470
385          465 PVEL = 0.
             470 CONTINUE
                 PIF = PVEL/PWVL
                 PF = 2. * PI * PIF
                 THRESH = BIAS(M,PFA)
390              XPN = FLOAT(M)
                 FA = FDB/10.
                 FB = 10. ** FA
                 RXN = 4.0E-15 * BMHZ * FB
                 YXN = 10. * ALOG10(RXN * 1000.0) * 0.025
395              VW = VBMWTH/DEG
                 HW = HBMWTH/DEG
                 DELR = V * 60./RPM
                 DESC = DESRT/1000.
                 DSL = DESC * DELR
400              DIST = SHORE * 1000.
                 SLST = SLOST * 1000.
                 SLF = SLOF * 1000.
                 TST = (THST - THF)/(SLF - SLST)
                 IF(THST.EQ.THF)          TSL = 0.
405              IF(IPOL - 2)          420,420,410
             410 SIGMA = SIGMA/5.
             420 CONTINUE
                 CDX = 1.
                 TIM = 0.
410              REC = (PWR * SPI * (AEGAIN * AMBDA) ** 2)/(((4. * PI) ** 3) *
                1TOTLOS)
                 IF(PJ - 1.)          715,720,720
             715 CJAM = 0.
                 FIJ = 0.
415              RJAM = 1.
                 GO TO 725
             720 CONTINUE
                 PJAM = PJ * 1.0E3
                 AERJAM = 10. ** (GJ/10.)
420              XLJAM = 10. ** (DJ/10.)
                 BJAM = BJ * 1.0E6
                 DECL = DECLJ/DEG
                 RJAM = RJ * 1000.
                 ZJAM = EXP(-1.15E-4 * ATM * RJAM)
425              AJAM = ANTJAM
                 HJAM = EARAD + AJAM
                 CJAM = PJAM * AERJAM * AEGAIN * ZJAM/(XLJAM * BJAM)
             725 CONTINUE
                 GRT = ZD
430              TRH = 0.
                 LM = LMAX/3
                 XLM = LM
                 DO 4444          IX = 1,3
                 IF(IX - 2)          1000,800,900
435          800 CALL PLOT(XX,YY,3)
                 YAP = - 200.
                 YAW = 40.
                 GO TO 2000
             900 CALL PLOT(XX,YY,3)
440              YAZ = - 200.
                 YAV = 40.
                  GO TO 3000
            1000 CONTINUE
                 DO 194          L = 1,LMAX
445              XL = L
           C     COMPUTE THE INSTANTANEOUS TARGET RANGE AND HEIGHT ON EACH AERIAL
           C     SCAN.
                 R = ZD - (XL - 1.) * DELR
                 HT = XIN - (XL - 1.) * DSL
450              A = EARAD + AAS
                 B = EARAD + TARG
                 AB = A - B
                 IF(AB)          530,530,520
             520 A = EARAD + TARG
455              B = EARAD + AAS
             530 CONTINUE
                 AX = AAS - TRH
                 BX = TARG - (XL - 1.) * DSL
                 XAB = AX - BX
460              IF(XAB)          225,225,220
             220 AX = TARG - (XL - 1.) * DSL
                 BX = AAS - TRH
             225 CONTINUE
           C     DETERMINE THE GEOMETRY OF THE DIRECT AND REFLECTED RAYS FOR THE
465        C     GIVEN TERRAIN PROFILE.
                 XI = SQRT((R * R - AB * AB)/(A * B))
                 D = (B * B - R * R - A * A)/(2. * A * R)
                 E = ATAN2(D,SQRT(1. - D * D))
                 BEAMLS = 1.
```

Fig. C.1 *Continued*

```
470           IF(RPM - 0.)          106,106,107
          106 BEAMLS = EXP(5.55 * (E/VW) ** 2)
          107 CONTINUE
              REC = REC/BEAMLS
              RXC = XI * EARAD
475           PAX = 2. * SQRT(EARAD * (AX + BX) + RXC * RXC/4.)/SQRT(3.)
              IF(BX - AX)          104,105,105
          104 CPH = 2. * EARAD * (AX - BX) * RXC/(PAX * PAX * PAX)
              GO TO 35
          105 CPH = 2. * EARAD * (BX - AX) * RXC/(PAX * PAX * PAX)
480        35 CONTINUE
              SPH = SQRT(1. - CPH * CPH)
              PHA = ATAN(SPH/CPH)
              RAX = RXC/2. + PAX * COS((PHA + PI)/3.)
              IF(BX - AX)          121,122,122
485       121 ALPHA = RAX/EARAD
              BETA = XI - ALPHA
              GO TO 123
          122 BETA = RAX/EARAD
              ALPHA = XI - BETA
490       123 CONTINUE
              X = SQRT(AX * AX + EARAD * A * ALPHA * ALPHA)
              Y = SQRT(BX * BX + EARAD * B * BETA * BETA)
              DELTA = ATAN((A - EARAD * COS(ALPHA))/(EARAD * SIN(ALPHA)))
              GRT = ALPHA * (EARAD + TRH)
495           IF(XAB)          570,570,560
          560 GRT = BETA * (EARAD + TRH)
          570 CONTINUE
              GAM = PI/2. - ABS(DELTA - ALPHA)
              IF(THST - THF)          212,214,212
500       212 TSL = ATAN(TST)
              GO TO 216
          214 TSL = 0.
          216 CONTINUE
              GDR = EARAD * ALPHA * ANTHIT/AAS
505           IF(GRT.GT.SLF)          TSL = 0.
              IF(GRT - SLST)          80,80,41
           41 IF(TSL - ABS(DELTA - ALPHA))          43,43,42
           42 IF(GDR - SLST)          80,102,102
           80 TRH = AERSEA
510       102 TSL = 0.
              GO TO 44
           43 TRH = (SLF - GRT) * TAN(TSL)
           44 CONTINUE
              GAM = GAM + TSL
515           DIELCO = PERM
              COND = RHC
              IF(GRT - DIST)          126,125,125
          125 DIELCO = WDLCO
              COND = WCOND
520           TRH = AERSEA
          126 CONTINUE
              QC = 60. * AMBDA * COND
              IF(IPOL - 2)          127,128,127
          127 CONTINUE
525     C     VERTICAL POLARIZATION
              AO = DIELCO * COS(GAM)
              OZ = SQRT((DIELCO - (SIN(GAM)) ** 2) ** 2 + QC * QC)
              OY = SQRT(0.5 + 0.5 * (DIELCO -(SIN(GAM)) ** 2)/OZ)
              OX = SQRT(0.5 - 0.5 * (DIELCO - (SIN(GAM)) ** 2)/OZ)
530           O = AO - OY * SQRT(OZ)
              P = QC * COS(GAM) - OX * SQRT(OZ)
              Q = AO + OY * SQRT(OZ)
              PA = QC * COS(GAM) + OX * SQRT(OZ)
              OA = O * Q + P * PA
535           OB = P * Q - O * PA
              OC = Q * Q + PA * PA
              RHOVER = SQRT((OA * OA + OB * OB)/(OC * OC))
              AC = OB/OA
              IF(ABS(AC) - 1.E-07)          721,240,240
540       721 AMU = PI
              IF(IPOL - 2)          255,255,717
          240 IF(AC)          246,246,244
          244 AMU = ATAN(AC)
              IF(IPOL - 2)          255,255,717
545       246 AMU = PI - ABS(ATAN(AC))
          717 AMUVER = AMU
              IF(IPOL - 2)          255,255,128
          128 CONTINUE
        C     HORIZONTAL POLARIZATION
550           QB = DIELCO - (SIN(GAM)) ** 2
              HOZ = SQRT(QB * QB + QC * QC)
              HOY = SQRT(0.5 + 0.5 * QB/HOZ)
              HOX = SQRT(0.5 - 0.5 * QB/HOZ)
              HO = COS(GAM) - HOY * SQRT(HOZ)
555           HP = HOX * SQRT(HOZ)
              HQ = COS(GAM) + HOY * SQRT(HOZ)
              HA = HO * HQ - HP * HP
              HB = HP * (HO + HQ)
              HC = HQ * HQ + HP * HP
560           RHOHOR = SQRT((HA * HA + HB * HB)/(HC * HC))
              HAC = HB/HA
              AC = HAC
              AMU = PI + ABS(ATAN(AC))
              AMUHOR = AMU
565           IF(IPOL - 2)          255,255,722
          722 CONTINUE
        C     CIRCULAR POLARIZATION
              RVT = RHOVER * RHOVER
              RHT = RHOHOR * RHOHOR
```

Fig. C.1 *Continued*

```
570              RVH = 2. * RHOVER * RHOHOR * COS(AMUHOR - AMUVER)
                 SICA = RHOVER * SIN(AMUVER)
                 SICB = RHOHOR * SIN(AMUHOR)
                 COCA = RHOVER * COS(AMUVER)
                 COCB = RHOHOR * COS(AMUHOR)
575              IF(ICIR - 1)        714,716,255
             714 RHOCIR = 0.5 * SQRT(RVT + RHT - RVH)
                 AMUCIR = PI + ATAN((SICA - SICB)/(COCA - COCB))
                 AMU = AMUCIR
                 GO TO 255
580          716 RHOCIR = 0.5 * SQRT(RVT + RHT + RVH)
                 AMUCIR = PI + ATAN((SICA + SICB)/(COCA + COCB))
                 AMU = AMUCIR
             255 CONTINUE
                 ZY = X + Y
585              S1 = (2. * X * Y)/(EARAD * ZY)
                 S2 = (1. + S1)/COS(GAM)
                 DA = SQRT(1. + S1 * (1. + S2))
                 DX = 1./DA
                 WAVE = SEA
590              IF(GRT - DIST)          130,131,131
             130 WAVE = UNDUL
             131 CONTINUE
                 WHT = WAVE
                 RUFS = EXP(-8. * (PI * WHT * COS(GAM)/(4. * AMBDA)) ** 2)
595              CALL AERPOD(E,FORM)
                 IF(AMBDA - 0.01)        108,110,110
             110 IF(AMBDA - 0.1)        119,119,108
             119 RACC = 27.47 * AMBDA * AMBDA - 4.2408 * AMBDA + 0.149679
                 RAIN = RACC * RRT
600          108 CONTINUE
                 RFA = (RAIN * RR + FOG * RF) * 1000. + ATM * R
                 ZR = EXP(-1.15E-4 * 2. * RFA)
                 E1 = FORM * ZR
                 CALL AERPLD(DELTA,FORM1)
605              IF(IPOL - 2)        132,133,129
             132 EX = FORM1 * RHOVER * DX * RUFS
                 GO TO 124
             133 EX = FORM1 * RHOHOR * DX * RUFS
                 GO TO 124
610          129 EX = FORM1 * RHOCIR * DX * RUFS
             124 CONTINUE
                 EX = EX * ZR
                 PADI = X * (1. - COS(DELTA + E))
           C     DETERMINE THE PATTERN PROPAGATION FACTOR FOR THE GIVEN AERIAL CO-
615        C     VERAGE DIAGRAM AND TERRAIN.
                 PATH = PADI/AMBDA
                 IF(PATH - 0.125)        692,692,697
             692 FIELD = 2. * FORM * SIN(PI * PATH)
                 GO TO 698
620          697 FIELD = SQRT(E1 * E1 + EX * EX + 2. * E1 * EX * COS(AMU + 2. * PI
                1* PATH))
             698 CONTINUE
                 IF(PJ.EQ.0.)          GO TO 1150
                 IF(JAM - 0)          194,1160,1165
625         1160 RJAM = R
                 FIJ = FIELD
                 GO TO 1150
            1165 CONTINUE
                 COXIJ = (A * A + HJAM * HJAM - RJAM * RJAM)/(2. * A * HJAM)
630              SIXIJ = SQRT(1. - COXIJ * COXIJ)
                 XIJ = ATAN(SIXIJ/COXIJ)
                 COEJ = HJAM * SIN(XIJ)/RJAM
                 SIEJ = SQRT(1. - COEJ * COEJ)
                 EJ = ATAN(SIEJ/COEJ)
635              RXCJ = XIJ * EARAD
                 PAXJ = 2.*SQRT(EARAD*(AAS + AJAM)+RXCJ*RXCJ/4.)/SQRT(3.)
                 COPHAJ = 2. * EARAD * (AAS - AJAM) * RXCJ/(PAXJ ** 3)
                 SIPHAJ = SQRT(1. - COPHAJ * COPHAJ)
                 PHAJ = ATAN(SIPHAJ/COPHAJ)
640              RAXJ = RXCJ/2. + PAXJ * COS((PHAJ + PI)/3.)
                 BETJ = RAXJ/EARAD
                 ALJ = XIJ - BETJ
                 DELJ = ATAN((A - EARAD * COS(ALJ))/(EARAD * SIN(ALJ)))
                 CALL SDLB(TLT,HW,VW,DECL,EJ,DELJ,HORZ,HORZ1)
645              XJ = SQRT(AAS * AAS + EARAD * A * ALJ * ALJ)
                 YJ = SQRT(AJAM * AJAM + EARAD * HJAM * BETJ * BETJ)
                 PDJ = (XJ + YJ - RJAM)/AMBDA
                 PHJ = COS(PI * (1. + 2. * PDJ))
                 FIJ = SQRT(HORZ * HORZ + HORZ1 * HORZ1 + 2. * HORZ * HORZ1 * PHJ)
650         1150 CONTINUE
                 CONJAM = CJAM * (AMBDA/(4. * PI * RJAM)) ** 2
                 XN = RXN + CONJAM * FIJ * FIJ
                 RECPWR = REC * SIGMA/(XN * (R ** 4))
                 SIG = RECPWR * XN * (FIELD ** 4)
655              IF(R - R2D)         710,710,700
             700 CALL DIFREG(EARAD,AMBDA,R,AERHIT,XIN,SS)
                 SIG = SIG * SS * SS
             710 CONTINUE
                 XIF = SLF/EARAD
660              BF = EARAD + THF
                 RTF = SQRT(A * A + BF * BF - 2. * A * BF * COS(XIF))
                 CEF = (A * A + RTF * RTF - BF * BF)/(2. * A * RTF)
                 SEF = SQRT(1. - CEF * CEF)
                 EF = ATAN(SEF/CEF)
665              IF(CEF)         526,528,528
             526 EF = PI + EF
             528 CONTINUE
                 ET = PI/2. + E
                 IF(SLF - R)         523,523,524
```

Fig. C.1 *Continued*

```
670          523 CONTINUE
                 IF(EF - ET)         524,521,522
             521 SIG = SIG * 0.5
                 GO TO 524
             522 SIG = 0.
675          524 CONTINUE
           C     SHADOWING OF SURFACE TARGETS BY SEA WAVES.
                 IF(PIA - 0.)         615,615,590
             590 CONTINUE
                 IF(WHT - TARG)         615,600,600
680          600 SWH = TARG + WHT * SIN(PF * TIM/2.)/(DEG * PI)
                 AU = EARAD + ANT
                 BU = EARAD + SWH
                 DU = (AU * AU + R * R - BU * BU)/(2. * AU * R)
                 EU = ATAN2(DU,SQRT(1. - DU * DU))
685              SWH1 = WHT * SIN((PF * TIM/2.) + PI/2.)/(DEG * PI)
                 R1 = R - PWVL/4.
                 BV = EARAD + SWH1
                 DV = (AU * AU + R1 * R1 - BV * BV)/(2. * AU * R1)
                 EV = ATAN2(DV,SQRT(1. - DV * DV))
690              IF(EV - EU)         615,610,605
             605 SIG = 0.
                 GO TO 615
             610 SIG = 0.5 * SIG
             615 CONTINUE
695              TIM = TIM + 60./RPM
           C     DETERMINE THE CLUTTER SIGNAL POWER AND THE SIGNAL TO CLUTTER PLUS
           C     NOISE RATIO.
                 CODE = (R * R + AX * (AX + 2. * EARAD))/(2. * A * R)
                 ADE = ATAN(SQRT(1. - CODE * CODE)/CODE)
700              ANU = PI/2. - ADE
                 SAPP = R * SIN(ADE)/EARAD
                 APP = ATAN(SAPP/SQRT(1. - SAPP * SAPP))
                 CALL CLUTTR(E,ANU,APP,XI,R,CNR)
                 SN = SIG/CNR
705        C     SN IS THE SINGLE PULSE SIGNAL TO CLUTTER-PLUS-NOISE RATIO.
           C     COMPUTE THE PROBABILITY OF DETECTION FOR THE CHOSEN TARGET TYPE.
                 IF(K.EQ.1)         GO TO 134
                 IF(K.EQ.2)         GO TO 137
                 IF(K.EQ.3)         GO TO 140
710              IF(K.EQ.4)         GO TO 149
                 IF(K.EQ.5)         GO TO 172
             134 CONTINUE
           C
           C     SWERLING CASE I
715        C
                 IF(M - 1)         135,135,136
             135 PRODET = EXP(-THRESH/(1. + SN))
                 GO TO 190
             136 XP = XPN * SN
720              IF(XP - 0.)         1004,1004,1005
            1004 XP = 1.0E-08
            1005 CONTINUE
                 TP = 1. + XP
                 XTP = ALOG(TP) - ALOG(XP)
725              PT = (XPN - 1.) * XTP - THRESH/TP
                 IF(PT - 88.029692)         1007,1007,1006
            1006 PT = 88.029692
            1007 CONTINUE
                 PR = 1. - XCGAM(THRESH,M-2,DUM)
730              TQ = 1. + 1./XP
                 TC = THRESH/TQ
                 OH = EXP(PT) * XCGAM(TQ,M-2,DUM)
                 PRODET = PR + OH
                 GO TO 190
735          137 CONTINUE
           C
           C     SWERLING CASE II
           C
                 TIX = THRESH/(1. + SN)
740              IF(M - 1)         138,138,139
             138 PRODET = EXP(- TIX)
                 GO TO 190
             139 PRODET = 1. - XCGAM(TIX,M-1,DUM)
                 GO TO 190
745          140 CONTINUE
           C
           C     SWERLING CASE III
           C
                 IF(M - 2)         141,142,143
750          141 TY = 1. + SN/2.
                 TZ = 1. + 2./SN
                 TV = THRESH/TY
                 PRODET = (1. + THRESH/(TY * TZ)) * EXP(-TV)
                 GO TO 190
755          142 TW = THRESH/(1. + SN)
                 PRODET = (1. + TW) * EXP(-TW)
                 GO TO 190
             143 UA = 2./(2. + XPN * SN)
                 UB = 1. - UA
760              UC = UB * THRESH
                 UD = UA * THRESH
                 IF(UC - XPN)         144,148,148
             144 SUM = 0.
                 TERM = 1.
765              J = M
             145 TEMP = SUM + TERM
                 IF(SUM - TEMP)         146,147,147
             146 SUM = TEMP
                 TERM = TERM * UC/FLOAT(J)
770              J = J + 1
```

Fig. C.1 *Continued*

```
              GO TO 145
          147 PRODET = 1. - XCGAM(THRESH,M-2,DUM) + UD * ELOG(THRESH,M-2) +
             1UB * ELOG(THRESH,M-1) * (1. + UD - (XPN - 2.) * UA/UB) * SUM
              GO TO 190
775       148 PRODET = 1. - XCGAM(THRESH,M-3,DUM) + THRESH * ELOG(THRESH,M-3) *
             1UA/UB + EXP(-UD -(XPN - 2.)*ALOG(UB))*(1. + UD - (XPN - 2.) *
             2UA/UB) * XCGAM(UC,M-3,DUM)
              GO TO 190
          149 CONTINUE
780     C
        C     SWERLING CASE IV
        C
              SUM = 0.
              UA = 2./(2. + SN)
785           UB = 1. - UA
              QU = UA/UB
              UC = UA * THRESH
              UD = UB * THRESH
              KS = (3. * XPN + UD)/2. - SQRT((XPN - 1. + UD) ** 2/4. + UD *
790          1(XPN + 1.))
              KS = MIN0(KS,M)
              KS = MAX0(KS,0)
              KL = KS
              J = M - KS
795           FKS = KS
              KL = MIN0(KS,M)
              KM = 2 * M - 1 - KS
              IF(THRESH - XPN * (1. + UB))          161,150,150
          150 GS = 1. - XCGAM(UC,KM,TN)
800           IF(GS)            160,160,151
          151 TS = EXP(FKS * ALOG(UA) + (XPN - FKS) * ALOG(UB) + SUMNL(M) -
             1SUMNL(KS) - SUMNL(J) + ALOG(GS))
              G = GS
              TERM = TS
805           TL = TN
          152 TEMP = SUM + TERM
              IF(SUM - TEMP)          153,155,155
          153 SUM = TEMP
              IF(KL)            155,155,154
810       154 TL = TL * UC/FLOAT(2 * M - KL)
              TERM = TERM * FLOAT(KL) * (G + TL)/(QU * FLOAT(M - KL + 1) * G)
              G = G + TL
              KL = KL - 1
              GO TO 152
815       155 IF(KS - M)            156,160,160
          156 TERM = TS * QU * FLOAT(M - KS) * (GS - TN)/(FLOAT(KS + 1) * GS)
              G = GS - TN
              TL = TN * FLOAT(KM)/UC
              KL = KS + 1
820       157 TEMP = SUM + TERM
              IF(SUM - TEMP)          158,160,160
          158 SUM = TEMP
              IF(KL - M)            159,160,160
          159 TERM = TERM * QU * FLOAT(M - KL) * (G - TL)/(FLOAT(KL + 1) * G)
825           G = G - TL
              KN = 2 * M - 1 - KL
              TL = TL * FLOAT(KN)/UC
              KL = KL + 1
              GO TO 157
830       160 PRODET = SUM
              GO TO 190
          161 KO = 2 * M - 1 - KS
              GS = XCGAM(UC,KO,TN)
              IF(GS)            171,171,162
835       162 TS = EXP(FKS * ALOG(UA) + (XPN - FKS) * ALOG(UB) + SUMNL(M) -
             1SUMNL(KS) - SUMNL(J) + ALOG(GS))
              G = GS
              TERM = TS
              TL = TN
840       163 TEMP = SUM + TERM
              IF(SUM - TEMP)          164,166,166
          164 SUM = TEMP
              IF(KL)            166,166,165
          165 TL = TL * UC/FLOAT(2 * M - K)
845           TERM = TERM * FLOAT(KL) * (G - TL)/(QU * FLOAT(M - KL + 1) * G)
              G = G - TL
              KL = KL - 1
              GO TO 163
          166 IF(KS - M)          167,171,171
850       167 TERM = TS * QU * FLOAT(M - KS) * (GS + TN)/(FLOAT(KS + 1) * GS)
              G = GS + TN
              TL = TN * FLOAT(KO)/UC
              KL = KS + 1
          168 TEMP = SUM + TERM
855           IF(SUM - TEMP)          169,171,171
          169 SUM = TEMP
              IF(KL - M)            170,171,171
          170 TERM = TERM * QU * FLOAT(M - KL) * (G + TL)/(FLOAT(KL + 1) * G)
              G = G + TL
860           KP = 2 * M - 1 - KL
              TL = TL * FLOAT(KP)/UC
              KL = KL + 1
              GO TO 168
          171 PRODET = 1. - SUM
865           GO TO 190
          172 CONTINUE
        C
        C     NON - FLUCTUATING TARGET
        C
```

Fig. C.1 *Continued*

```
870              SUM = 0.
                 XP = XPN * SN
                 IF(THRESH - XP - XPN)        180,173,173
           173   KS = -(XPN + 1.)/2. + SQRT(((XPN - 1.)/2.) ** 2 + XP * THRESH)
                 KS = MAX0(KS,0)
875              GS = 1. - XCGAM(THRESH,KS + M - 1,TN)
                 TS = ELOG(XP,KS) * GS
                 G = GS
                 KL = KS
                 TERM = TS
880              TL = TN
           174   TEMP = SUM + TERM
                 IF(SUM - TEMP)        175,177,177
           175   SUM = TEMP
                 IF(KL)        177,177,176
885        176   TERM = TERM * FLOAT(KL) * (G - TL)/(XP * G)
                 G = G - TL
                 KL = KL - 1
                 TL = TL * FLOAT(KL + M)/THRESH
                 GO TO 174
890        177   TL = TN * THRESH/FLOAT(KS + M)
                 KL = KS + 1
                 G = GS + TL
                 TERM = TS * XP * G/(GS * FLOAT(KL))
           178   TEMP = SUM + TERM
895              IF(SUM - TEMP)        179,189,189
           179   SUM = TEMP
                 TL = TL * THRESH/FLOAT(KL + M)
                 KL = KL + 1
                 TERM = TERM * XP * (G + TL)/(G * FLOAT(KL))
900              G = G + TL
                 GO TO 178
           180   KS = - 1. - XPN/2. + SQRT(XPN ** 2/4. + XP * THRESH)
                 KS = MAX0(KS,0)
                 KQ = KS + M - 1
905              GS = XCGAM(THRESH,KQ,TN)
                 IF(GS)        188,188,181
           181   TS = ELOG(XP,KS) * GS
                 G = GS
                 TERM = TS
910              KL = KS
                 TL = TN
           182   TEMP = SUM + TERM
                 IF(SUM - TEMP)        183,185,185
           183   SUM = TEMP
915              IF(KL)        185,185,184
           184   TERM = TERM * FLOAT(KL) * (G + TL)/(XP * G)
                 G = G + TL
                 TL = TL * FLOAT(KL + M - 1)/THRESH
                 KL = KL - 1
920              GO TO 182
           185   TL = TN * THRESH/FLOAT(KS + M)
                 KL = KS + 1
                 G = GS - TL
                 TERM = TS * XP * G/(GS * FLOAT(KL))
925        186   TEMP = SUM + TERM
                 IF(SUM - TEMP)        187,188,188
           187   SUM = TEMP
                 TL = TL * THRESH/FLOAT(KL + M)
                 TERM = TERM * XP * (G - TL)/(G * FLOAT(KL + 1))
930              G = G - TL
                 KL = KL + 1
                 GO TO 186
           188   SUM = 1. - SUM
           189   PRODET = SUM
935        190   CONTINUE
                 IF(IPROB - 1)        6030,6035,194
          6030   DEPRO(L) = 100. * PRODET
                 GO TO 6040
          6035   PMISS = 1. - PRODET
940              CDX = CDX * PMISS
                 CUDEPR = 1. - CDX
                 DEPRO(L) = 100. * CUDEPR
          6040   CONTINUE
                 IF(0.005 - DEPRO(L))        192,192,191
945        191   DEPRO(L) = 0.0
           192   CONTINUE
                 RANGE(L) = R/1000.
                 HFT = HT
                 RT = ANTHIT
950              IF(DHT)        550,550,540
           540   HFT = AERHIT
                 RT = HT
                 RT = RT - AERSEA
           550   CONTINUE
955              CNDBM(L) = 10. * ALOG10(CNR * 1000.)
                 IF(SIG - 0.)        730,730,735
           730   SIGDBM(L) = - 250.
                 GO TO 740
           735   SIGDBM(L) = 10. * ALOG10(SIG * 1000.)
960        740   CONTINUE
                 SNDB = SIGDBM(L) - CNDBM(L)
                 XXP = RANGE(L)
                 XPRO(L) = XXP
                 YPRO(L) = DEPRO(L)
965              SIGLEV = SIGDBM(L)
                 IF(SIGLEV + 200.)        2100,2100,2200
          2100   SIGLEV = - 200.
          2200   CONTINUE
                 ZZY = ZD - 2. * XLM * DELR
```

Fig. C.1 *Continued*

```
970               XLO = XL - 2. * XLM
                  IF(ABS(XLO) - DELR * 1.0E-08)        8550,8560,8560
             8550 XO = ZZY * XUNIT
                  YO = DEPRC(L)/10. + 0.05
             8560 CONTINUE
975               ZZZ = ZD - XLM * DELR
                  XLN = XL - XLM
                  IF(ABS(XLN) - DELR * 1.0E-08)        8520,8540,8540
             8520 XD = ZZZ * XUNIT
                  YD = 0.1 + ABS(SIGLEV/40.)
980               IF(SIGLEV + 200.)        8530,8530,8540
             8530 YD = 0.1
             8540 CONTINUE
                  XSIG(L) = XXP
                  YSIG(L) = SIGLEV
985               XPAR(L) = XXP
                  YPAR(L) = CNDBM(L)
                  IF(HFT)          194,210,210
              210 WRITE(6,193) RANGE(L),RT,HFT,SIGDBM(L),CNDBM(L),SNDB,DEPRC(L)
              193 FORMAT(1X,F7.2,5X,F8.2,8X,F8.2,5X,F9.3,2X,F9.3,8X,F7.2,19X,F6.2)
990           194 CONTINUE
                  XZ = 0.
                  XV = 25.
                  YZ = 0.
                  YV = 10.
995               CALL AXIS(0.,0.,8HRANGE KM,-8,20.,0.,XZ,XV,0)
                  CALL AXIS(0.,0.,28HPROBABILITY OF DETECTION     ,28,10.,90.,YZ,YV,
                 1-1)
                  CALL SYMBOL(-2.15,6.6,0.3,21,90.,-1)
                  IF(IPROB - 1)       6025,6020,196
1000         6020 CALL SYMBOL(-3.,2.,0.3,10HCUMULATIVE,90.,10)
             6025 CONTINUE
                  CALL SYMBOL(1.,15.,1.,TITLE,0.,30)
                  IF(IPOL - 2)       116,117,640
              116 CALL SYMBOL(1.,14.,0.3,21HVERTICAL POLARIZATION,0.,21)
1005              GO TO 118
              117 CALL SYMBOL(1.,14.,0.3,23HHORIZONTAL POLARIZATION,0.,23)
                  GO TO 118
              640 CALL SYMBOL(1.,14.,0.3,21HCIRCULAR POLARIZATION,0.,21)
              118 CONTINUE
1010              CALL SYMBOL(8.5,14.,0.3,20HPROB. OF FALSE ALARM,0.,20)
                  CALL SYMBOL(14.,14.,0.3,2H10,0.,2)
                  CALL SYMBOL(14.5,14.3,0.2,2H-5,0.,2)
                  CALL SYMBOL(1.,13.5,0.3,26HPULSE WIDTH        MICROSEC,0.,26)
                  CALL NUMBER(4.1,13.5,0.3,TAU,0.,4HF5.1)
1015              CALL SYMBOL(1.,13.,0.3,20HRADAR HEIGHT       M,0.,20)
                  CALL NUMBER(4.4,13.,0.3,ANTHIT,0.,4HF5.1)
                  CALL SYMBOL(1.,12.5,0.3,18HR.C.A.        SQ.M.,0.,18)
                  CALL NUMBER(2.8,12.5,0.3,SIGMA,0.,4HF5.1)
                  CALL SYMBOL(8.5,13.5,0.3,28HAERIAL SCAN RATE       R.P.M.,0.,28)
1020              CALL NUMBER(12.9,13.5,0.3,RPM,0.,4HF3.0)
                  CALL SYMBOL(1.,12.,0.3,24HTARGET HEIGHT          M,0.,24)
                  CALL NUMBER(4.7,12.,0.3,TARGHT,0.,4HF6.0)
                  CALL SYMBOL(1.,11.5,0.3,9HSEA STATE,0.,9)
                  CALL NUMBER(4.,11.5,0.3,STATE,0.,4HF2.0)
1025              IF(K.EQ.1)       GO TO 400
                  IF(K.EQ.2)       GO TO 402
                  IF(K.EQ.3)       GO TO 404
                  IF(K.EQ.4)       GO TO 406
                  IF(K.EQ.5)       GO TO 408
1030          400 CALL SYMBOL(7.,12.5,0.3,15HSWERLING CASE I,0.,15)
                  GO TO 300
              402 CALL SYMBOL(7.,12.5,0.3,16HSWERLING CASE II,0.,16)
                  GO TO 300
              404 CALL SYMBOL(7.,12.5,0.3,17HSWERLING CASE III,0.,17)
1035              GO TO 300
              406 CALL SYMBOL(7.,12.5,0.3,16HSWERLING CASE IV,0.,16)
                  GO TO 300
              408 CALL SYMBOL(7.,12.5,0.3,22HNON-FLUCTUATING TARGET,0.,22)
              300 CONTINUE
1040              CALL SYMBOL(XO,YO,0.3,2HPD,0.,2)
                  CALL SYMBOL(XD,YD,0.3,1HS,0.,1)
                  XPRO(LMAX + 1) = XZ
                  XPRO(LMAX + 2) = XV
                  YPRO(LMAX + 1) = YZ
1045              YPRO(LMAX + 2) = YV
                  CALL LINE(XPRO,YPRO,LMAX,1,0,0)
                  GO TO 4444
             2000 CONTINUE
                  CALL AXIS(21.,0.,30HSIGNAL AND CLUTTER + NOISE DBM,-30,10.,90.,YA
1050             1,YAW,-1)
                  XPAR(LMAX + 1) = XZ
                  XPAR(LMAX + 2) = XV
                  YPAR(LMAX + 1) = YAP
                  YPAR(LMAX + 2) = YAW
1055              CALL LINE(XPAR,YPAR,LMAX,1,0,0)
                  GO TO 4444
             3000 CONTINUE
                  XSIG(LMAX + 1) = XZ
                  XSIG(LMAX + 2) = XV
1060              YSIG(LMAX + 1) = YAZ
                  YSIG(LMAX + 2) = YAV
                  CALL LINE(XSIG,YSIG,LMAX,1,0,0)
             4444 CONTINUE
                  XC = XX - 2.
1065              YYZ = ABS(10. * ALOG10(XN * 1000.))/48.5
                  IF(PJ)         1185,1185,1180
             1180 XC = XX - 3.2
                  CALL SYMBOL(XC,YYZ,0.3,7HJAMMING,0.,7)
                  GO TO 1190
```

Fig. C.1 *Continued*

```
1070          1185 CALL SYMBOL(XC,YYZ,0.3,5HC + N,0.,5)
              1190 CONTINUE
                   CALL PLOT(XX,YY,3)
                   CALL SYMBOL(XX,1.2,0.2,13HRADIC HORIZON,90.,13)
                   CALL PLOT(XX,0.,3)
1075               YP = 2.
                   CALL PLOT(XX,YP,2)
                   CALL PLOT(50.,0.,-3)
               196 CONTINUE

                   STOP
1080               END
```

Fig. C.1 *Continued*

```
 1                 SUBROUTINE ARRAY(NA,ZD,RPM,V,RANGE,SIGDBM,CNDBM,DEPRO,XPRO,YPRO,
                  1DHT,TITLE,LMAX)
                   DIMENSION  RANGE(LMAX),SIGDBM(LMAX),CNDBM(LMAX),DEPRO(LMAX)
                   DIMENSION  XPRO(NA),YPRO(NA)
 5                 DIMENSION  TITLE(3)
           C       SUBROUTINE ARRAY DETERMINES THE NUMBER OF VALUES TO BE COMPUTED.
                   XLMAX = (ZD - ABS(DHT)) * RPM/(60. * V)
                   LMAX = XLMAX
                   RETURN
10                 END
```

Fig. C.2 *Listing of subroutine ARRAY*

```
 1                 SUBROUTINE CLUTTR(E,ANL,APP,XI,R,CNR)
                   COMMON  AGEX,AMBDA,BKB,BMHZ,DEC,DIST,FASTEP,GAM,GRT,HW,PCR,PFA,PI,
                  1REC,RRT,SCV,SEA,TAU,TLT,VW,XAB,XN,ZP,ZR,ZX,
                  2ICIR,IPOL,M,MM,MTI
 5         C       SUBROUTINE CLUTTR CONSIDERS IMPROVEMENTS DUE TO EITHER FREQUENCY
           C       AGILITY,MTI,PULSE COMPRESSION OR ANY OF THEIR COMBINATIONS AND
           C       COMPUTES THE CLUTTER PLUS NOISE SIGNAL POWER.
           C       FREQUENCY AGILE CLUTTER PERFORMANCE IMPROVEMENT - ACCORDING TO
           C       J.R.MARTIN --THE FREQUENCY AGILE MAGNETRON STORY - VARIAN-EASTERN
10         C       TUBE DIVISION BROCHURE 2836,AUGUST,1972.
                   PCIMP = PCR
                   IF(FASTEP - 0.)         100,100,90
                90 FAGIMP = SQRT(FASTEP)
                   GO TO 7
15             100 CONTINUE
                   IF(AGEX - 0.)       2,2,1
                 1 XNP = FLOAT(MM)
                   IF(XNP - 20.)       3,4,4
                 4 XNP = 20.
20               3 PPFS = AGEX/XNP
                   IF(PPFS - 1./TAU)       2,6,6
                 2 FAGIMP = 1.
                   GO TO 7
                 6 FAGIMP = SQRT(XNP)
25               7 CONTINUE
           C       MTI IMPROVEMENT ACCORDING TO BARTON - RADAR EQUATIONS FOR JAMMING
           C       AND CLUTTER,IEEE TRANSACTIONS VOL,AES-3,NO.6,NOV,1967.
                   TZ = FLOAT(MM)
                   ZT = TZ * TZ
30                 IF(MTI - 1)       20,25,30
                20 XMTIMP = 1.
                   GO TO 35
                25 XMTIMP = 1./((1.386/ZT) * (1. - (0.693/ZT)))
                   GO TO 35
35              30 IF(MTI - 3)       31,32,33
                31 XMTIMP = 1./((3.844/(ZT * ZT)) * (1. - (2.3/ZT)))
                   GO TO 35
                32 XMTIMP = 1./((1.386/ZT) * (1. - (1.386/ZT)))
                   GO TO 35
40              33 XMTIMP = 1./((7.687/(ZT * ZT)) * (1. - (4.62/ZT)))
                35 CONTINUE
                   CV = 299.793 * TAU/2.
                   ARC = PI * HW * VW * CV * (R * R)/4.
                   CCR = 6. * (RRT ** 1.6)/((10. ** 14) * (AMBDA ** 4))
45                 IF(IPOL - 2)       18,18,5
                 5 IF(ICIR - 1)       17,18,18
                17 CCR = 0.01 * CCR
                18 CONTINUE
                   CALL AERPCD(E,FORM)
50                 RCA = ARC * CCR * (FORM ** 4)
                   RCL = REC * RCA * ZR/(R ** 4)
                   SINC = 0.303 * AMBDA/SEA
                   TSC = SINC/SQRT(1. - SINC * SINC)
                   GMAX = ATAN(TSC)
```

Fig. C.3 *Listing of subroutine CLUTTR*

```
 55            IF(BKB - 0.)        9,8,9
             8 CCS = 0.
               ACL = 0.
               GO TO 14
             9 CONTINUE
 60            IF(R - ZB)        74,74,72
            72 SCL = 0.
               GO TO 77
            74 IF(XAB)        40,40,45
            40 PSI = ABS(ANU - APP)
 65            CALL AERPLD(ANU,FORM1)
               FORM = FORM1
               GO TO 50
            45 PSI = E - XI
               IF(PSI - 0.)        15,15,16
 70         15 PSI = 1.0E-08
            16 CONTINUE
               CALL AERPOD(E,FORM)
            50 CONTINUE
      C        GROUND CLUTTER ACCORDING TO D.K.BARTON,RADAR EQUATIONS FOR JAMMING
 75   C        AND CLUTTER,IEEE TRANSACTIONS VOL.AES-3,NO.6,NOV.1967.
               IF(GRT - DIST)        10,22,22
            10 CCS = 3.1601E-04/AMBDA
               IF(R - ZB)        65,60,60
            60 CCS = CCS/((R/ZB) ** 4)
 80         65 CONTINUE
               CCSDB = 10. * ALOG10(CCS)
               GO TO 12
            22 CONTINUE
      C        SEA CLUTTER ACCORDING TO D.K.BARTON,RADAR EQUATIONS FOR JAMMING
 85   C        AND CLUTTER,IEEE TRANSACTIONS VOL.AES-3,NO.6,NOV.2967.
      C        SKOLNIK,RADAR HANDBOOK,CHAPTER 1.
      C        NATHANSON,RADAR DESIGN PRINCIPLES,PAGES 66-67
               CCSDB = - 64. + 6.*BKB + 10.*ALOG10(SIN(PSI)) - 10.*ALOG10(AMBDA)
               IF(PSI - GMAX)        11,11,12
 90         11 CCSDB = CCSDB - 40. * ALOG10(GMAX/PSI)
            12 CONTINUE
               CLC = R * TAN(VW)/CV
               CCS = 10. ** (CCSDB/10.)
               ACL = 2. * R * CV * TAN(HW/2.)/COS(PSI)
 95            IF(TAN(PSI) - CLC)        14,13,13
            13 ACL = PI * (R * R) * TAN(HW/2.) * TAN(VW/2.)/SIN(PSI)
            14 CONTINUE
               SCA = ACL * CCS * (FORM ** 4)
               SCL = REC * SCA * ZR/(R ** 4)
100         77 CONTINUE
               CIMTI = 10. ** (SCV/10.)
               CLIMP = CIMTI * XMTIMP * FAGIMP * FCIMP
               CNA = RCL + SCL/CLIMP
               CNR = XN + CNA
105            RETURN
               END
```

Fig. C.3 *Continued*

```
  1            FUNCTION BIAS(N,FAP)
               DOUBLE PRECISION  YM,YN,YO
      C        THIS SUBROUTINE COMPUTES THE DETECTION THRESHOLD FOR THE COMPUTED
      C        NUMBER OF HITS/SCAN AND THE SPECIFIED PROBABILITY OF FALSE ALARM.
  5            YM = 50.
            10 CALL DXCGAM(N,YM,GM)
               IF(GM.NE.FAP)        GO TO 20
               BIAS = YM
               RETURN
 10         20 IF(GM.GT.FAP)        GO TO 30
               YM = YM/2.
               GO TO 10
            30 YN = 500.
            40 CALL DXCGAM(N,YN,GN)
 15            IF(GN.NE.FAP)        GO TO 50
               BIAS = YN
               RETURN
            50 IF(GN.LT.FAP)        GO TO 60
               YN = 2. * YN
 20            GO TO 40
            60 CONTINUE
               KMAX = 1000
               TOLER = 1.0E-05
               K = 0
 25         70 YO = (YN + YM)/2.
               CALL DXCGAM(N,YO,GO)
               IF(GO.NE.FAP)        GO TO 80
               BIAS = YO
               RETURN
 30         80 IF(GO.LT.FAP)        GO TO 90
               YM = YO
               GO TO 100
            90 YN = YO
           100 IF(((YN - YM)/YM).LT.TOLER)        GO TO 110
 35            K = K + 1
               IF(K.GT.KMAX)        GO TO 120
               GO TO 70
           110 BIAS = (YN + YM)/2.
               RETURN
 40        120 BIAS = 0.
               RETURN
               END
```

Fig. C.4 *Listing of subroutine BIAS*

```
 1           FUNCTION XCGAM(A,N,RT)
       C     XCGAM IS THE INCOMPLETE GAMMA FUNCTION REQUIRED FOR THE COMPUTA-
       C     TION OF THE PROBABILITY OF DETECTION.
             SUM = 0.
 5           L = A
             IF(L - N)        1,2,2
           1 J = N + 1
             GL = ELOG(A,J)
             RT = GL * (FLOAT(J))/A
10         3 GR = SUM + GL
             IF(SUM - GR)       4,5,5
           4 SUM = GR
             J = J + 1
             GL = GL * A/FLOAT(J)
15           GO TO 3
           5 XCGAM = SUM
             RETURN
           2 J = N
             GL = ELOG(A,J)
20           RT = GL
           6 GR = SUM + GL
             IF(SUM - GR)        7,8,8
           7 SUM = GR
             IF(J - 1)        8,9,9
25         9 GL = GL * FLOAT(J)/A
             J = J - 1
             GO TO 6
           8 XCGAM = 1. - SUM
             RETURN
30           END
```

Fig. C.5 *Listing of subroutine XCGAM*

```
 1           FUNCTION ELOG(YB,N)
       C     ELOG IS AN AUXILIARY FUNCTION NEEDED FOR THE COMPUTATION OF XCGAM.
             XPNT = - YB
             IF(N)        2,2,1
 5         1 PN = N
             XPNT = XPNT + PN * ALOG(YB) - SUMNL(N)
           2 ELOG = EXP(XPNT)
             RETURN
             END
```

Fig. C.6 *Listing of subroutine ELOG*

```
 1           FUNCTION SUMNL(N)
       C     SUMNL IS AN AUXILIARY FUNCTION NEEDED FOR THE COMPUTATION OF THE
       C     ELOG FUNCTION.
             SUM = 0.
 5           IF(N - 1)        1,1,2
           2 DO 3       J = 1,N
             SUM = SUM + ALOG(FLOAT(J))
           3 CONTINUE
           1 SUMNL = SUM
10           RETURN
             END
```

Fig. C.7 *Listing of subroutine SUMNL*

```
 1           FUNCTION DXCGAM(N,Y,G)
             DOUBLE PRECISION  G,GA,GM,QG,SA,SM,Y,YM
       C     DXCGAM COMPUTES THE THRESHOLD BY ITERATIONS USING THE BISECTION
       C     METHOD.
 5           IF(N.EQ.1)        GO TO 60
             YM = Y
             GA = - Y
             IF(YM.GT.175.)        GO TO 10
             GM = DEXP(GA)
10           GO TO 20
          10 GM = 0.0
          20 QG = DLOG(YM)
             M = N - 1
             DO 50     K = 1,M
15           SA = K
             GA = GA + QG - DLOG(SA)
             IF(GA.LE.-175.)        GO TO 30
             SM = DEXP(GA)
             GO TO 40
20        30 SM = 0.0
          40 GM = GM + SM
          50 CONTINUE
             G = GM
             GO TO 80
25        60 IF(Y.GT.175.)        GO TO 70
             G = DEXP(- Y)
             GO TO 80
          70 G = 0.0
          80 DXCGAM = G
30           RETURN
             END
```

Fig. C.8 *Listing of subroutine DXCGAM*

```
  1           SUBROUTINE DIFREG(EAR,WL,R,AHT,THT,SS)
        C     ACCORDING TO DOMB AND PRYCE,J.I.E.E.,VOL.94 (1947),PART III
        C     AND KERR - PROPAGATION OF SHORT RADIO WAVES - PP 122 - 125.
              PI = 3141592.653589E-06
  5           YYK = EAR * EAR * WL/PI
              YK = ALOG10(YYK)/3.
              YL = 10. ** YK
              XXK = EAR * ((WL/PI) ** 2)
              XK = ALOG10(XXK)/3.
 10           YH = (10. ** XK)/2.
              Z1 = AHT/YH
              Z2 = THT/YH
              YX = R/YL
              U = 0.
 15           DO 65        L = 1,2
              IF(L - 1)         25,25,30
           25 X = ALOG10(Z1)
              ZT = Z1
              GO TO 35
 20        30 X = ALOG10(Z2)
              ZT = Z2
           35 CONTINUE
              IF(ZT - 0.6)        40,40,45
           40 U = 20. * X + U
 25           GO TO 60
           45 IF(ZT - 1.)        50,55,55
           50 ZZ = 1.
              U = POLAGR(X,ZZ) + U
              GO TO 60
 30        55 ZZ = 2.
              U = POLAGR(X,ZZ) + U
           60 CONTINUE
           65 CONTINUE
              V = 10. * (ALOG10(4. * PI) + ALOG10(YX) - 175.45497E-02 * YX)
 35           SDB = U + V
              SS = 10. ** (SDB/10.)
              RETURN
              END
```

Fig. C.9 *Listing of subroutine DIFREG*

```
  1           FUNCTION POLAGR(X,ZZ)
        C     LAGRANGE INTERPOLATION POLYNOMIAL.
              IF(ZZ - 1.)         1,1,2
            1 X0 = ALOG10(0.6)
  5           X1 = ALOG10(0.7)
              X2 = ALOG10(0.8)
              X3 = ALOG10(0.9)
              X4 = ALOG10(1.)
              Y0 = - 4.436975
 10           Y1 = - 3.141403
              Y2 = - 1.524849
              Y3 = - 0.186871
              Y4 = 1.984728
              GO TO 3
 15         2 X0 = ALOG10(1.)
              X1 = ALOG10(25.)
              X2 = ALOG10(50.)
              X3 = ALOG10(75.)
              X4 = ALOG10(100.)
 20           Y0 = 1.984728
              Y1 = 72.239110
              Y2 = 106.938113
              Y3 = 133.138437
              Y4 = 155.000030
 25         3 CONTINUE
              A = X - X1
              B = X - X2
              C = X - X3
              D = X - X4
 30           E = X0 - X1
              F = X0 - X2
              G = X0 - X3
              H = X0 - X4
              O = X1 - X2
 35           P = X1 - X3
              Q = X1 - X4
              R = X - X0
              S = X2 - X3
              T = X2 - X4
 40           W = X3 - X4
              D0 = (A * B * C * D)/(E * F * G * H)
              D1 = (R * B * C * D)/((-E) * O * P * Q)
              D2 = (R * A * C * D)/((-F) * (-O) * S * T)
              D3 = (R * A * B * D)/((-G) * (-P) * (-S) * W)
 45           D4 = (R * A * B * C)/((-H) * (-Q) * (-T) * (-W))
              U = Y0 * D0 + Y1 * D1 + Y2 * D2 + Y3 * D3 + Y4 * D4
              POLAGR = U
              RETURN
              END
```

Fig. C.10 *Listing of subroutine POLAGR*

List of used symbols

a	modified earth radius
A	aerial aperture
	length
	signal amplitude
$A(x)$	amplitude distribution across an aerial's aperture
b	video bandwidth
B	bandwidth
	length
B_ϕ	3 dB beamwidth in the direction of ϕ
c	velocity of propagation of light
C	capacitance
C_{00}	cost of correct decision when $x(t) = n(t)$
C_{11}	cost of correct decision when $x(t) = s(t) + n(t)$
C_m	cost of miss
C_{fa}	cost of false alarm
d	ground range
	diameter
ds	elementary length
D	directivity of an aerial
E	electric field strength
	electric force
$\bar{E}$	electric field vector
E_i	incident electric field
E_s	scattered electric field
f	frequency
f_r	pulse repetition frequency
$f_1(t)$	transmitted signal in the time domain
$f_2(t)$	received signal in the time domain
F	noise figure
$F(\epsilon)$	radiation in the direction ϵ relative to maximum radiation
$F_1(j\omega)$	transmitted signal in the frequency domain

$F_2(j\omega)$	received signal in the frequency domain
G	aerial gain
$G(j\omega)$	voltage gain of an amplifier
G_0	maximum aerial gain
h	aerial height
	altitude
$h(t)$	network transfer function
H	target height
$\bar{H}$	magnetic field vector
$H(j\omega)$	network transfer function transform
$H_1(\#)$	first order Struve function
$I_n(x)$	modified Bessel function of the first kind and order n
$J_n(x)$	Bessel function of the first kind and order n
k	Boltzmann's constant
	earth radius correction factor
	number of hits per scan
	number of integrated pulses
	number of variates of signal plus noise
	wavelength constant
	wave number
K	sea wave height
K_B	Beaufort wind scale
L	inductance
	length
	losses
L_c	collapsing loss
m	complex index of refraction of water
	average number of successes in a given interval
m_n	nth moment
M	water content of a cloud
MDR	maximum displayed range
MTI	moving target indicator
n	integer
	number of lobes
	modified index of refraction
$n(t)$	noise voltage
n_{fa}	false alarm number
N	mean squared noise voltage
	refractivity
N_s	surface refractivity
p	transformation variable
$p(\#)$	probability density
	probability distribution function
$p(a\|b)$	conditional probability density function

P	power
P_c	cumulative probability of detection
P_d	probability of detection
P_{fa}	probability of false alarm
P_m	probability of miss
P_R	received power
P_T	transmitted power
P_0	probability that noise alone will exceed the bias
$P(\#)$	probability
$P(a\|b)$	conditional probability
PPI	plan position indicator display
PRF	pulse repetition frequency
q	Booker variable
Q	circuit magnification factor
r	earth radius
	rainfall rate
r_i	radius of raindrops
R	radar range
	resistance
R_D	dynamic resistance
R_u	maximum unambiguous range
R_0	range at unity signal-to-noise ratio
s	sweep speed
$s(t)$	signal voltage
S	signal amplitude
SCV	sub-clutter visibility
SN	peak signal to average noise ratio
t	dwell-time of aerial beam on target
t_r	rise time
T_a	noise temperature of an aerial
T_0	absolute temperature
u	integer
v	instantaneous illuminated volume
V	total surveyed volume of space
w	angular frequency
	angular velocity
w_0	angular frequency at resonance
W	wind velocity
W.R.E.	Weapons Research Establishment
x	signal-to-noise ratio
$\bar{x}$	average signal-to-noise ratio
$x(t)$	signal voltage
Y	threshold
	bias

Y_b	bias level
Z	impedance
α	angle
	attenuation coefficient
β	angle
γ	angle of incidence
	absorption coefficient
γ_{oo}	absorption coefficient of atmospheric oxygen
γ_{wo}	absorption coefficient of uncondensed water vapour
θ	angle
ϵ	angle of radiation
	dielectric constant
	ray elevation angle
ϵ_0	dielectric constant of free space
ϵ_R	complex dielectric constant
ξ	phase shift
η	number of pulse intervals per sweep
η_0	operator efficiency factor
θ	angle of incidence
	horizontal 3 dB beamwidth
	wind aspect angle
κ	lapse rate of the refractive index
λ	wavelength
$\Lambda(\#)$	likelihood ratio
μ	modified index of refraction
	phase shift on reflection
μ_0	modified index of surface refraction
ρ	collapsing ratio
	correlation coefficient
	radius of curvature
	reflection coefficient
ρ_{eff}	effective collapsing ratio
σ	r.m.s. value
	standard deviation
	target echoing area
σ^2	variance
τ	pulse width
	time interval
τ_{fa}	false alarm time
Ψ	phase difference
	angle
	vertical 3 dB beamwidth
ϕ	vertical 3 dB beamwidth

ψ	phase difference
	angle
	grazing angle
ψ_c	critical grazing angle
ω	angular frequency
	angular velocity

References

ABBOTT, R. – *see* BEAN, B. R.

ALBRECHT, H. J.: 'On the relationship between electrical ground parameters', *P.I.E.E.E.* **53**, No. 5, May 1965, p. 554

ALLER, L. H.: *Astrophysics, the Atmospheres of the Sun and Stars*, The Ronald Press Co., New York, 1953

AMENT, W. S.: 'Toward a theory of reflection by a rough surface', *P.I.R.E.* **41**, No. 1, Jan. 1953, pp. 142–146

AMENT, W. S.: 'Forward and back scattering by certain rough surfaces', *I.R.E. Trans.* **AP-4**, July 1956, pp. 369–373

AMENT, W. S.: 'Reciprocity and scattering by certain rough surfaces', *I.R.E. Trans.* **AP-8**, March 1960, pp. 167–174

ANANASSO, F.: 'Coping with rain above 11 GHz', *MSN* **10**, No. 3, March 1980, pp. 58–72

BAKER, C. H.: *Man and Radar Displays*, Agardograph No. 60, The MacMillan Co., New York, 1962

BARLOW, H. M. and CULLEN, A. L.: *Micro-Wave Measurements*, Constable & Co. Ltd., London, 1950

BARRICK, D. E. – *see* RUCK, G. T. *et al.*

BARTHOLOMA, J.: 'Die Antenne der Mittelbereich – Radaranlage', *Telefunken Zeitung* **34**, No. 131, March 1961, pp. 33–41

BARTON, D. K.: 'Radar equations for jamming and clutter', *I.E.E.E. Trans.* **AES-3**, No. 6, Nov. 1967, pp. 340–355

BARTON, D. K. and WARD, H. R.: *Handbook of Radar Measurement*, Prentice-Hall Inc., New Jersey, 1969

BEAN, B. R. and ABBOTT, R.: 'Oxygen and water vapour absorption of radio waves in the atmosphere', *Geofisica Pura e Applicata* **37**, II, 1957, pp. 127–144

BEAN, B. R. and DUTTON, E. J.: 'Radio meteorology', *NBS Monograph No. 92*, Washington, 1966

BEAN, B. R., DUTTON, E. J. and WARNER, B. D.: 'Weather effects on radar', *see* SKOLNIK, M. I., 1970

BEARD, C. I.: 'Coherent and incoherent scattering of microwaves from the ocean', *I.R.E. Trans.* **AP-9**, No. 5, Sept. 1961, pp. 470–483

BEARD, C. I., KATZ, I. and SPETNER, L. M.: 'Phenomenological vector model of microwave reflection from the ocean', *I.R.E. Trans.* **AP-4**, No. 2, Apr. 1956, pp. 162–167

BECKENBACH, E. F.: *Modern Mathematics for the Engineer*, McGraw Hill Book Co., New York, 1956

BECKMANN, P.: *Probability in Communication Engineering*, Harcourt, Brace & World Inc., New York, 1967

BECKMANN, P.: *The Depolarization of Electromagnetic Waves*, The Golem Press, Boulder, Colorado, 1968

BECKMANN, P.: 'Scattering by non-Gaussian surfaces', *I.E.E.E. Trans.* **AP-21**, March 1973, pp. 169–175

BECKMANN, P. and SPIZZICHINO, A.: *The Scattering of Electromagnetic Waves from Rough Surfaces*, Pergamon Press, New York, 1963

BELKINA, M. G. – *see* FOK, V. A. *et al.*

BIGGS, A. W.: 'Terrain influences on effective ground conductivity', *I.E.E.E. Trans.* **GE-8**, No. 2, Apr. 1970, pp. 106–114

BLAKE, L. V. : 'A guide to basic pulse-radar maximum range calculations. Part I – Equations, definitions and aids to calculation', *NRL Report 5868*, Dec. 1963

BLAKE, L. V.: 'Recent advancements in basic radar range calculation techniques', *I.R.E. Trans.* **MIL-5**, No. 2, Apr. 1967, pp. 154–164

BLAKE, L. V.: 'Machine plotting of radio/radar vertical – plane coverage diagram', *NRL Report 7098*, June 1970

BLAKE, L. V.: 'A FORTRAN computer program to calculate the range of a pulse radar', *NRL Report 7448*, Aug. 1972

BLASBALG, H.: 'Experimental results in sequential detection', *I.R.E. Trans.* **IT-5**, No. 2, June 1959, pp. 41–51

BLOCH, F. – *see* VAN VLECK, J. H.

BLOMQUIST, A. – *see* JOSEPHSON, B.

BLUTHGEN, J.: *Allgemeine Klimatographie*, Walter de Gruyter & Co., Berlin, 1964

BOOKER, H. G.: 'Elements of radio meteorology. How weather and climate cause unorthodox radar vision beyond the geometrical horizon', *J.I.E.E.* **93**, Pt IIIa, 1946, pp. 69–78

BORING, J. G., FLYNT, E. R., LONG, M. W. and WIDERQUIST, V. R.: 'Sea return study', *Final Report Project No. 157-96*, Engineering Experiment Station, Georgia Institute of Technology, 1957 AD 246180

BREMMER, H. – *see* VAN DER POL, B.

BRENNAN, L. E. – *see* MALLET, J. D.

BUDDEN, K. G.: *Radio Waves in the Ionosphere*, Cambridge University Press, 1961*a*

BUDDEN, K. G.: *The Wave-Guide Mode Theory of Wave Propagation*, Logos Press Ltd., London, 1961b

BUREAU OF METEOROLOGY: *Selected Tables of Australian Rainfall and Related Data*, Melbourne, 1961

BUSSGANG, J. J. and JOHNSON, N.: 'Design of sequential scan with frame time constraints', *I.E.E.E. Internat. Conv. Record*, Pt. 4, 1965, pp. 153–157

BUSSGANG, J. J. and MIDDLETON, D.: 'Optimum sequential detection of signals in noise', *I.R.E. Trans.* **IT-1**, No. 3, Dec. 1955, pp. 5–18

CAMPBELL, G. A. and FOSTER, R. M.: *Fourier Integrals for Practical Applications*, Van Nostrand Reinhold, Princeton, New Jersey, 1947

CARTER, C. J. – *see* REBER, E. E. *et al.*

C.C.I.R. Documents of the Xth Plenary Assembly, Geneva 1963, Vol. II, propagation

CHU, T. S.: 'Rain induced cross-polarization at cm and mm wavelength', *B.S.T.J.* **53**, No. 8, Oct. 1974, pp. 1557–1580

CRAIG, A. T.: See HOGG, R.V.

CRISPIN, J. W. Jr, GOODRICH, R. F. and SIEGEL, K. M. A.: 'A theoretical method for the calculation of the radar cross section of aircraft and missiles', University of Michigan – Report No. 2591-1-M, July 1959

CRISPIN, J. W. and MAFFETT, A. L.: I: 'Radar cross-section estimation for simple shapes', *P.I.E.E.E.* **53**, No. 8, Aug. 1965, pp. 833–848. II: 'Radar cross-section estimation for complex shapes', *P.I.E.E.E.* **53**, No. 8, Aug. 1965, pp. 972–982

CRISPIN, J. W. and SIEGEL, K. M. (editors): 'Methods of radar cross-section analysis', Academic Press, N.Y. 1968

CRONEY, J.: 'Improved radar visibility of small targets in sea clutter', *Radio and Electronic Engineer* **32**, No. 3, Sept. 1966, pp. 135–148

DAVENPORT, W. B., Jr.: 'Signal-to-noise ratios in band-pass limiters', *J.A.P.* **24**, No. 6, June 1953, pp. 720–727

DAVENPORT, W. B., Jr. and ROOT, W.L.: 'Random signals and noise', McGraw Hill Book Co., N.Y. 1958

DAVID, P. and VOGE, J.: 'Propagation of waves', Pergamon Press, London 1969

DiFRANCO, J. V. and RUBIN, W. L.: 'Radar detection', Prentice Hall, Englewood Cliffs, N.J., 1968

DOLUCHANOW, M. P.: *Die Ausbreitung von Funkwellen*, VEB Verlag Technik, Berlin, 1956

DUTTON, E. J. – *see* BEAN, B. R.

EMDE, F. – *see* JAHNKE, E.

·FEHLNER, L. F.: 'Marcum's and Swerling's data on target detection by a pulsed radar', *A.P.L. Report TG-451*, Johns Hopkins University, 1962

FLYNT, E. R. – *see* BORING, J. G. *et al.*

FOK, V. A., VAINSHTEIN, L. A. and BELKINA, M. G.: 'Radiowave propagation in surface tropospheric ducts', *Radiotechnika i elektronika* **3**, No. 12, Dec. 1958, pp. 1411–1429

FOWLER, C. A. – *see* WARD, H. R.

FRIIS, H. T.: 'Noise figures of radio receivers', *P.I.R.E.* **32**, July 1944, pp. 419–422

FRY, T. C.: *Probability and Its Engineering Uses*, The MacMillan Co., London, 1928

GLOVER, G. H. – *see* TAYLOR, R. C.

GOLDSTEIN, H. – *see* KERR, D. E.

GOODRICH, R. F. – *see* CRISPIN, J. W. Jr.

GRANT, C. R. and YAPLEE, B. S.: 'Back scattering from water and land at centimeter and millimeter wavelengths', *P.I.R.E.* **45**, July 1957, pp. 972–982

GRISETTI, R. S., SANTA, M. M. and KIRKPATRICK, G. M.: 'Effect of internal fluctuations and scanning on clutter attenuation in MTI radar', *I.R.E. Trans.* **ANE-2**, No. 1, March 1955, pp. 37–41

HALL, W. M.: 'Prediction of pulse radar performance', *P.I.R.E.* **44**, No. 2, Feb. 1956, pp. 224–231

HALL, W. M.: 'General radar equation', *Space/Aeronautics R & D Handbook*, 1962–1963

HALL, W. M. and WARD, H. R.: 'Signal-to-noise loss in moving target indicator', *P.I.E.E.E.* **56**, No. 2, Feb. 1968, pp. 233–234

HAMERMESH, M. – *see* VAN VLECK, J. H.

HANCOCK, J. C.: *An Introduction to the Principles of Communication Theory*, McGraw Hill Book Co., Inc., New York, 1961

HANSEN, V. G. and WARD, H. R.: 'Detection performance of the cell averaging LOG/CFAR receiver', *I.E.E.E. Trans.* **AES-8**, No. 5, Sept. 1972, pp. 648–652

HELSTROM, C. W.: 'A Range Sampled Sequential Detection System', *I.R.E. Trans.* **IT-8**, No. 1, Jan. 1962

HELSTROM, C. W.: *Statistical Theory of Signal Detection*, 2nd edition, Pergamon Press, London, 1968

HOGG, R. V. and CRAIG, A. T.: *Introduction to Mathematical Statistics*, The MacMillan Co., New York, 1959

HOLLIS, R.: 'False alarm time in pulse radar', *P.I.R.E.* **42**, No. 7, July 1954, p. 1189

I.E.E.E. Standard Dictionary of Electrical and Electronics Terms, John Wiley & Sons, Inc., New York, 1972

IRVING, J. and MULLINEUX, N.: *Mathematics in Physics and Engineering*, Academic Press, New York, 1959

JAHNKE, E. and EMDE, F.: *Tables of Functions*, Dover Publications, New York, 1945

JASIK, N.: *Antenna Engineering Handbook*, McGraw Hill Book Co., Inc., New York, 1964

JOHNSON, N.: 'Cumulative detection probability for Swerling III and IV targets', *P.I.E.E.E.* **54**, No. II, Nov. 1966, pp. 1583–1584

JOHNSON, N. – *see* BUSSGANG, J. J.

JORDAN, E. C.: *Electromagnetic Waves and Radiating Systems*, Constable & Co. Ltd., London, 1953

JOSEPHSON, B. and BLOMQUIST, A.: 'The influence of moisture in the ground, temperature and terrain on ground wave propagation in the VHF band', *I.R.E. Trans.* **AP-6**, No. 2, Apr. 1958, pp. 169–172

KAPLAN, S. M. and McFALL, R. W.: 'The statistical properties of noise applied to radar performance', *P.I.R.E.* **39**, No. 1, Jan. 1951, pp. 56–60

KAPUR, J. N. and SAXENA, H. C.: *Mathematical Statistics*, S. Chand & Co. Ltd., New Delhi, 1972

KATZ, I. – *see* BEARD, C. I. *et al.*: 'Radar reflectivity of the ocean surface for circular polarization', *I.E.E.E. Trans.* **AP-11**, No. 4, July 1963, pp. 451–453

KAY, I. W. – *see* KLINE, M.

KELL, R. E.: 'On the derivation of bistatic RCS from monostatic measurements', *P.I.E.E.E.* **53**, No. 8, Aug. 1965, pp. 983–988

KERR, D. E. (editor): *Propagation of Short Radio Waves*, Vol. 13, M.I.T. Radiation Lab. Series, McGraw-Hill Book Co., New York, 1951

KIRKPATRICK, G. M. – *see* GRISETTI, R. S. *et al.*

KLINE, M. and KAY, I. W.: *Electromagnetic Theory and Geometrical Optics*, Interscience Publishers, New York, 1965

KO, H. C.: 'The distribution of cosmic radio background radiation', *P.I.R.E.* **48**, No. 1, Jan. 1958, pp. 208–215

KRICHBAUM, C. K. – *see* RUCK, G. T. *et al.*

KUHRDT, G.: 'Grenzen der Ziel-Wahrnembarkeit in MTI – Radarsystemen', *Telefunken – Zeitung* **34**, No. 131, March 1961, pp. 42–50

LAWSON, J. L. and UHLENBECK, C. E.: *Threshold Signals*, Vol. 24, M.I.T. Radiation Lab. Series, McGraw Hill Book Co., New York, 1950

LEE, Y. W.: *Statistical Theory of Communication*, J. Wiley & Sons. Inc., New York, 1964

LILLEY, A. E. – *see* MEEKS, M. L.

LIPSON, H. I. – *see* WARD, H. R. *et al.*

LIVINGSTON, M. L.: 'The effect of antenna characteristics on antenna noise temperature and system SNR', *I.R.E. Trans.* **SET-7**, No. 3, Sept. 1961, pp. 71–79

LONG, M. W. – *see* BORING, J. G. *et al.*

LUNDIEN, J. R.: 'Terrain analysis by electromagnetic means: radar responses to laboratory prepared soil samples', *U.S. Army Waterways Expt. Sta. Tech. Rept. 3-639*, Rept. 2, 1966

MacDONALD, D.: 'The uses and potential of radio modelling', *R.R.E. Tech. Note 761*, 1972

MAFFETT, A. L. – *see* CRISPIN, J. W.

MALLETT, J. S. and BRENNAN, L. E.: 'Cumulative probability of detection of targets approaching a uniformly scanning search radar', *P.I.E.E.E.* **51**, No. 4, Apr. 1963, pp. 596–601

MALLINCKRODT, A. J. and SOLLENBERGER, T. E.: 'Optimum pulse-time determination', *I.R.E. Trans.* **IT-3**, No. 2, March 1954, pp. 151–159

MANNASSE, R.: 'Maximum angular accuracy of tracking a radio star by lobe comparison', *I.R.E. Trans.* **AP-8**, No. 1, Jan. 1960, pp. 50–56

MARCUM, J. I.: 'A statistical theory of target detection by pulsed radar', with 'Mathematical appendix', *I.R.E. Trans.* **IT-6**, No. 2, Apr. 1960

MARCUS, M. B. and SWERLING, P.: 'Sequential detection in radar with multiple resolution elements', *I.R.E. Trans.* **IT-8**, No. 2, April 1962, pp. 237–245

MARTIN, J. R.: *The Frequency Agile Magnetron Story*, Varian Eastern Tube Dvn. – Brochure 2836, Aug. 1972

MATTHEWS, P. A.: *Radio Wave Propagation: VHF and Above*, Chapman & Hall Ltd., London, 1965

MEEKS, M. L. and LILLEY, A. E.: 'The microwave spectrum of oxygen in the earth's atmosphere', *J. of Geophys. Res.* **68**, No. 6, March 15, 1963, pp. 1683–1703

MENZEL, D. H. (editor): *The Radio Noise Spectrum*, Harvard University Press, Cambridge, Mass. 1960

MERRITT, S. M.: *Mathematics Manual*, McGraw-Hill Book Co. Inc., New York, 1962

MIDDLETON, D. – BUSSGANG, J. J.

MIDDLETON, D.: *An Introduction to Statistical Communication Theory*, McGraw Hill Book Co., New York, 1960

MITCHELL, R. L. – *see* REBER, E. E.

MÜCKE, J. and RÖHRICH, K.: 'Mittelbereich – Radaranlage für den Flugsicherungs-Kontrolldienst', *Telefunken Zeitung* **34**, No. 131, March 1961, pp. 5–12

MULLINEUX, N. – *see* IRVING, J.

NATHANSON, F. E.: *Radar Design Principles – Signal Processing and the Environment*, McGraw Hill Book Co., New York, 1969

NORTON, K. A. and OMBERG, A. C.: 'The maximum range of a radar set', *P.I.R.E.* **35**, No. 1, Jan. 1947, pp. 4–24

OGUCHI, T.: 'Attenuation of electromagnetic waves due to rain with distorted raindrops', *J. Radio Res. Lab. (Tokyo)* 7, No. 33, Sept. 1960, pp. 467–485

OMBERG, A. C. – *see* NORTON, K. A.

OTTERSEN, H.: 'Radar Angels and their relationship to meteorological factors', *F.O.A. Report* **4**, No. 2, Apr. 1970

PACHARES, J.: 'A table of bias levels useful in radar detection problems', *I.R.E. Trans.* **IT-4**, No. 1, Jan. 1958, pp. 38–45

PAPOULIS, A.: *The Fourier Integral and Its Applications*, McGraw Hill Book Co., Inc., New York, 1962

PAVEL'YEV, A. G. – *see* PROSIN, A. V.

PAYNE-SCOTT, R.: 'The visibility of small echoes on radar PPI displays', *P.I.R.E.* **36**, No. 2, Feb. 1948, pp. 180–196

PEARSON, K.: *Table of Incomplete Gamma-Functions*, Cambridge University Press, London, 1946

PETERS, J.: *Einfuehrung in die Allgemeine Informations Theorie*, Springer-Verlag, Berlin, 1967

PRESTON, G. W.: 'The search efficiency of the probability ratio sequential search radar', *I.E.E.E. Internat. Conv. Record*, Pt. 4, 1960, pp. 116–124

PROSIN, A. V. and PAVEL'YEV, A. G.: 'Calculation of the power of a radio signal scattered by a statistically uneven surface', *Telecommunication and Radio Engineering* **26/27**, Apr. 1972, pp. 78–87

RAPPAPORT, S. S.: 'On practical setting of the detection threshold', *P.I.E.E.E.* **57**, No. 8, Aug. 1969, pp. 1420–1421

REBER, E. E., MITCHELL, R. L. and CARTER, C. J.: 'Attenuation of the 5 mm wavelength band in a variable atmosphere', *I.E.E.E. Trans.* **AP-18**, No. 4, July, 1980, pp. 472–479

RICE, S. O.: 'The mathematical analysis of random noise', *B.S.T.J.* **23**, July 1944, pp. 282–332 and **24**, Jan. 1945, pp. 46–156

RICE, S. O.: 'Reflection of electromagnetic waves from slightly rough surfaces', *Com. Pure Appl. Mat.* **4**, 1951, pp. 351–378

ROHAN, P.: 'Umísťování ultrakrátkovlnných stanic v terénu vzhledem k šíření vln', *Věstník Spoj, Vojska* **1**, No. 2, 1948

ROHAN, P.: 'Noise factor, noise figure and noise temperature', *Weapons Research Establishment (W.R.E.) Tech. Memo. PAD 176*, 1964

ROHAN, P.: 'Graphs for cumulative detection probability calculations,' *W.R.E. Tech. Memo 444 (AP)*, 1971

ROHAN, P.: 'A computer programme for the computation of Bessel functions of the first kind and orders 0, 1, 2 and 3', *W.R.E. Tech. Memo. 807 (AP)*, 1973*a*

ROHAN, P.: 'A FORTRAN II programme for the evaluation of the error function by a digital computer', *W.R.E. Tech. Memo. 858 (AP)*, 1973*b*

ROHAN, P.: 'Machine computation of the sine integral, of its complement and of the cosine integral', *W.R.E. Tech. Memo. 965 (AP)*, 1973*c*

ROHAN, P.: 'Surveillance radar performance assessment by mathematical modelling', Ph.D. Thesis, Dept. of Electrical Engineering, The University of Adelaide, 1981

ROHAN, P. and SODEN, L. B.: 'A survey of the southern sky at 55 MHz', *Aust. J. of Physics* **23**, 1970, pp. 223–225

RÖHRICH, K. – *see* MÜCKE, J.

ROOT, W. L. – *see* DAVENPORT, W. B. Jr.

ROTHERAM, S.: 'Propagation theory for the evaporation duct', *Paper presented at NATO Advanced Study Institute*, Sorento, 1973

RUCK, G. T., BARRICK, D. E., STUART, W. D. and KRICHBAUM, C. K.: *Radar Cross-Section Handbook*, Plenum Press, New York, 1970

SANTA, M. M. – *see* GRISETTI, R. S.

SAXENA, H. C. – *see* KAPUR, J. H.

SCHWARTZ, M.: 'Statistical decision theory and detection of signals in noise', in BERKOWITZ, R. S. *Modern Radar*, J. Wiley & Sons, New York, 1965

SIEGEL, K. M. A. – *see* CRISPIN, J. W. Jr.

SILVER, S. (editor): *Microwave Antenna Theory and Design*, Vol. 12, M.I.T. Radiation Lab. Series, McGraw Hill Book Co., New York, 1949

SKOLNIK, M. I.: *Introduction to Radar Systems*, McGraw Hill Book Co., New York, 1962

SKOLNIK, M. I.: 'An empirical formula for the radar cross section of ships at grazing incidence', *I.E.E.E. Trans.* **AES-10**, No. 2, March 1974, p. 292

SKOLNIK, M. I.: *Radar Handbook*, McGraw Hill Book Co., New York, 1970

SLATER, L. J.: *Confluent Hypergeometric Functions*, Cambridge University Press, 1960

SMITH, A. A. and SODEN, L. B.: 'Field study of ducting and tropospheric radar propagation around Darwin, Australia', *P.I.R.E.E.* **35**, No. 12, Dec. 1974, pp. 390–395

SODEN, L. B. – *see* ROHAN, P.

SODEN, L. B. – *see* SMITH, A. A.

SOLLENBERGER, T. E. – *see* MALLINCKRODT, A. J.

SPETNER, L. M. – *see* BEARD, C. I. *et al.*

SPIZZICHINO, A. – *see* BECKMANN, P.

STARR, A. T.: *Radio and Radar Technique*, Sir Isaac Pitman & Sons, Ltd., London, 1953

STRATTON, J. A.: *Electromagnetic Theory*, McGraw-Hill Book Co., New York, 1941

STUART, W. D. – *see* RUCK, G. T. *et al.*

SWERLING, P.: 'Maximum angular accuracy of a pulsed search radar', *P.I.R.E.* **44**, No. 9, Sept. 1956, pp. 1146–1155

SWERLING, P.: 'Probability of detection for fluctuating target', *I.R.E. Trans.* **IT-6**, No. 2, Apr. 1960

SWERLING, P. – *see* MARCUS, M. B.

TAYLOR, R. G. and GLOVER, C. H.: 'Low-level radar propagation in coastal waters', *Electronic Letters* **11**, No. 11, pp. 246–248

TENNE-SENS, A. U.: 'A computer program for plotting radar detection probabilities', *C.R.C., Radar Section, Tech. Memo. No. 22*, Ottawa, 1971

TWERSKY, V.: 'On scattering and reflection of electromagnetic waves by rough surfaces', *I.R.E. Trans.* **AP-5**, No. 1, Jan. 1957, pp. 81–90

UHLENBECK, G. E. – *see* LAWSON, J. L.

VAINSHTEIN, L. A. – *see* FOK, V. A. *et al.*

VALLEY, G. E. and WALLMAN, H.: *Vacuum Tube Amplifiers*, Vol. 18, M.I.T. Radiation Lab. Series, McGraw Hill Book Co., New York, 1948

VAN der POL, B. and BREMMER, H.: 'Further note on the propagation of radio waves over a finitely conducting spherical earth', *Phil. Mag.* **27**, Series 7, No. 182, March 1939, pp. 261–275

VANNICOLA, V.: 'Fluctuation loss and diversity gain for in-phase systems with post-detection integration', *I.E.E.E. Trans.* **AES-9**, No. 2, March 1973, pp. 290–295
VAN VLECK, J. H.: 'The absorption of microwaves by oxygen', *Phys. Review* **71**, No. 7, April 1947*a*, pp. 413–424
VAN VLECK, J. H.: 'The absorption of microwaves by uncondensed water vapor', *Phys. Review* **71**, No. 7, April 1947*b*, pp. 425–433
VAN VLECK, J. H., BLOCH, F. and HAMERMESH, M.: 'Theory of radar reflection from wires of thin metallic strips', *J.A.P.* **18**, No. 3, March 1947, pp. 274–294
VOCE, J. – *see* DAVID, P.
WALLMAN, H. – *see* VALLEY, G. E.
WARD, H. R. – *see* BARTON, D. K.
WARD, H. R. – *see* HALL, W. M.
WARD, H. R. – *see* HANSEN, V. G.
WARD, H. R., FOWLER, C. A. and LIPSON, H. I.: 'GCA radars. Their history and state of development', *P.I.E.E.E.* **62**, No. 6, June 1974, pp. 705–716
WARNER, B. D. – *see* BEAN, B. R. *et al.*
WATSON, G. N.: *A Treatise on the Theory of Bessel Functions*, Cambridge University Press, 1922
WEINSTOCK, W. W. – *see* BERKOWITZ, R. S.: *Modern Radars*, J. Wiley & Sons. New York, 1965
WENISCH, G.: 'Prévision des échos radar du sol d'après la topographie du terrain', *L'Onde Electrique* **51**, No. 8, Sept. 1971, pp. 696–703
WESTON, V. H.: 'Theory of absorbers in scattering', *I.E.E.E. Trans.* **AP-11**, Sept. 1963, pp. 578–584
WHITE, D. M.: 'Radar simulation and analysis by digital computer', *A.P.L. Tech. Memo. TG-952*, The Johns Hopkins University, 1968
WIDERQUIST, V. R. – *see* BORING, J. G. *et al.*
WOODWARD, P. M.: 'A method of calculating the field over a plane aperture required to produce a given polar diagram', *J.I.E.E.* **93**, Pt. IIIA, No. 10, 1946, pp. 1554–1558
WOZENCRAFT, J. M. and JACOBS, I. M.: *Principles of Communication Engineering*, J. Wiley & Sons, New York, 1955
YAPLEE, B. S. – *see* GRANT, C. R.

Index